Emerging Hydro-Climatic Patterns, Teleconnections and Extreme Events in Changing World at Different Timescales

Emerging Hydro-Climatic Patterns, Teleconnections and Extreme Events in Changing World at Different Timescales

Editors

Ankit Agarwal
Naiming Yuan
Kevin K.W. Cheung
Roopam Shukla

MDPI • Basel • Beijing • Wuhan • Barcelona • Belgrade • Manchester • Tokyo • Cluj • Tianjin

MDPI

Editors
Ankit Agarwal
Department of Hydrology
Indian Institute of Technology
Roorkee
Roorkee
India

Naiming Yuan
Institute of Atmospheric Physics
Chinese Academy of Sciences
Beijing
China

Kevin K.W. Cheung
E3-Complexity Consultant
Complexity Consultant
Sydney
Australia

Roopam Shukla
Adaptation in Agriculture
Systems, RD2: Climate
Resilience
Potsdam Institute for Climate
Impact Research (PIK)
Potsdam
Germany

Editorial Office
MDPI
St. Alban-Anlage 66
4052 Basel, Switzerland

This is a reprint of articles from the Special Issue published online in the open access journal *Atmosphere* (ISSN 2073-4433) (available at: www.mdpi.com/journal/atmosphere/special_issues/Hydro-Climatic).

For citation purposes, cite each article independently as indicated on the article page online and as indicated below:

LastName, A.A.; LastName, B.B.; LastName, C.C. Article Title. *Journal Name* **Year**, *Volume Number*, Page Range.

ISBN 978-3-0365-2953-0 (Hbk)
ISBN 978-3-0365-2952-3 (PDF)

Contents

About the Editors

Ankit Agarwal

Dr Ankit Agarwal is a hydro-climatologist interested in interdisciplinary research and teaching. His research aims to understand multi-scale interactions between different components of the Atmosphere and Hydrosphere, and in particular between climate patterns and extreme events. He gained training as a civil engineer and hydrologist during his master degree at IIT-Delhi and Technical University Dresden. He diversified his expertise in climatology and nonlinear dynamics during his PhD at the University of Potsdam. Currently, Dr Ankit is an Assistant Professor at the Department of Hydrology, Indian Institute of Technology Roorkee, responsible for developing new methods and applying them in hydrology and climatology to advance understanding. He also enriches master and PhD students with his expertise and knowledge. Dr Ankit is currently serving as an Associate Editor in many well-known journals such as *Scientist Reports-Nature*, *Hydrological Sciences Journal* (IAHS) and *Weather, Climate and Society* (AMS). He also served as a guest editor for *Atmosphere (MDPI)*, in *European Physics Journal Special Topics* (Springer), Water and Build Environment, a section in *Frontiers*.

Naiming Yuan

Dr Naming Yuan is currently employed at the Institute of Atmospheric Physics, Chinese Academy of Sciences located in Be, China. He has expertise in climate variability, climate change and monsoon understanding.

Kevin K.W. Cheung

Dr Kevin Cheung has expertise on the dynamics of severe weather systems and how these systems change in behaviour under climate change, with applications of nonlinear techniques such as complex networks and concepts from criticality. Currently, he is a consultant at the E3-Complexity Consultant, Sydney, Australia.

Roopam Shukla

Roopam Shukla is presently a postdoctoral researcher at the Adaptation in Agricultural Systems working group. Roopam Shukla has her research interest in vulnerability and adaptation in smallholder farming systems. She is also a member of Geo.X Young Academy as a member of the Geo-Society group. Roopam was contributing author for the Intergovernmental Panel on Climate Change (IPCC) AR6 WGII (Chapter 16), where she contributes specifically to the synthesis of evidence on aspects of equity in adaptation responses. She serves as an associate editor for the *Weather, Climate, and Society* journal and a review editor for the Climate Risk Management section of *Frontiers in Climate* journal. Roopam pursued her Ph.D. from TERI School of Advanced Sciences, India, focusing on vulnerability, resilience, and adaptation of agricultural communities. She holds a master's degree in Climate Science and Policy.

Preface to "Emerging Hydro-Climatic Patterns, Teleconnections and Extreme Events in Changing World at Different Timescales"

This is a book brought together with the support of all our authors' unconventional research to understand the complex climate system. The book is intended for professionals, researchers, and academicians who in addition to their subject knowledge on climate systems, would like to go deeper into the domain of climatic teleconnections, climatic patterns and their interactions with hydro-climatic variables. It presents many who have come to realise that the complexity, randomness and teleconnections are the critical blueprints of the hydro-climatic systems. The 13 papers in the book converge and create a holistic understanding of essential tools for regional hydro-meteorological predictions and highlight the potential explanatory role in understanding the spatiotemporal variability of hydrological variables.

Prof. Ankit Agarwal has expertise in complex networks, multi-scale analysis, teleconnections and patterns understanding. Prof. Naming Yuman has expertise in climate variability, climate change and Monsoon understanding. Dr Kevin Cheung has expertise on the dynamics of severe weather systems and how these systems change in behaviour under climate change, with applications of nonlinear techniques such as complex networks and concepts from criticality. Dr Roopam Shukla works on the intersection of climate impact analysis and linking to vegetation dynamics across multiple scales. We, the editorial team, are grateful for the trust and permission of the authors to share such critical pieces of research with everyone.

Ankit Agarwal, Naiming Yuan, Kevin K.W. Cheung, Roopam Shukla
Editors

atmosphere

MDPI

Editorial

Emerging Hydro-Climatic Patterns, Teleconnections, and Extreme Events in Changing World at Different Timescales

Ankit Agarwal [1,*], Naiming Yuan [2], Kevin K. W. Cheung [3] and Roopam Shukla [4]

1 Department of Hydrology, Indian Institute of Technology, Roorkee 247667, India
2 Institute of Atmospheric Physics, Chinese Academy of Sciences, Hua Yan Li #40, Beijing 100029, China; mingyuan@hotmail.com
3 E^3-Complexity Consultant, Sydney, NSW 2122, Australia; kkwc09@gmail.com
4 Adaptation in Agriculture Systems, RD2: Climate Resilience, Potsdam Institute for Climate Impact Research (PIK), Leibniz Association, Telegrafenberg, 14476 Potsdam, Germany; shukla@pik-potsdam.de
* Correspondence: ankit.agarwal@hy.iitr.ac.in

Citation: Agarwal, A.; Yuan, N.; Cheung, K.K.W.; Shukla, R. Emerging Hydro-Climatic Patterns, Teleconnections, and Extreme Events in Changing World at Different Timescales. *Atmosphere* **2022**, *13*, 56. https://doi.org/10.3390/atmos13010056

Received: 27 December 2021
Accepted: 27 December 2021
Published: 30 December 2021

Publisher's Note: MDPI stays neutral with regard to jurisdictional claims in published maps and institutional affiliations.

The *Atmosphere* Special Issue, entitled "Emerging Hydro-Climatic Patterns, Teleconnections and Extreme Events in Changing World at Different Timescales", comprises thirteen original papers.

Climate is a complex system regulated by interactions among components of Earth at different spatial and temporal scales. Unravelling spatiotemporal patterns and interactions among climate variables, especially those related to hydroclimate, has always been an important task for geoscientists in general and climatologists in particular, mostly because it contributes significantly to better prediction and forecasting. Teleconnection is a fundamental component of the climate system that refers to the climate variability links between geographically separated regions. Teleconnections, such as El Niño Southern Oscillation (ENSO), Indian Ocean Dipole (IOD), North Atlantic Oscillation (NAO), and the Madden Julian Oscillation (MJO), are often analyzed in their mature phase of variability with the recognition that the teleconnections have an evolving spatiotemporal scale. However, complexities are intrinsic to natural systems, and for this reason, the task of identifying patterns and interactions has always been challenging. Coupled with the existing challenges of global-warming-induced climate change, these patterns and interactions become further unusual, unexpected, and unpredictable. With these challenging realities, climate science studies globally recognize that climatic and other geophysical processes are intrinsically nonlinear and carry multiscale features and influences that are generally time-varying. This Special Issue is primarily prompted by these realizations and an implied aspiration to develop a collection of advanced studies addressing the abovementioned issues.

Teleconnections serve as an essential tool for regional hydrometeorological predictions and highlight the potential explanatory role in understanding the spatiotemporal variability of hydrological variables. There are six papers in this Special Issue related to this topic. Harry West et al. [1] focused on the spatio-temporal understanding of the rainfall signatures in Great Britain based on the North Atlantic Oscillation (NAO) rainfall response variability. They showed a stronger and more consistent NAO rainfall response, with a greater probability of more extreme wet/dry conditions. However, greater NAO rainfall variability during winter was found in the SE. A more spatially consistent rainfall response marks the summer months and finds variability in wet/dry magnitude and directionality. Wayne Yuan-Huai Tsai [2] reported the cause for the Northern Queensland Floods during February 2019 as a record-breaking sub-seasonal peak rainfall event. The event was induced by an anomalously strong monsoon depression moderated by the convective phases of an MJO and an equatorial Rossby wave. This study reported that the equatorial Rossby wave is the leading cause for the lower forecast skill. Chao Wang [3] focused on the evolution characteristics of the daily-scale Silk Road pattern and its effect on heatwaves in the Yangtze River Valley based on multi-reanalysis daily datasets and station data. The study reported that sinking adiabatic warming and clear-sky radiation warming can be considered the possible causes of the YRV heat waves. Xiaoduo Pan [4] described the refined characteristics of moisture recycling over the Heihe River Basin (HRB) using the Weather Research and Forecasting model. The study concluded that the wind dominantly transported water vapor of the HRB from the west and north directions, and the west one was much larger than the north one. In addition, precipitation over the HRB was triggered mainly by the vapor transport from the west to the east. Nachiketa Acharya [5] explored the onset of the rainy season for seven agroclimatic zones over Vietnam. The spatiotemporal characteristics of zonal onset date are analyzed using the teleconnection with Niño 3.4 anomalies. The interannual variation in the rainy season onset date is approximately two weeks across all agroclimatic zones. Central Highlands and South zones were found to have a potential for onset prediction, as both regions were linked with Nino 3.4 anomalies. Md Wahiduzzaman [6] investigated the spatiotemporal variabilities and trends of thunderstorms days over Bangladesh and their teleconnections with El Niño Southern Oscillation (ENSO) and Indian Ocean Dipole (IOD). The Mann–Kendall test reveals an increasing trend in thunderstorms for May and June, particularly in north and northeastern regions. The connection between thunderstorms and ENSO/IOD indicates a decrease in thunderstorm activities in Bangladesh during the El Niño and positive IOD years.

The emerging spatiotemporal characteristics of extreme events due to climate variability and change helps to understand the propagation of the extreme events. There are three papers related to this topic. Kevin K. W. Cheung [7] analyzed the spatial distribution of the daily precipitation concentration index inside the Greater Sydney Metropolitan Area using Moran's spatial autocorrelation. The spatial distribution has revealed the nature and mechanisms underlying the distribution of torrential rains over space within the metropolis of Sydney under the warming climate. AVS Kalyan [8] conducted a multiscale spatiotemporal analysis of extreme events in India's Gomati River Basin. Using Sen's slope estimator and Hurst exponent analysis, the authors computed the present and future spatial variation of sixteen extreme climate indices. The study has shown that the number of dry days increased, highlighting that the basin is getting drier. Next, the periodicities and non-stationary features were estimated using the continuous wavelet transform and find a dominant two-year period in D95P has changed to the four years after 1984 and remains in the past two decades. Furthermore, the joint probability estimation using the copula theory highlights the underestimation of the return period due to the ignorance of mutual dependence. Changjun Wan [9] explored the spatiotemporal aggregation characteristics of extreme precipitation over China using the local spatial autocorrelation and spatiotemporal scanning models. Integrating both models can effectively detect the accumulation of extreme events.

Several precipitation products are available to model the hydrological behavior of watersheds, primarily derived from gauge-based observations, remote sensing, and reanalysis. This Focus Issue touches on this exciting research; for instance, Setti et al. [10] evaluate the influence of different precipitation products on model parameters and streamflow predictive uncertainty using a soil water assessment tool (SWAT) model. They showed that bias-corrected TRMM data could be a good alternative to ground observations for driving the hydrological model of forest dominated catchment over India.

Jianfeng Wang [11] presented the impacts of harsh weather conditions on the transportation system using a stratified Cox model and a heterogeneous Markov chain model, and showed that weather variables, including temperature, humidity, snow depth, and ice/snow precipitation, have a significant impact on train performance.

To improve the simulation of tropical cyclones over the North Indian Ocean (NIO), Gundapuneni Venkata Rao [12] investigated the impact of seven microphysical parameterization schemes using the ARW model. From sensitivity analysis, the WSM3 scheme simulated the cyclones Nilofar, Kyant, Daye, and Phethai well, whereas the cyclones Hudhud, Titli, and Ockhi are best simulated by WSM6. The study suggests that the WSM3 scheme can be used as the first best scheme to predict post-monsoon tropical cyclones over the NIO.

To disentangle the devastating landslides and widespread flooding over Central Chile, Piero Mardones [13] explored the freezing level distribution along the western slope of the subtropical Andes (30°–38° S) for the present climate. They estimated the changes from the 21st century using the free tropospheric height of the 0 °C isotherm (H0) as a proxy. The mean value under wet conditions toward the end of the century (under RCP8.5) is close to, or higher than, the upper quartile of the H0 distribution in the current climate. Under RCP8.5, even moderate daily precipitation can increase river flow to levels that are considered hazardous for central Chile.

Findings reported in the SI highlight and support the urgent need to consider the remote teleconnections, extreme events, multiscale variability, and dynamics of the geophysical process in general and climate process in particular. All published 13 studies contributed significantly to our existing knowledge and will appeal to the broader society of Earth scientists and modellers given the problems they face in understanding the spatio-temporal variability of the climate variables, coupled between the teleconnections and extreme events. This Special Issue advanced our understanding of these emerging patterns, teleconnections, and extreme events in a changing world for more accurate prediction or projection of their changes, especially on different spatial–time scales.

Author Contributions: A.A. conceptualized the theme of the SI and prepared the original draft. N.Y., K.K.W.C. and R.S. supported the idea, reviewed and edited the SI extensively. All editor would like to thank the authors for their contributions to this Special Issue, and the reviewers for their constructive and helpful comments to improve the manuscripts. The editor is grateful to Alicia Wang for her kind support in processing and publishing this Special Issue. All authors have read and agreed to the published version of the manuscript.

Funding: This research received no external funding.

Conflicts of Interest: The authors declare no conflict of interest.

References

1. West, H.; Quinn, N.; Horswell, M. Spatio-Temporal Variability in North Atlantic Oscillation Monthly Rainfall Signatures in Great Britain. *Atmosphere* **2021**, *12*, 763. [CrossRef]
2. Tsai, W.Y.-H.; Lu, M.-M.; Sui, C.-H.; Cho, Y.-M. Subseasonal Forecasts of the Northern Queensland Floods of February 2019: Causes and Forecast Evaluation. *Atmosphere* **2021**, *12*, 758. [CrossRef]
3. Wang, C.; Wen, Y.; Wang, L.; Kang, X.; Liu, Y. The Evolution Characteristics of Daily-Scale Silk Road Pattern and Its Relationship with Summer Temperature in the Yangtze River Valley. *Atmosphere* **2021**, *12*, 747. [CrossRef]
4. Pan, X.; Ma, W.; Zhang, Y.; Li, H. Refined Characteristics of Moisture Cycling over the Inland River Basin Using the WRF Model and the Finer Box Model: A Case Study of the Heihe River Basin. *Atmosphere* **2021**, *12*, 399. [CrossRef]
5. Acharya, N.; Bennett, E. Characteristic of the Regional Rainy Season Onset over Vietnam: Tailoring to Agricultural Application. *Atmosphere* **2021**, *12*, 198. [CrossRef]
6. Wahiduzzaman, M.; Islam, A.R.M.T.; Luo, J.; Shahid, S.; Uddin, M.J.; Shimul, S.M.; Sattar, M.A. Trends and Variabilities of Thunderstorm Days over Bangladesh on the ENSO and IOD Timescales. *Atmosphere* **2020**, *11*, 1176. [CrossRef]
7. Cheung, K.K.W.; Rasuly, A.A.; Ji, F.; Chang, L.T.-C. Spatial Characteristics of Precipitation in the Greater Sydney Metropolitan Area as Revealed by the Daily Precipitation Concentration Index. *Atmosphere* **2021**, *12*, 627. [CrossRef]
8. Kalyan, A.; Ghose, D.K.; Thalagapu, R.; Guntu, R.K.; Agarwal, A.; Kurths, J.; Rathinasamy, M. Multiscale Spatiotemporal Analysis of Extreme Events in the Gomati River Basin, India. *Atmosphere* **2021**, *12*, 480. [CrossRef]
9. Wan, C.; Cheng, C.; Ye, S.; Shen, S.; Zhang, T. Recognizing the Aggregation Characteristics of Extreme Precipitation Events Using Spatio-Temporal Scanning and the Local Spatial Autocorrelation Model. *Atmosphere* **2021**, *12*, 218. [CrossRef]
10. Setti, S.; Maheswaran, R.; Sridhar, V.; Barik, K.K.; Merz, B.; Agarwal, A. Inter-Comparison of Gauge-Based Gridded Data, Reanalysis and Satellite Precipitation Product with an Emphasis on Hydrological Modeling. *Atmosphere* **2020**, *11*, 1252. [CrossRef]
11. Wang, J.; Yu, J. Train Performance Analysis Using Heterogeneous Statistical Models. *Atmosphere* **2021**, *12*, 1115. [CrossRef]
12. Venkata Rao, G.; Venkata Reddy, K.; Sridhar, V. Sensitivity of Microphysical Schemes on the Simulation of Post-Monsoon Tropical Cyclones over the North Indian Ocean. *Atmosphere* **2020**, *11*, 1297. [CrossRef]
13. Mardones, P.; Garreaud, R.D. Future Changes in the Free Tropospheric Freezing Level and Rain–Snow Limit: The Case of Central Chile. *Atmosphere* **2020**, *11*, 1259. [CrossRef]

atmosphere

MDPI

Article

Spatio-Temporal Variability in North Atlantic Oscillation Monthly Rainfall Signatures in Great Britain

Harry West *, **Nevil Quinn** and **Michael Horswell**

Centre for Water, Communities and Resilience, Department of Geography and Environmental Management, University of the West of England, Bristol BS16 1QY, UK; Nevil.Quinn@uwe.ac.uk (N.Q.); michael.horswell@uwe.ac.uk (M.H.)

* Correspondence: harry.west@uwe.ac.uk

Abstract: The North Atlantic Oscillation (NAO) is the primary atmospheric-oceanic circulation/ teleconnection influencing regional climate in Great Britain. As our ability to predict the NAO several months in advance increases, it is important that we improve our spatio-temporal understanding of the rainfall signatures that the circulation produces. We undertake a high resolution spatio-temporal analysis quantifying variability in rainfall response to the NAO across Great Britain. We analyse and map monthly NAO-rainfall response variability, revealing the spatial influence of the NAO on rainfall distributions, and particularly the probability of wet and dry conditions/extremes. During the winter months, we identify spatial differences in the rainfall response to the NAO between the NW and SE areas of Britain. The NW area shows a strong and more consistent NAO-rainfall response, with greater probability of more extreme wet/dry conditions. However, greater NAO-rainfall variability during winter was found in the SE. The summer months are marked by a more spatially consistent rainfall response; however, we find that there is variability in both wet/dry magnitude and directionality. We note the implications of these spatially and temporally variable NAO-rainfall responses for regional hydrometeorological predictions and highlight the potential explanatory role of other atmospheric-oceanic circulations.

Keywords: North Atlantic Oscillation; NAO; rainfall signatures; spatio-temporal analysis

check for updates

Citation: West, H.; Quinn, N.; Horswell, M. Spatio-Temporal Variability in North Atlantic Oscillation Monthly Rainfall Signatures in Great Britain. *Atmosphere* **2021**, *12*, 763. https://doi.org/10.3390/atmos12060763

Academic Editors: Ankit Agarwal, Naiming Yuan, Kevin K. W. Cheung and Roopam Shukla

Received: 15 May 2021
Accepted: 11 June 2021
Published: 13 June 2021

1. Introduction

Weather in Great Britain can be highly variable, fluctuating between wet and dry extremes. The North Atlantic Oscillation (NAO) atmospheric-oceanic circulation has long been cited as the leading mode of climate variability in the North Atlantic region [1–3] due to its influence on the location and amplitude of the North Atlantic Jet Stream [4]. The NAO teleconnection is commonly defined by the sea level pressure (SLP) variation between two meridional dipoles: the Icelandic low-pressure action point and the Azores anticyclone. Fluctuations in the difference in SLP between Iceland and the Azores leads to the occurrence of NAO positive (NAO+) phases, representing a greater than normal difference in SLP between the two dipoles, or NAO negative (NAO−) phases representing a weaker than normal difference in SLP. The strength and phase of the NAO can be quantified by the North Atlantic Oscillation Index (NAOI) [3].

Previous work has explored the influence of sub-annual variability and phase of the NAO on weather and climate in Great Britain. Strong positive correlations are often reported between the NAOI and winter rainfall in the north-western areas of the country [5–7], whilst weaker negative winter correlations have been found in the south-east and central areas [8,9]. Previous work has quantified these opposing regional rainfall responses to NAO+ and NAO− phases relative to when the NAO is in a weak neutral state, finding that in winter, average monthly rainfall increases under NAO+ and decreases under NAO− conditions can be as much as 200–300 mm in the north-west [10]. Negative correlations between the winter NAOI and modelled snow cover have also been found in the north-west of Scotland [11], suggesting that winter NAO− phases are associated with higher snowfall in these regions.

Whilst the magnitude of the NAOI is weaker in the summer months [12], significant NAOI-rainfall correlations have been reported [13]. However, the correlation coefficients are generally weaker in summer than during the winter months [10]. In winter, the NAO-rainfall response is characterised by a north-west/south-east spatial divide as described above, whilst in summer the spatial signature is more homogenous across the country [12,13]. In summer months, average monthly rainfall increases under NAO− conditions and decreases under NAO+ conditions are approximately 50–100 mm, compared to NAO neutral conditions [10].

These NAO rainfall signatures propagate through the hydrological cycle, with NAO responses observed across Great Britain in catchment runoff [14–16], groundwater [17,18] and fluvial water temperatures [19], although this propagation can be moderated by catchment characteristics such as topography, landcover and geology [15,16]. Ongoing research is exploring whether this understanding of NAO-rainfall-flow propagation can be incorporated into seasonal streamflow modelling and forecasting [20,21].

As discussed above, the NAO has been found to influence hydrometeorology across the country, and its spatial signature is evident in monthly average rainfall datasets [10]. However, several studies have reported inconsistency in seasonal NAO-rainfall responses across Great Britain. Hall and Hanna [13] present three winter rainfall anomaly maps produced by the UK Met Office for 2013/2014, 2014/2015 and 2015/2016 (see their Figure 1). The reported NAOI values all indicate an NAO+ phase of varying magnitudes. However, only in the winter of 2014/2015 was the typical north-west/south-east winter NAO-rainfall response (as described above) observed. This study concluded by foregrounding the role of other North Atlantic teleconnections as secondary modes of climate variability, such as the East Atlantic Pattern (EA) and Scandinavian Pattern, in influencing regional rainfall (and temperatures) in Britain [13]. Variability in winter NAO-rainfall signatures, both in terms of strength and spatiality, has also been attributed to other atmospheric teleconnections, in particular positive and negative phases of the EA, in other work [22,23].

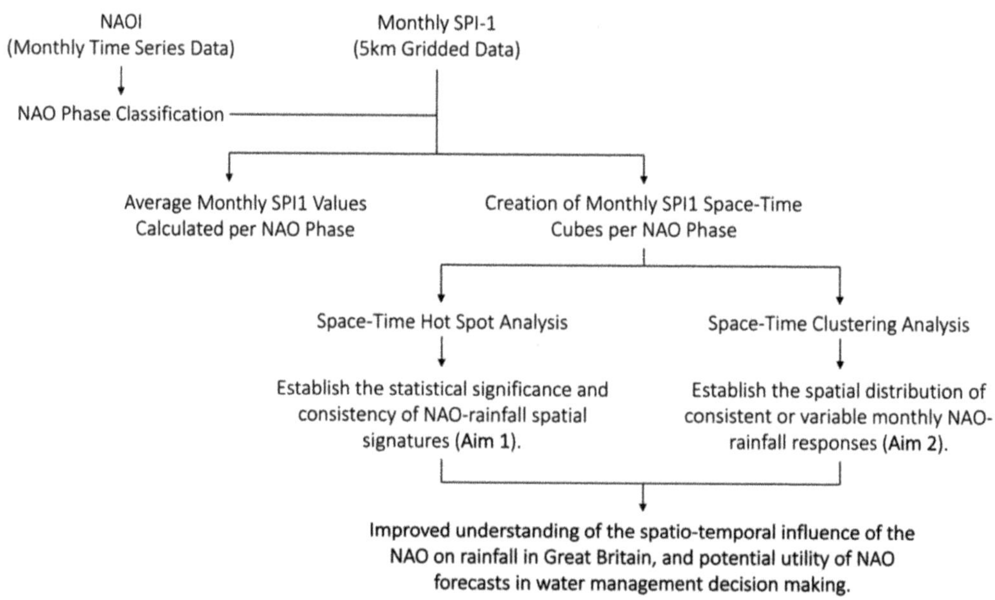

Figure 1. Flow chart of the analytical stages of this study.

Our ability to predict the NAO, especially during its stronger phases during the winter months [24,25], has improved. In particular, recent research from the National Center for Atmospheric Research (NCAR) and the UK Met Office suggests that winter NAO prediction several months in advance may be increasingly possible [26,27]. As NAO forecasting skill improves, it is important that we continue to develop our spatial and temporal understanding of the rainfall signatures associated with the circulation, especially as evidence of the propagation of NAO rainfall deviations to other hydrometeorological variables, such as runoff and groundwater, continues to emerge.

Whilst previous studies have explored average monthly and seasonal NAO-rainfall responses across Great Britain [1,10,13], as far as we are aware, no study has quantified the spatial influence of the NAO on rainfall distributions, and particularly the probability of wet and dry conditions/extremes. Understanding NAO-rainfall variability is important as it underpins the potential application of NAO forecasts in water management decision making. This study responds to this need through the novel application of spatio-temporal statistics and high spatial and temporal resolution Standardised Precipitation Index (SPI) data. Specifically, we aim to:

1. Assess the statistical significance and consistency of NAO-rainfall spatial signatures.
2. Establish the spatial distribution of consistent or variable monthly NAO-rainfall responses.

2. Materials and Methods

Figure 1 summarises the key stages of this analysis to explore space-time variability in monthly NAO-rainfall response across Great Britain, which are explained in full below.

2.1. Data

A range of indices have been used in previous work to quantify the phase and strength of the NAO [1,10,12,18], and the choice of NAOI can have a notable influence on subsequent analyses [28]. As this study focuses on quantifying monthly NAO-rainfall response variability across a full year, an NAOI calculated using a principal components (PC) analysis of the leading empirical orthogonal function of sea level pressure anomalies in the North Atlantic region was used [3]. These NAO indices avoid the limitation of indices directly derived using station-measured data in the summer months when the NAO dipoles can move away from the monitoring stations [28]. As a result, the use of station-based NAO indices in summer months can produce weaker and non-significant NAOI-rainfall correlation analyses [10]. Monthly PC-based NAOI data were downloaded for the period January 1900–December 2015 from NCAR [29].

Previous studies have used different approaches to defining NAO+ and NAO− phases using the NAOI. In this study, the NAO phase was defined as half the standard deviation plus/minus the long-term mean of the NAOI dataset [30]. By identifying NAO phases in this way, we could remove months where the NAO signal is weak, allowing only clear NAO+ and NAO− rainfall responses to be considered. NAO+ phases were defined as having a NAOI > 0.502 and NAO− < -0.503 (data between this range were classified as NAO neutral and removed). Table 1 shows the distribution of NAO phase and months for the period January 1900–December 2015.

We used the 5 km gridded Standardised Precipitation Index (SPI) dataset from the UK Centre for Ecology and Hydrology (CEH) [31] to represent precipitation. Standardised indices have been widely used in research exploring the hydrometeorological response to atmospheric-oceanic teleconnections [10,32,33]. The SPI dataset from CEH was calculated by fitting a gamma distribution to modelled historical rainfall using a standard period of 1961–2010 [31]. For this study, the SPI calculated with a one-month accumulation period (SPI-1) was used to give a relative indication of wetness/dryness compared to the standard period at a monthly scale. The SPI-1 data for the same period as the monthly NAOI from NCAR (January 1900–December 2015) were downloaded from CEH.

Table 1. Frequency Distribution of NAO Phases per Month (January 1900–December 2015).

Period	NAO+ Frequency	NAO− Frequency	NAO Neutral Frequency
Dec	47	40	29
Jan	52	36	28
Feb	47	40	29
WINTER	**146**	**116**	**86**
Mar	45	38	33
Apr	29	32	55
May	28	24	64
SPRING	**102**	**94**	**152**
Jun	26	24	66
Jul	16	16	84
Aug	16	19	81
SUMMER	**58**	**59**	**231**
Sep	19	21	76
Oct	31	33	52
Nov	29	31	56
AUTUMN	**79**	**85**	**184**

The SPI-1 values are normally distributed and can range from −5 to 5, although approximately 95% of values occur within the range of −2 (extremely dry) to 2 (extremely wet), and 68% within the range of −1 to 1 [31]. We use a qualitative classification of SPI-1 thresholds to indicate the relative degree of wetness/dryness (Table 2) in the interpretation of our results.

Table 2. Qualitative Descriptors for SPI-1 Values adapted from [34].

Qualitative Descriptor	SPI-1 Value Range
Extremely Wet	2.0–5.0
Very Wet	1.5–1.99
Moderately Wet	1.0–1.49
Near Normal	−0.99–0.99
Moderately Dry	−1.49–−1.0
Severely Dry	−1.99–−1.5
Extremely Dry	−5.0–−2.0

2.2. *Calculation of Monthly Average SPI-1 Values*

To provide a point of comparison, before undertaking the NAO-rainfall space-time variability analysis described below, we mapped the average monthly SPI-1 values under NAO+ and NAO− phases over the period January 1900–December 2015. The average monthly SPI-1 value under each phase was calculated for each 5 km pixel. This analysis provided a dataset of average rainfall conditions (represented by the SPI-1) for each calendar month under NAO+ and NAO− conditions.

2.3. *Space-Time Data Array*

To undertake the spatio-temporal analyses, the monthly SPI-1 data were structured into a series of data arrays (a space-time cube). Each array contained the monthly 5 km gridded SPI-1 data stacked in time ascending order for each NAO phase. The number of time steps in each space-time cube equalled the monthly-phase frequency values in Table 1. A space-time cube was created for each month under NAO+ and NAO− conditions using Esri ArcGIS Pro software (Version 2.6).

2.4. *Space-Time Hot Spot Analysis (Getis Ord Gi* Statistic)*

The Getis-Ord Gi* statistic [35] identifies clusters of significantly high and low values within a spatial dataset—identified as hot or cold spots with varying significance levels (90, 95 or 99%). The statistic indicates whether the spatial clustering of high/low values is more pronounced than would be expected in a random spatial distribution of those same values [36]. The Getis-Ord Gi* statistic is commonly used in spatial statistical analyses in health and crime applications [37,38]. This study's application of it in the field of hydrometeorology/climatology is novel.

As the Getis-Ord Gi* statistic has a spatial component, its calculation requires consideration of how pixels are spatially related, so that high/low value clusters can be identified (i.e., a conceptualisation of spatial relationships between the 5 km SPI-1 pixels). Because of the spatially continuous gridded SPI-1 data, we used the 'edges and corners' or 'queens' contiguity rule, where each pixel's statistical neighbourhood included the eight pixels which share a boundary or point of contact in the four cardinal and diagonal directions. Hot spots in this context represent clusters of pixels where the SPI-1 values are significantly high, whilst cold spots represent the inverse (low SPI-1 values). The Getis-Ord Gi* statistic therefore allows for the identification of statistically significant wet or dry (high SPI-1/low SPI-1) spatial patterns.

The Getis-Ord Gi* statistic was calculated for each time step (SPI-1 dataset) in the space–time cubes for NAO+ and NAO− conditions. The output is a multi-variate dataset, indicating the percentage time each pixel is in either a significant hot (high SPI-1 value—wet) or cold spot (low SPI-1 value—dry). This enables an evaluation of the statistical significance and consistency of the spatial rainfall (wet/dry) response to the NAO across Great Britain and provides an assessment as to the spatial probability of significant wet/dry conditions under NAO+ and NAO− phases.

2.5. *Space-Time Clustering Analysis*

The second spatio-temporal analysis was a space-time clustering process [39], which grouped 5 km pixels with similar SPI-1 values across space, time, and NAO phase for each month. The clustering used a k-means algorithm with spatially random starting seeds. During the cluster calculation, SPI-1 time series similarity was assessed using the Euclidean distance between the SPI-1 time series values (the square root of the sum of squared differences in SPI-1 values across time) [39]. The algorithm produced 90 space-time clustering solutions, with potential results between two and ten output clusters. The optimal clustering solution was identified as that which had the highest pseudo-F statistic. This statistic describes the within-cluster SPI-1 time series similarity and between-cluster SPI-1 time series difference—the larger the pseudo-F statistic of the clustering solution, the greater the distinctiveness of each individual cluster in space and time [39].

The range of the average SPI-1 values within each space-time cluster was investigated using box plots, frequency histograms and descriptive statistics, quantifying the distribution of SPI-1 within clusters and providing a measure of the consistency/variability in space-time rainfall response to the phase of the NAO.

3. Results

3.1. *Average Monthly SPI-1 Values*

Figures 2 and 3 present the average monthly SPI-1 values under NAO+ and NAO− conditions. Whilst these are average monthly SPI-1 values and mask extreme values, clear spatial and temporal signatures of the NAO can be detected. In the winter months, the NAO rainfall response described in the introduction can be observed [1,7,12,13]. Under NAO+ conditions, the north-west areas have higher/positive average SPI-1 values, indicating wet conditions, and under NAO− conditions, the region is typically dry (negative SPI-1 values). The inverse average wet/dry response, albeit weaker, is seen in the southern and eastern areas.

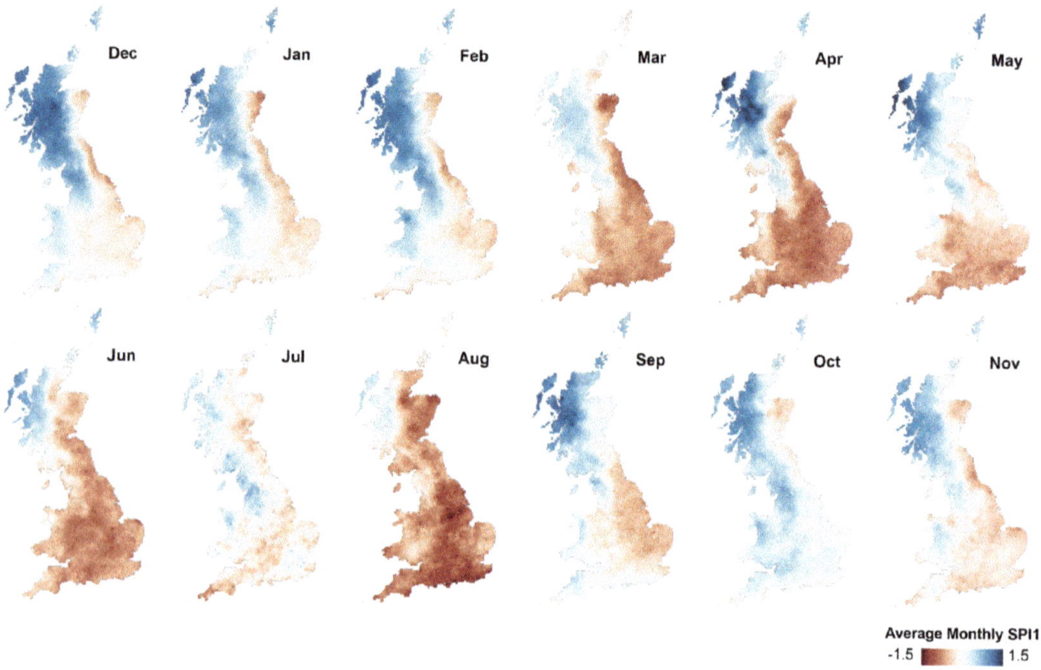

Figure 2. Average monthly SPI-1 values under NAO+ conditions (January 1900–December 2015).

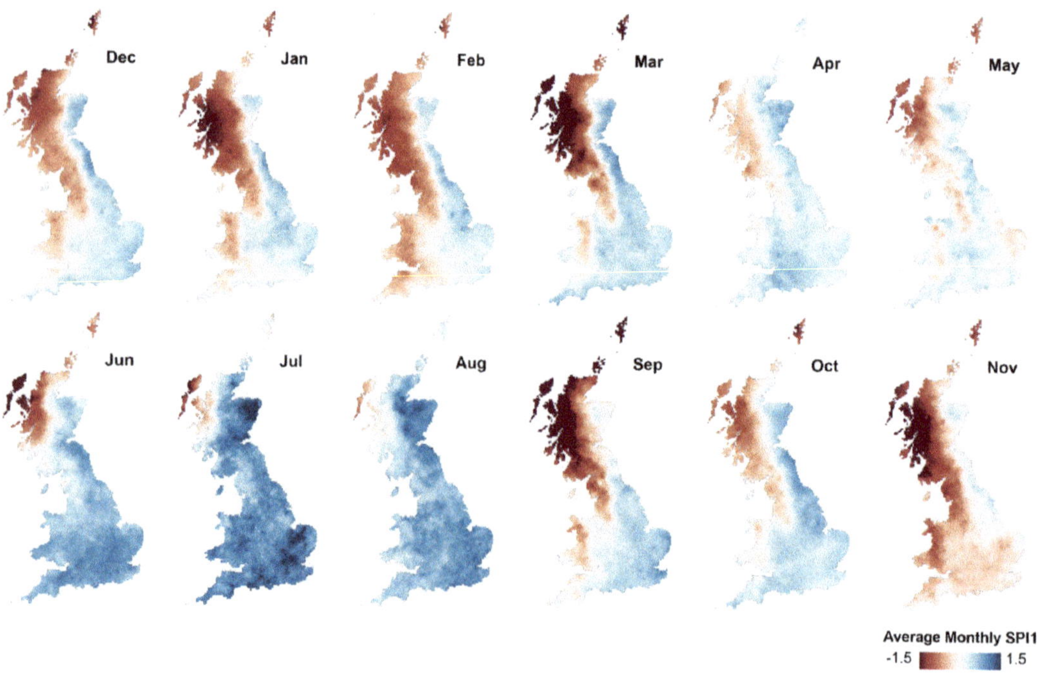

Figure 3. Average monthly SPI-1 values under NAO− conditions (January 1900–December 2015).

Moving through to the summer months, the average NAO-rainfall response becomes more spatially homogeneous (i.e., there is less difference between the different areas of the country), and the wet/dry response in the north-west during winter is inverted; a spatio-temporal pattern also noted in other studies [10,12,13].

3.2. Space-Time Hot Spot Results

Figures 4 and 5 show the results of the space-time hot spot analysis. This analysis involved calculating the Getis-Ord Gi* for each time step (SPI-1 dataset) of the month/phase space–time cubes as described in Section 2.3, allowing us to examine the statistical significance and consistency of the patterns discussed in Section 3.1 above. The mapped results in Figures 4 and 5 below show the percentage of time each 5 km pixel was in a statistically significant hot spot (i.e., a cluster of pixels with high/wet SPI-1 values) or cold spot (i.e., a cluster of pixels low/dry SPI-1 values). These maps indicate the spatial probability of significant wet/dry conditions under NAO+ and NAO− phases. Averages of these results for the nine Met Office Climate Districts for Great Britain are shown in Figure 6.

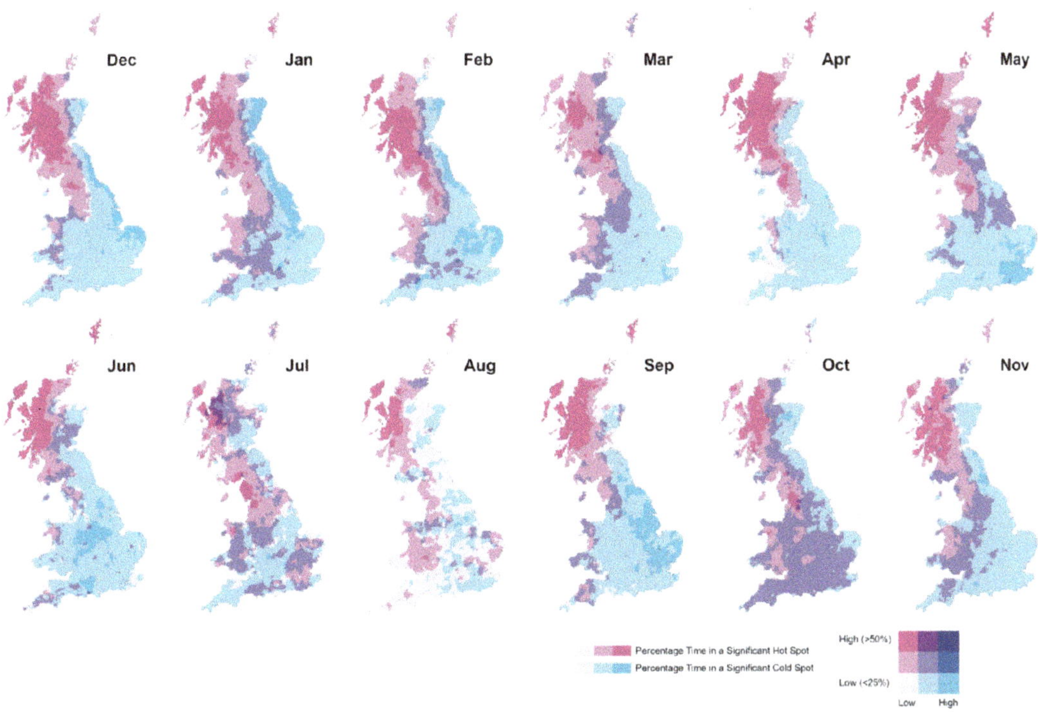

Figure 4. Space-time hot spot results under NAO+ conditions. Pinks indicate where the percentage of time in a significant hot spot is high (indicating wetter conditions), and light blues where the percentage of time in a significant cold spot is high (indicating drier conditions). Light grey indicates low occurrence of both hot and cold spots, whilst mixed shades indicate occurrence of both hot and cold spots. Disaggregated versions of these maps can be found in Appendix A.

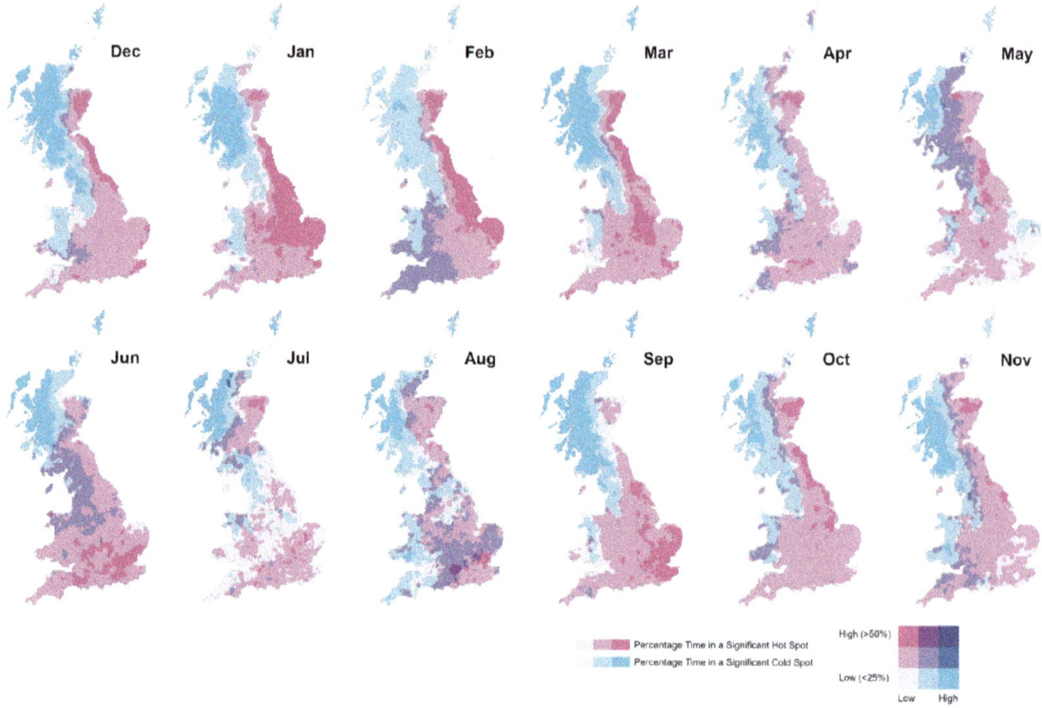

Figure 5. As per Figure 4 but under NAO− conditions.

The space-time hot spot analysis reveals statistically significant spatio-temporal patterns in the NAO-rainfall response across Great Britain. The winter months are marked by the previously noted north-west/south-east spatial divide of opposing wet/dry responses to the NAO, as identified in other studies [1,10,18]. In our analysis (Figures 4 and 5), this pattern is shown by the location of widespread hot and cold spots in the NW and SE areas, indicating significant clusters of pixels with high/low (wet/dry) SPI-1 values. This suggests that the wet/dry spatial pattern we detect in winter rainfall associated with the phase of the NAO is statistically significant.

Looking across the record, this NW/SE opposing response appears to have a relatively high degree of consistency in some areas, for example, Scotland North and West, and to a slightly lesser extent England South West and Central South and East Anglia are in a statistically significant wet/dry cluster (hot/cold spot) for a relatively high proportion of the analysis period (Figure 6). This suggests that there is a higher probability that the NAO+ and NAO− phases will result in this statistically significant winter NW/SE spatial pattern. However, it should be noted that in winter, some areas, for example Scotland East and England North West and Wales North, show little or no difference between the occurrence of significant clusters of wet/dry SPI-1 values (hot/cold spots). Therefore, the effect of the NAO in producing significant wet/dry spatial patterns in these Climate Districts is more limited.

Figures 2 and 3 show that the spatial rainfall response to the NAO is on average more homogeneous in the summer months [10,12,13]. As a result, less discernible patterns are found in the hot spot analysis, and the occurrence of statistically significant hot/cold spots (i.e., significant clusters of high/low SPI-1 values) is more variable in space and time (Figures 4 and 5). Most Climate Districts show minor differences in the occurrence of significant clusters of wet/dry SPI-1 values in summer, except for Scotland North and West (Figure 6).

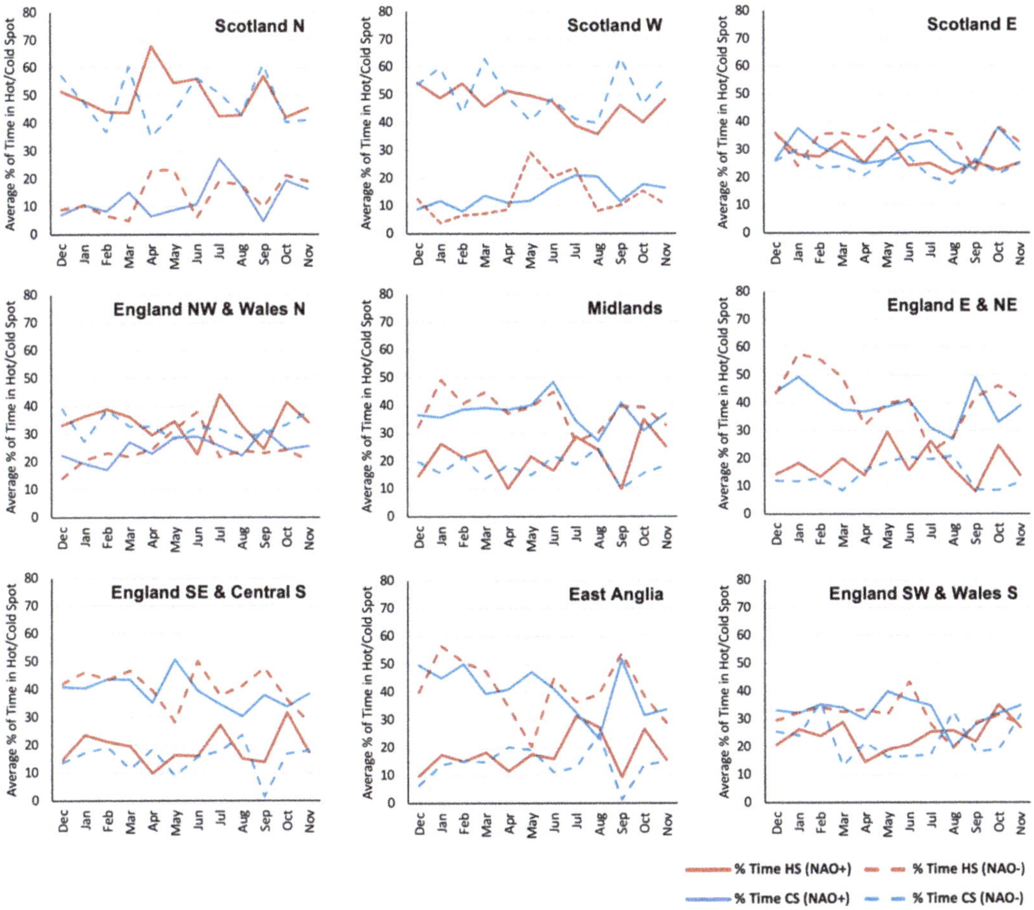

Figure 6. Regional average percentage of time in a significant hot/cold spot (based on the mapped results in Figures 4 and 5). Regions based on the Met Office Climate Districts for Great Britain.

3.3. *Space-Time Clustering Results*

Figures 7 and 8 present the space-time clustering results. This analysis explored the variability around the average NAO responses identified in Figures 2 and 3 (discussed in Section 3.1 above). Each cluster represents spatial groupings of 5 km pixels which have a similar response to the NAO phase during that month across the temporal record. The optimal number of clusters based on the pseudo-F statistic was consistently three. In Figures 7 and 8, the more saturated the blue/red colour, the more distinctive the space-time cluster wet/dry response to the NAO (i.e., the space-time median value is wetter/drier). Less saturated clusters indicate a space-time median SPI-1 value closer to 0, suggesting a less distinctive wet/dry average rainfall response to the NAO across the temporal record. The distributions of cluster median SPI-1 values per time step are plotted in Figure 9, and the associated descriptive statistics are shown in Figure 10. Figures 11 and 12 show frequency histograms of the cluster median SPI-1 values and the percentage of time each cluster experiences wet/dry conditions.

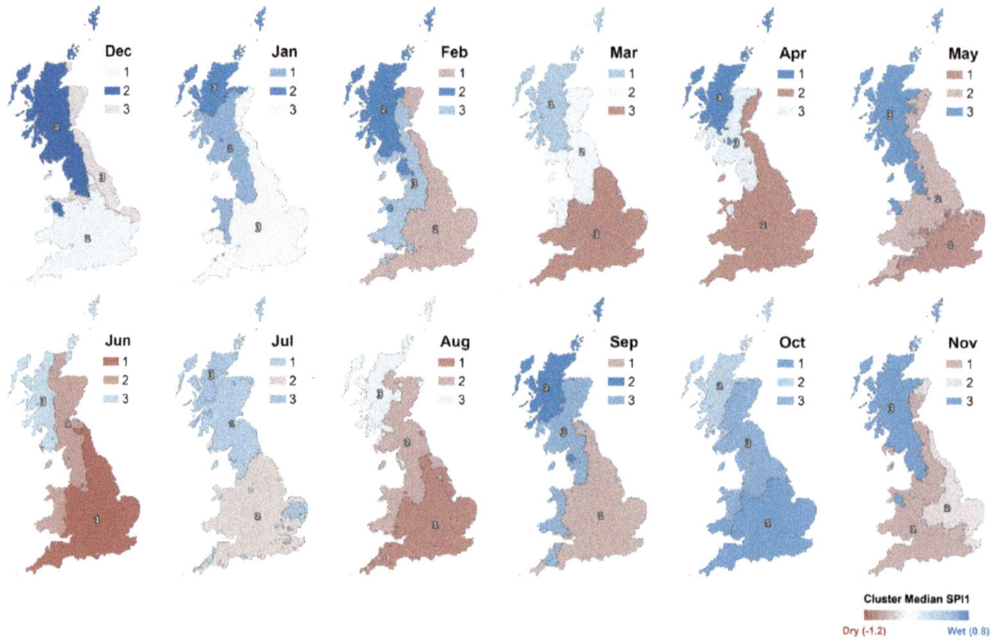

Figure 7. Monthly space-time clusters of SPI-1 values under NAO+ conditions. The clusters are coloured based on the median SPI-1 value of the cluster in space and time. Blue indicates a wetter median, whereas red indicates a drier median. Less saturated clusters indicate median SPI-1 values closer to 0.

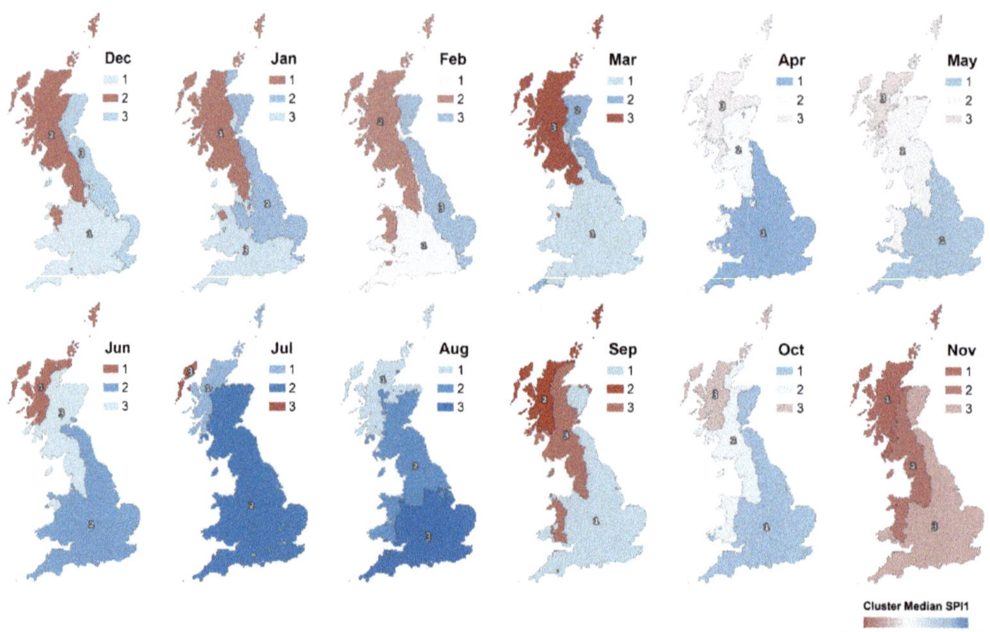

Figure 8. As per Figure 7 but showing the monthly space-time clusters of SPI-1 values under NAO− conditions.

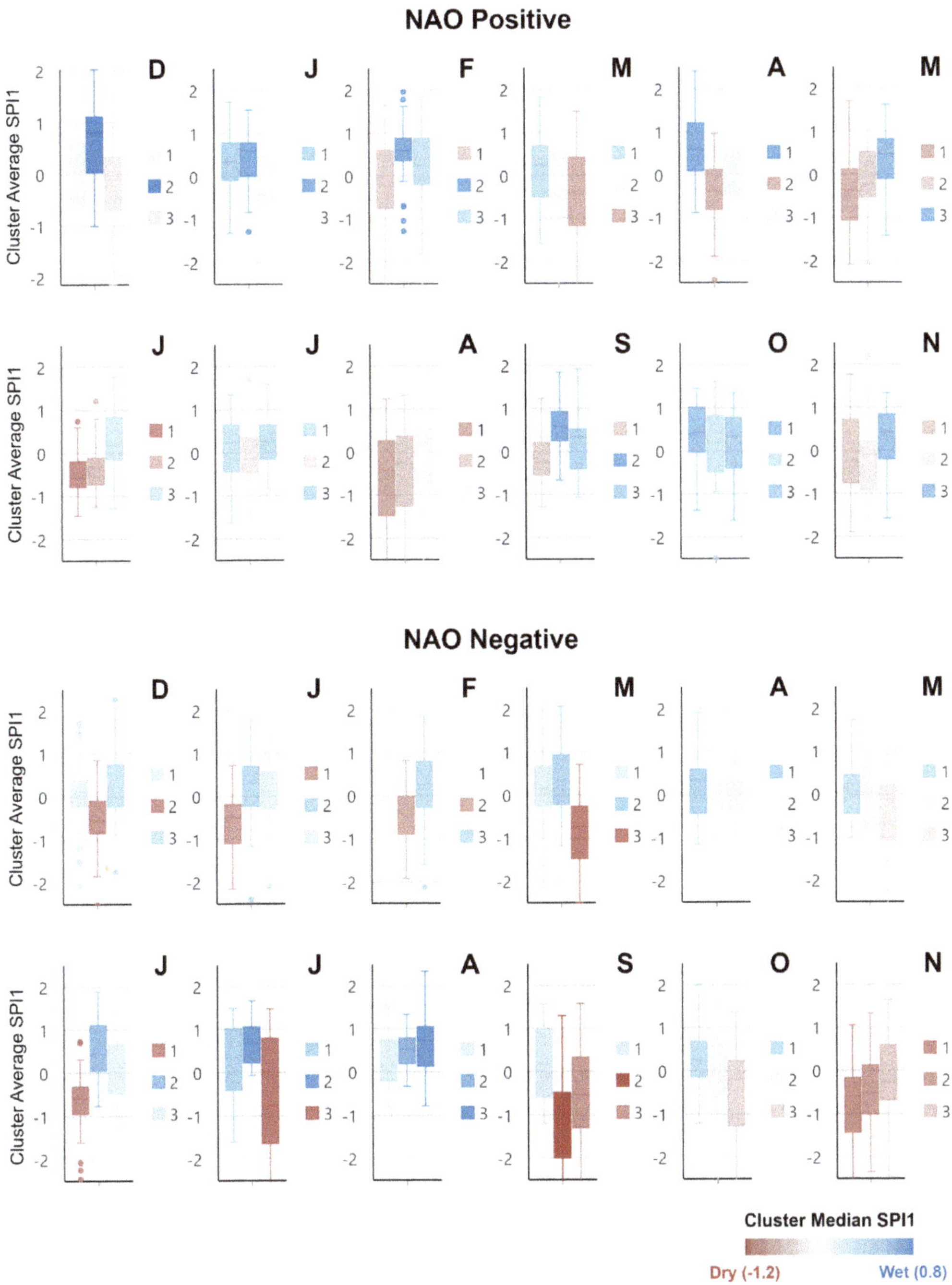

Figure 9. Box plots representing the distribution of cluster median SPI-1 values for the space-time clusters mapped in Figures 7 and 8. As with Figures 7 and 8 the box plots are coloured based on the median SPI-1 value of the cluster in space and time. Individual points represent outlier values.

	Dec NAO Positive			Dec NAO Negative		
	1	2	3	1	2	3
Median	0.044	0.802	-0.114	0.115	-0.551	0.219
Mean	-0.018	0.567	-0.101	0.072	-0.518	0.250
Minimum	-1.891	-1.009	-2.134	-0.208	-2.516	-1.755
Maximum	2.010	2.028	1.868	1.733	0.852	2.263
Range	3.901	3.037	4.002	1.941	3.367	4.018
First Quartile	-0.667	0.037	-0.711	-0.219	-0.844	-0.229
Third Quartile	0.631	1.114	0.335	0.400	-0.093	0.757
IQR	1.298	1.078	1.046	0.619	0.751	0.986
STD	0.966	0.740	0.920	0.905	0.742	0.822
Variance	0.933	0.547	0.846	0.819	0.551	0.675

	Jan NAO Positive			Jan NAO Negative		
	1	2	3	1	2	3
Median	0.345	0.597	-0.010	-0.497	0.304	0.130
Mean	0.318	0.450	-0.093	-0.646	0.215	-0.037
Minimum	-1.321	-1.289	-1.821	-2.655	-2.395	-2.797
Maximum	1.736	1.541	1.305	0.719	1.771	1.373
Range	3.057	2.830	3.126	3.374	4.166	4.170
First Quartile	-0.089	0.004	-0.823	-1.097	-0.237	-0.301
Third Quartile	0.793	0.784	0.658	-0.179	0.707	0.571
IQR	0.882	0.780	1.481	0.918	0.944	0.872
STD	0.678	0.601	0.852	0.751	0.833	0.914
Variance	0.459	0.362	0.726	0.565	0.695	0.835

	Feb NAO Positive			Feb NAO Negative		
	1	2	3	1	2	3
Median	-0.240	0.598	0.250	0.005	-0.421	0.274
Mean	-0.117	0.568	0.286	-0.112	-0.541	0.232
Minimum	-2.610	-1.279	-0.257	-2.980	-3.149	-2.123
Maximum	1.642	1.950	1.835	1.682	0.813	1.864
Range	4.252	3.230	2.092	4.663	3.962	3.987
First Quartile	-0.758	0.359	-0.198	-0.667	-0.891	-0.261
Third Quartile	0.608	0.881	0.886	0.611	-0.006	0.804
IQR	1.367	0.522	1.084	1.279	0.885	1.065
STD	0.966	0.656	0.910	1.102	0.877	0.875
Variance	0.933	0.430	0.829	1.214	0.768	0.765

	Mar NAO Positive			Mar NAO Negative		
	1	2	3	1	2	3
Median	0.246	0.028	-0.394	0.153	0.342	-0.729
Mean	0.166	-0.190	-0.396	0.152	0.342	-0.819
Minimum	-1.587	-2.561	-2.899	-2.119	-1.180	-2.505
Maximum	1.826	1.313	1.495	2.198	2.071	0.716
Range	3.413	3.874	4.393	4.317	3.250	3.220
First Quartile	-0.496	-0.971	-1.164	-0.259	-0.218	-1.479
Third Quartile	0.704	0.607	0.426	0.677	0.941	-0.259
IQR	1.200	1.578	1.590	0.936	1.159	1.220
STD	0.803	0.995	1.097	0.848	0.721	0.818
Variance	0.644	0.990	1.204	0.718	0.520	0.669

	Apr NAO Positive			Apr NAO Negative		
	1	2	3	1	2	3
Median	0.590	-0.380	0.062	0.349	0.010	-0.057
Mean	0.606	-0.459	0.052	0.215	0.007	-0.036
Minimum	-0.875	-2.754	-1.543	-1.150	-1.171	-1.375
Maximum	2.414	0.958	1.622	1.911	1.237	1.350
Range	3.289	3.712	3.165	3.061	2.408	2.725
First Quartile	0.100	-0.813	-0.456	-0.439	-0.392	-0.356
Third Quartile	1.215	0.134	0.632	0.593	0.514	0.352
IQR	1.116	0.947	1.088	1.032	0.906	0.708
STD	0.826	0.862	0.796	0.764	0.624	0.589
Variance	0.682	0.743	0.634	0.583	0.389	0.347

	May NAO Positive			May NAO Negative		
	1	2	3	1	2	3
Median	-0.377	-0.223	0.475	0.243	-0.027	-0.117
Mean	-0.384	-0.163	0.347	0.090	-0.046	-0.485
Minimum	-2.096	-2.072	-1.421	-1.007	-1.474	-2.544
Maximum	1.708	1.040	1.634	1.717	1.547	0.935
Range	3.804	3.111	3.056	2.724	3.021	3.479
First Quartile	-1.073	-0.534	-0.094	-0.452	-0.841	-1.081
Third Quartile	0.130	0.532	0.826	0.452	0.628	0.216
IQR	1.202	1.066	0.920	0.904	1.468	1.297
STD	0.901	0.852	0.705	0.648	0.881	0.985
Variance	0.811	0.725	0.497	0.421	0.776	0.971

	Jun NAO Positive			Jun NAO Negative		
	1	2	3	1	2	3
Median	-0.581	-0.326	0.184	-0.591	0.441	0.119
Mean	-0.468	-0.268	0.269	-0.684	0.530	0.172
Minimum	-1.453	-1.249	-1.286	-2.449	-0.762	-1.301
Maximum	0.742	1.213	1.770	0.726	1.895	1.897
Range	2.195	2.461	3.055	3.175	2.657	3.198
First Quartile	-0.788	-0.732	-0.150	-0.943	0.051	-0.479
Third Quartile	-0.183	-0.101	0.842	-0.312	1.110	0.658
IQR	0.605	0.631	0.992	0.631	1.060	1.137
STD	0.540	0.636	0.763	0.840	0.757	0.841
Variance	0.291	0.405	0.583	0.705	0.574	0.707

	Jul NAO Positive			Jul NAO Negative		
	1	2	3	1	2	3
Median	0.233	-0.133	0.268	0.338	0.683	-0.758
Mean	0.025	-0.072	0.087	0.265	0.656	-0.772
Minimum	-1.624	-1.603	-3.140	-1.606	-0.065	-4.827
Maximum	1.347	1.696	1.593	1.493	1.662	1.474
Range	2.971	3.298	4.733	3.099	1.727	6.300
First Quartile	-0.452	-0.484	-0.158	-0.413	0.223	-1.646
Third Quartile	0.655	0.363	0.652	1.023	1.060	0.796
IQR	1.107	0.848	0.809	1.436	0.836	2.442
STD	0.846	0.849	1.089	0.933	0.534	1.826
Variance	0.716	0.720	1.186	0.871	0.285	3.335

	Aug NAO Positive			Aug NAO Negative		
	1	2	3	1	2	3
Median	-0.422	-0.250	0.049	0.156	0.540	0.732
Mean	-0.568	-0.441	-0.087	0.242	0.487	0.556
Minimum	-2.837	-2.565	-2.554	-0.792	-0.330	-0.782
Maximum	1.238	1.314	1.739	1.539	1.337	2.344
Range	4.075	3.879	4.293	2.331	1.667	3.125
First Quartile	-1.478	-1.260	-0.744	-0.206	0.199	0.132
Third Quartile	0.271	0.379	0.588	0.756	0.796	1.050
IQR	1.749	1.639	1.333	0.963	0.597	0.918
STD	1.085	1.055	1.163	0.618	0.484	0.732
Variance	1.178	1.114	1.352	0.382	0.234	0.536

	Sep NAO Positive			Sep NAO Negative		
	1	2	3	1	2	3
Median	-0.246	0.590	0.343	0.172	-1.219	-0.537
Mean	-0.169	0.581	0.161	0.193	-1.257	-0.502
Minimum	-1.280	-0.665	-1.064	-1.201	-3.027	-2.651
Maximum	1.238	1.840	1.923	1.575	1.295	1.574
Range	2.518	2.505	2.988	2.776	4.323	4.226
First Quartile	-0.556	0.255	-0.407	-0.589	-2.002	-1.306
Third Quartile	0.214	0.935	0.527	0.980	-0.483	0.334
IQR	0.770	0.680	0.934	1.569	1.518	1.640
STD	0.684	0.565	0.733	0.864	1.029	1.127
Variance	0.467	0.319	0.537	0.746	1.058	1.270

	Oct NAO Positive			Oct NAO Negative		
	1	2	3	1	2	3
Median	0.435	0.206	0.344	0.269	0.042	-0.197
Mean	0.428	0.106	0.161	0.306	-0.128	-0.503
Minimum	-1.375	-2.658	-1.610	-1.209	-2.646	-3.936
Maximum	1.467	1.617	1.355	2.014	1.550	1.368
Range	2.842	4.275	2.965	3.224	4.196	5.304
First Quartile	-0.028	-0.495	-0.406	-0.137	-0.591	-1.289
Third Quartile	1.023	0.829	0.781	0.691	0.621	0.240
IQR	1.051	1.323	1.186	0.828	1.213	1.529
STD	0.718	0.993	0.844	0.724	0.899	1.165
Variance	0.515	0.985	0.713	0.524	0.808	1.356

	Nov NAO Positive			Nov NAO Negative		
	1	2	3	1	2	3
Median	-0.249	-0.101	0.434	-0.587	-0.512	-0.268
Mean	-0.075	-0.225	0.286	-0.815	-0.537	-0.131
Minimum	-1.899	-1.952	-1.568	-3.504	-3.628	-2.448
Maximum	1.766	2.201	1.344	1.058	1.327	1.634
Range	3.664	4.152	2.913	4.563	4.955	4.083
First Quartile	-0.763	-0.927	-0.209	-1.434	-1.013	-0.686
Third Quartile	0.723	0.216	0.849	-0.162	0.129	0.593
IQR	1.486	1.143	1.059	1.273	1.141	1.279
STD	0.902	0.937	0.734	1.032	1.115	0.972
Variance	0.814	0.879	0.538	1.066	1.244	0.944

Figure 10. Descriptive statistics for the space-time cluster median SPI-1 box plots in Figure 9 (IQR = interquartile range; STD = standard deviation). Formatting of colours is applied per statistical measure. Blue indicates wetter mean/median SPI-1 values, red indicates drier mean/median SPI-1 values. Green graduated shading indicates greater range/IQR values.

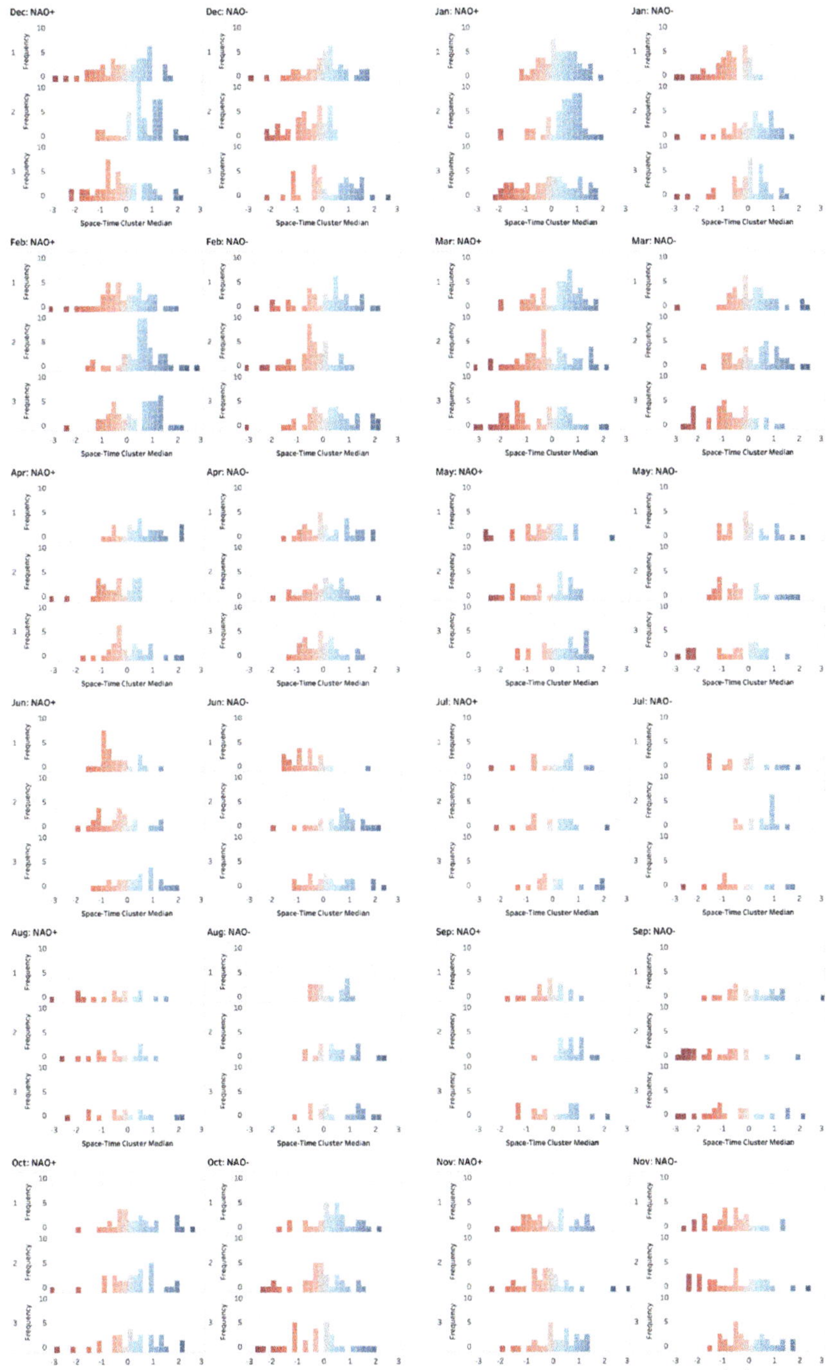

Figure 11. Frequency histograms of median SPI-1 values for the space–time clusters mapped in Figures 7 and 8.

	Cluster	Dry Conditions (%)	Wet Conditions (%)	Difference (%)
Dec NAO-	1	42.5	57.5	-15.0
	2	72.5	27.5	45.0
	3	45.0	55.0	-10.0
Dec NAO+	1	50.0	50.0	0.0
	2	17.4	82.6	-65.2
	3	60.9	39.1	21.7
Jan NAO-	1	83.3	16.7	66.7
	2	38.9	61.1	-22.2
	3	41.7	58.3	-16.7
Jan NAO+	1	30.8	69.2	-38.5
	2	21.2	78.8	-57.7
	3	51.9	48.1	3.8
Feb NAO-	1	40.0	60.0	-20.0
	2	70.0	30.0	40.0
	3	40.0	60.0	-20.0
Feb NAO+	1	53.2	46.8	6.4
	2	19.1	80.9	-61.7
	3	36.2	63.8	-27.7
Mar NAO-	1	47.4	52.6	-5.3
	2	36.8	63.2	-26.3
	3	78.9	21.1	57.9
Mar NAO+	1	33.3	66.7	-33.3
	2	57.8	42.2	15.6
	3	60.0	40.0	20.0
Apr NAO-	1	46.9	53.1	-6.3
	2	43.8	56.3	-12.5
	3	56.3	43.8	12.5
Apr NAO+	1	27.6	72.4	-44.8
	2	65.5	34.5	31.0
	3	55.2	44.8	10.3
May NAO-	1	50.0	50.0	0.0
	2	58.3	41.7	16.7
	3	54.2	45.8	8.3
May NAO+	1	60.7	39.3	21.4
	2	46.4	53.6	-7.1
	3	32.1	67.9	-35.7

	Cluster	Dry Conditions (%)	Wet Conditions (%)	Difference (%)
Jun NAO-	1	87.5	12.5	75.0
	2	29.2	70.8	-41.7
	3	41.7	58.3	-16.7
Jun NAO+	1	76.9	23.1	53.8
	2	73.1	26.9	46.2
	3	38.5	61.5	-23.1
Jul NAO-	1	50.0	50.0	0.0
	2	18.8	81.3	-62.5
	3	62.5	37.5	25.0
Jul NAO+	1	43.8	56.3	-12.5
	2	50.0	50.0	0.0
	3	50.0	50.0	0.0
Aug NAO-	1	50.0	50.0	0.0
	2	33.3	66.7	-33.3
	3	27.8	72.2	-44.4
Aug NAO+	1	68.8	31.3	37.5
	2	62.5	37.5	25.0
	3	62.5	37.5	25.0
Sep NAO-	1	47.6	52.4	-4.8
	2	90.5	9.5	81.0
	3	71.4	28.6	42.9
Sep NAO+	1	63.2	36.8	26.3
	2	5.3	94.7	-89.5
	3	42.1	57.9	-15.8
Oct NAO-	1	27.3	72.7	-45.5
	2	60.6	39.4	21.2
	3	51.5	48.5	3.0
Oct NAO+	1	45.2	54.8	-9.7
	2	38.7	61.3	-22.6
	3	38.7	61.3	-22.6
Nov NAO-	1	79.3	20.7	58.6
	2	62.1	37.9	24.1
	3	55.2	44.8	10.3
Nov NAO+	1	56.7	43.3	13.3
	2	66.7	33.3	33.3
	3	43.3	56.7	-13.3

Figure 12. Percentage of time the cluster median SPI-1 value is wet (positive SPI-1 values) and dry (negative SPI-1 values) based on the frequency histograms in Figure 11. The difference value represents the difference between the percentage of time under dry conditions and wet conditions.

3.4. *Examples of Consistent Monthly NAO-Rainfall Responses*

In the space-time clustering analysis, the NAO-rainfall response observed in the average monthly SPI-1 analysis (Figures 2 and 3) comes through clearly. For example, the north-west/south-east spatial divide in winter rainfall response [10] can be seen in Figures 4 and 5. During the winter months (DJF), the north-western areas experience the greatest change in rainfall under NAO+ and NAO− phases. In December, for example, the extremes of this response can be seen in NAO+ Cluster 2, which has a maximum cluster median SPI-1 value of 2 (extremely wet), and NAO− Cluster 2 which has a minimum cluster median SPI-1 value of −2.5 (extremely dry) (Figures 9 and 10).

Clusters covering these north-western areas in winter also show a relatively more consistent NAO-rainfall response compared to other parts of the country. For example, Dec NAO+ Cluster 2 has a 65% chance of being relatively wetter than drier, whilst the similarly located Dec NAO− Cluster 2 has a 45% probability of experiencing drier rather than wetter conditions (Figures 11 and 12). The interquartile range (IQR) of these two clusters also only include wet/dry cluster median SPI-1 values (Figure 9). Similar results for the north-western areas can be seen in January, for example the IQR for NAO− Cluster 1 only includes negative SPI-1 (dry) values (Figure 9).

In spring (MAM), similar spatial patterns to those described above can be seen, but with a north/south gradient in the clustering. As in winter, the median responses (Figures 7 and 8) match with the average monthly values mapped in Figures 2 and 3. For example, March NAO− Cluster 3 in north-western Scotland typically experiences notably dry conditions (Figure 8), with a 57% probability of experiencing drier rather than wetter conditions (Figures 11 and 12). This cluster also has a notably low minimum cluster median value of −2.5 representing extremely dry conditions.

In summer (JJA), the more spatially homogeneous wet/dry NAO−/NAO+ responses in Figures 2 and 3 can be seen in the cluster median values. Notably, the wet/dry directionality is the opposite to the NAO-rainfall response seen in the north-western areas during winter. For example, in June, NAO+ Clusters 1 and 2 which cover most of Great Britain (except for a cluster in the far north-west-Figure 7) show a relatively consistent dry response, with interquartile ranges (IQR) covering negative/dry SPI-1 values (Figure 9) and a 53% and 46% probability of drier rather than wetter conditions (Figures 11 and 12).

3.5. Examples of Variable Monthly NAO-Rainfall Responses

The space-time clustering analysis also reveals that whilst typical and relatively consistent NAO response signals can be observed, there is also significant NAO-rainfall response variability within some of the space-time clusters.

Even in areas that show relative consistency, variability in the cluster median SPI-1 values can still be observed (Figure 9). For example, in December, NAO− Cluster 2 in the north-western region (Figure 8), the minimum cluster median SPI-1 value is −2.5 which represents extremely dry conditions; however, the maximum is an SPI-1 value representing near-normal conditions (0.8). Whilst drier conditions are more likely in this cluster (72.5%), wet conditions were present for 27.5% of the time period analysed (Figures 11 and 12). In February, NAO+ Cluster 2, similarly covering the north-western area, the maximum cluster median SPI-1 value is 1.9 (severely wet) and the IQR covers positive values only. Whilst wetter than average conditions were found in this cluster for the majority of the time period (80%), the cluster did experience relatively dry conditions for 20% of the time period (Figures 11 and 12). These findings suggest that whilst the typical winter NAO responses can be observed in the north-western area, there is also some variability in both the magnitude of the NAO-rainfall response and the directionality (i.e., positive/wet, or negative/dry SPI-1 values).

The rainfall response in clusters covering the central, southern and eastern areas of Great Britain support the average winter NAO-rainfall response mapped in Figures 2 and 3. However, in comparison to the north-west, the winter rainfall response to NAO+ and NAO− phases is much more variable in clusters spanning these areas (Figures 7 and 8). For example, NAO+ Cluster 1 in December, has a notably large value range, with cluster median SPI-1 values ranging from a maximum of 2 (extremely wet) to −1.8 (severely dry) (Figures 9 and 10). The cluster median SPI-1 histograms also show a more normal distribution with more equal probability of relatively wet and dry conditions in this cluster (Figure 11). Similar variability in these areas can also be seen in January and February. Clusters spanning the central and southern areas of Britain (Figures 7 and 8) also show more variability in comparison to the north-west in spring.

In summer, some of the clusters covering the north-west exhibit more variability than similarly located clusters during the winter months. For example, June NAO+ Cluster 3 has a large cluster SPI-1 value range of −1.2 (moderately dry) to 1.7 (severely wet). Compared to the winter months in clusters covering the north-western area, the difference between the probability of wet and dry conditions associated with the phase of the NAO is reduced. In the example of June NAO+ Cluster 3, there is a 23% likelihood of experiencing wetter rather than drier conditions (Figures 11 and 12). Some clusters during the summer months have large SPI-1 value ranges (Figure 9), and the likelihood of relative wet/dry conditions is equally likely (Figure 12), for example July NAO+ Cluster 2 and August NAO+ Cluster 1, which span the central, southern and eastern areas of Great Britain.

In late summer, the average NAO-rainfall response across Great Britain can be notably strong in comparison to June and July [10]. Whilst this can be observed in the clustering analysis, there is variability in the magnitude of the NAO-rainfall response. For example, August NAO− Cluster 3 shows a 45% likelihood of wetter rather than drier conditions (Figure 12). However, the magnitude of these wet events can vary—the cluster average SPI-1 range varies from −0.7 (slightly dry, but near-normal) to 2.3 (extremely wet) (Figure 10), and the frequency histograms show a wider spread across positive SPI-1 values (Figure 11). This demonstrates that even when there are clear average signals, as mapped in Figures 2 and 3, there can be significant spatio-temporal variability in the NAO-rainfall response.

4. Discussion

This study sought to evaluate the variability in NAO-rainfall response across Great Britain at high spatial and temporal scales. Average 5 km gridded SPI-1 values were mapped under NAO+ and NAO− conditions for the period January 1900–December 2015 (Figures 2 and 3). This revealed distinctive spatial signatures of the NAO in average monthly rainfall, such as the winter north-west/south-east spatial divide and more spatially homogeneous summer rainfall responses also observed in other studies [1,7,10,12,13]. The key spatio-temporal differences in NAO-rainfall response and relative consistency/variability between winter and summer revealed by our space-time analyses are summarised in Figure 13.

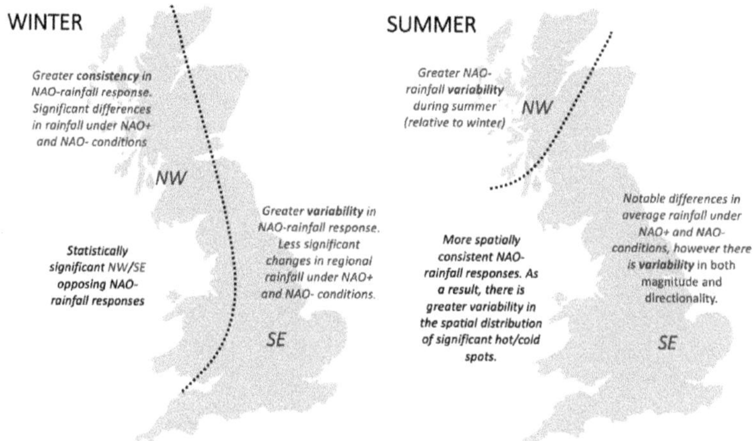

Figure 13. Schematic representation of the key spatio-temporal differences observed in NAO-rainfall response during the winter and summer months.

In the winter months, the results of the Getis-Ord Gi* space-time hot spot analysis confirm that the NW/SE spatial pattern (i.e., the spatial distribution of SPI-1 values) is statistically significant and consistent over the temporal record analysed (Figures 4 and 5). The space-time clustering analysis (Figures 7 and 8) also show these clear spatial patterns in the mean and median cluster average values, with clear differences as well in the frequency distribution of SPI-1 values related to NAO phase and region (Figures 11 and 12). This indicates a more spatially reliable estimate of monthly rainfall volume under NAO+ and NAO− phases may be possible during the winter months.

In the north-western areas of Great Britain, the space-time clustering and hot spot analyses reveal rainfall is very responsive to the phase of the NAO. There are clear differences in rainfall response between the two NAO phases, with values markedly fluctuating between wet (positive SPI- values) and dry (negative SPI-1 values) (Figure 9). Significant wet conditions occur under NAO+ conditions and dry conditions under NAO−, which supports the significant NAOI-rainfall correlations found in other studies [1,6,7,10]. Our space-time analyses show that the NAO-rainfall response in the north-western area during winter also shows greater consistency in these significant NAO+/NAO− wet/dry rainfall deviations. In some space-time clusters, the frequency histograms show that the probability of relative wetness/dryness differs significantly in the north-west during winter, for example Jan NAO− Cluster 1 has an 83% likelihood of experiencing dry conditions, whilst the similarly located NAO+ Clusters 1 and 2 have a 69% and 78% likelihood of wetter than average conditions (Figures 11 and 12). This indicates that we can have greater confidence in how the monthly rainfall volume in the north-western area will change with the phase of the NAO during the winter months. Improved winter NAO forecasting skill [25–27] may therefore allow for effective water management decisions to be taken if we are able utilise this forecasting skill to predict an upcoming period of rainfall surplus (NAO+) or deficit (NAO−). However, it is important to note that even with these more consistent NAO-rainfall responses, our analysis shows the wet/dry response magnitude can still vary in the north-west (Figures 9 and 11).

The southern, eastern and central areas of Great Britain have a consistent opposing (wet/dry) NAO-rainfall response to the north-west during the winter months (Figures 4 and 5). However, the relative change in monthly average rainfall under NAO+ and NAO− conditions is notably less (Figures 2 and 3), with median space-time cluster values in these areas being closer to 0 (Figures 9 and 10). There is also greater variability in the NAO-rainfall response—the median SPI-1 histograms for clusters in these areas are typically more distributed across wet/dry values (Figure 11). The differences in the likelihood of relative wet/dry conditions associated with the phase of the NAO are notably reduced, being within approximately 20–30% (Figure 12). These findings suggest clear variability in both wet/dry event magnitude and directionality. As a result, it can be concluded that the NAO has a weaker and more variable influence on rainfall in the southern, eastern, and central areas, and as such, NAO forecasts might be of less practical use in water management decision making in comparison to the north-western area.

Differences in average monthly SPI-1 values during the summer months were found between the two NAO phases (Figures 2 and 3), and less distinctive spatial differences between the north-western and southern/central areas of the country were found [10,12,13] (Figures 4 and 5). The more spatially consistent rainfall patterns during summer may be associated with greater convective rainfall generation [10], compared to orographic rainfall during winter [16]. However, an area of future research would be to explore the physical processes resulting in the difference between NAO winter and summer rainfall patterns. On average, NAO+ conditions result in drier summer months, and NAO− wetter summer months (Figures 2 and 3) aligning with negative NAOI-rainfall correlations observed in other studies [10,12,13]. The median space-time cluster values corroborate this (Figure 10), and in some cases clusters show more distinctive wet/dry summer responses, with the differences in the likelihood of relative wetness/dryness being approximately 50% for some clusters over the time period analysed (Figure 12). However, other clusters across the country also show notable variability in terms of magnitude and directionality in the NAO-rainfall response (Figure 9), with the relative probability of relative wetness/dryness being equal in some clusters (Figure 12). These findings demonstrate that even with clear average monthly signals (Figures 2 and 3), there can be significant variability in the NAO-rainfall (wet/dry) response, which may limit the practicality of NAO forecasts for water management decision making during summer.

In summary, our space-time analyses reveal that whilst typical NAO-rainfall signatures can be observed across the year, there is also significant NAO-rainfall response variability in space and time. This variability in both rainfall magnitude and wet/dry directionality may be a limiting factor in the utility of incorporating NAO forecasts into water management decision making [40], even though the accuracy of these NAO forecasts has improved in recent years [25–27]. The exception being in the north-western area during winter, where significant and more consistent changes in rainfall [10,18], and subsequently catchment hydrology [16], can be found relatable to the phase and strength of the NAO.

Variability in NAO-rainfall response in space and time across Britain might be explained by other North Atlantic and European atmospheric-oceanic circulations (teleconnections) moderating or enhancing the rainfall effect of the NAO and/or being potentially more dominant in driving regional rainfall in areas where the NAO's effect is weaker or when the NAO is in a neutral phase. As discussed above, in our analysis, the central, southern and eastern regions of Great Britain frequently had relatively variable NAO-rainfall responses under NAO+ and NAO− phases. The East Atlantic pattern in particular has been found to be positively correlated with rainfall these regions [13,41] and depending on its phase and strength may moderate or enhance the effect of the NAO on rainfall distribution and magnitude [22,23].

Our research supports the findings of Hall and Hanna [13], who suggest that even highly accurate NAOI forecasts might not provide enough information on their own to predict regional rainfall in Britain, and potentially subsequently catchment hydrology, several months in advance, without also considering the phase and magnitude of other atmospheric-oceanic circulations and climatic variables. As far as we are aware, no study has yet mapped at a high spatial and temporal (monthly) resolution the signature of these other North Atlantic/European circulations in regional rainfall across Great Britain.

In this study, the NAO was quantified using a PC-based NAOI, with phases defined using an approach adopted in previous work [10,30]. However, it is important to note that there is no universal approach to defining the NAO [42], and there is an opportunity for future work to explore the sensitivity of these results to the chosen NAOI and phase definition method. We have demonstrated the effectiveness of high resolution spatio-temporal analytical methods in exploring the meteorological impact of atmospheric circulations and revealing spatio-temporal climatic patterns. For example, the Getis-Ord Gi* statistic allowed for the identification of statistically significant spatial wet/dry patterns in the SPI-1 data in the winter months, although we acknowledge its limitation in detecting significant high/low value clusters in summer due to the more spatially consistent rainfall response. The space-time clustering analysis allowed us to look beyond average conditions and explore spatial and temporal consistency of the average and well-established NAO rainfall signatures in Great Britain. However, it is important to note that due to the random nature of the initial seed locations for the space-time clusters, modestly different results may be produced with a re-running of the analysis. This may be the case for locations (5 km pixels) where the spatial differences in the SPI-1 time series values are smaller and so may switch cluster membership between model runs, for example in the Midlands area between the more distinctive NW/SE zones during winter.

5. Conclusions

This study presents a novel application of space-time analyses to understand the variability in NAO-rainfall signatures at a high spatial (5 km) and temporal (monthly) resolution in Great Britain. Our analyses confirm that statistically significant NAO-rainfall signatures can be observed, and some regions show relatively high consistency in rainfall response to the phase of the NAO over time. However, our analyses also reveal that there is significant spatio-temporal variability in the rainfall response to the NAO, especially in the central, southern, and eastern areas of Great Britain. This has implications for the practical application of the NAOI in regional hydrometeorological forecasting as it is important to consider the variability in regional NAO-rainfall response under positive and negative phases of the NAO across Great Britain.

We suggest that such spatio-temporal variability might be explained by also considering the phase and magnitude of other atmospheric-oceanic teleconnections such as the East Atlantic Pattern. There is a need for high spatial and temporal resolution exploration of the hydrometeorological impact of these secondary modes of climate variability/teleconnections on rainfall in Great Britain, and in particular, the extent to which they might moderate or enhance the regional rainfall response to the NAO.

Author Contributions: Conceptualisation, H.W., N.Q. and M.H.; methodology, H.W., N.Q. and M.H.; software, H.W.; validation, H.W., N.Q. and M.H.; formal analysis, H.W.; investigation, H.W.; resources, H.W.; data curation, H.W.; writing—original draft preparation, H.W.; writing—review and editing, H.W., N.Q. and M.H.; visualisation, H.W., N.Q. and M.H.; supervision, N.Q. and M.H.; project administration, H.W. and N.Q. All authors have read and agreed to the published version of the manuscript.

Funding: This research received no external funding, and the APC was funded by the Department of Geography and Environmental Management at the University of the West of England, Bristol, UK.

Institutional Review Board Statement: Not applicable.

Informed Consent Statement: Not applicable.

Conflicts of Interest: The authors declare no conflict of interest.

Appendix A

Figure A1. Disaggregated space-time hot spot results (DJFM). Values indicate percentage of time in a statistically significant hot/cold spot.

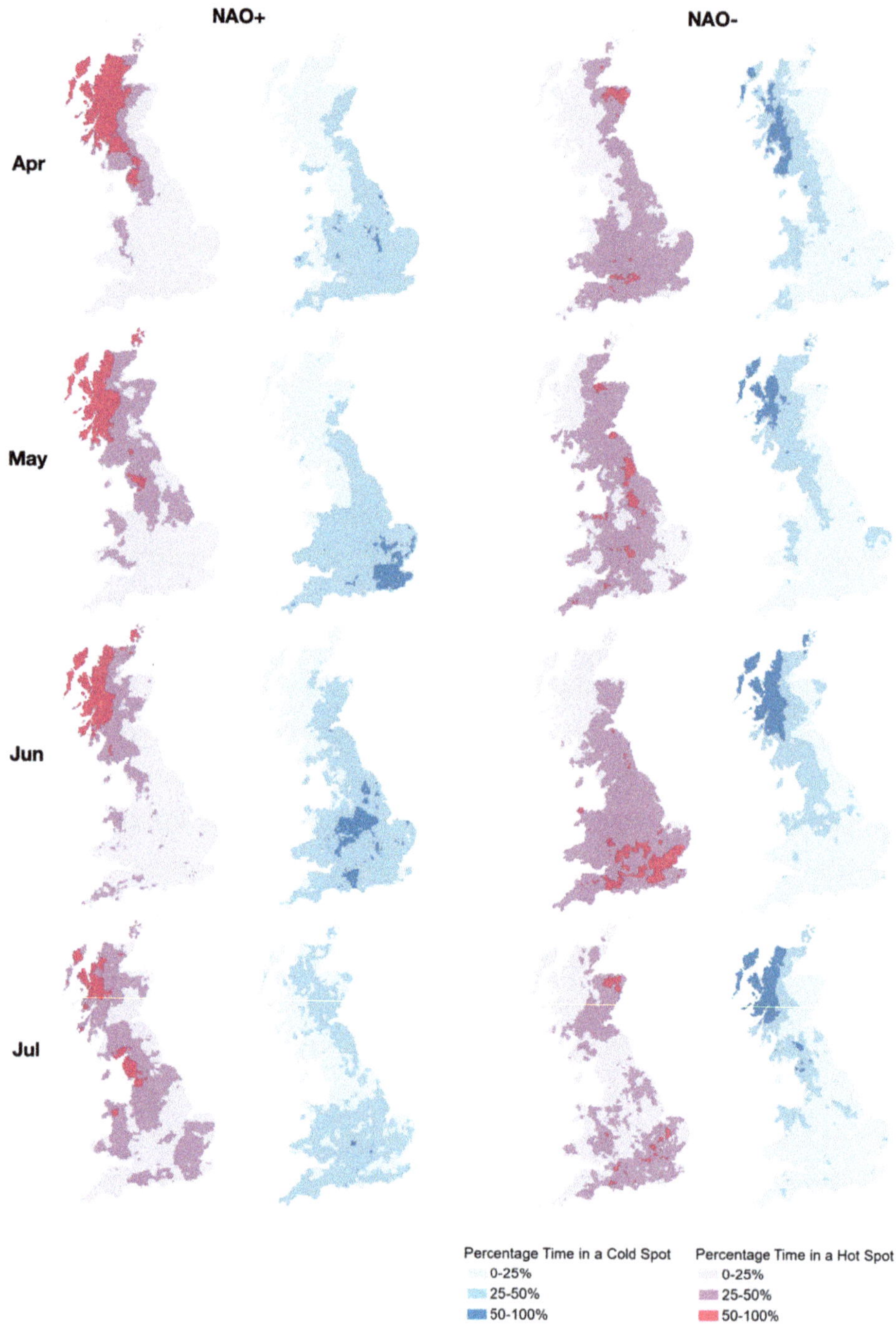

Figure A2. Disaggregated space-time hot spot results (AMJJ). Values indicate percentage of time in a statistically significant hot/cold spot.

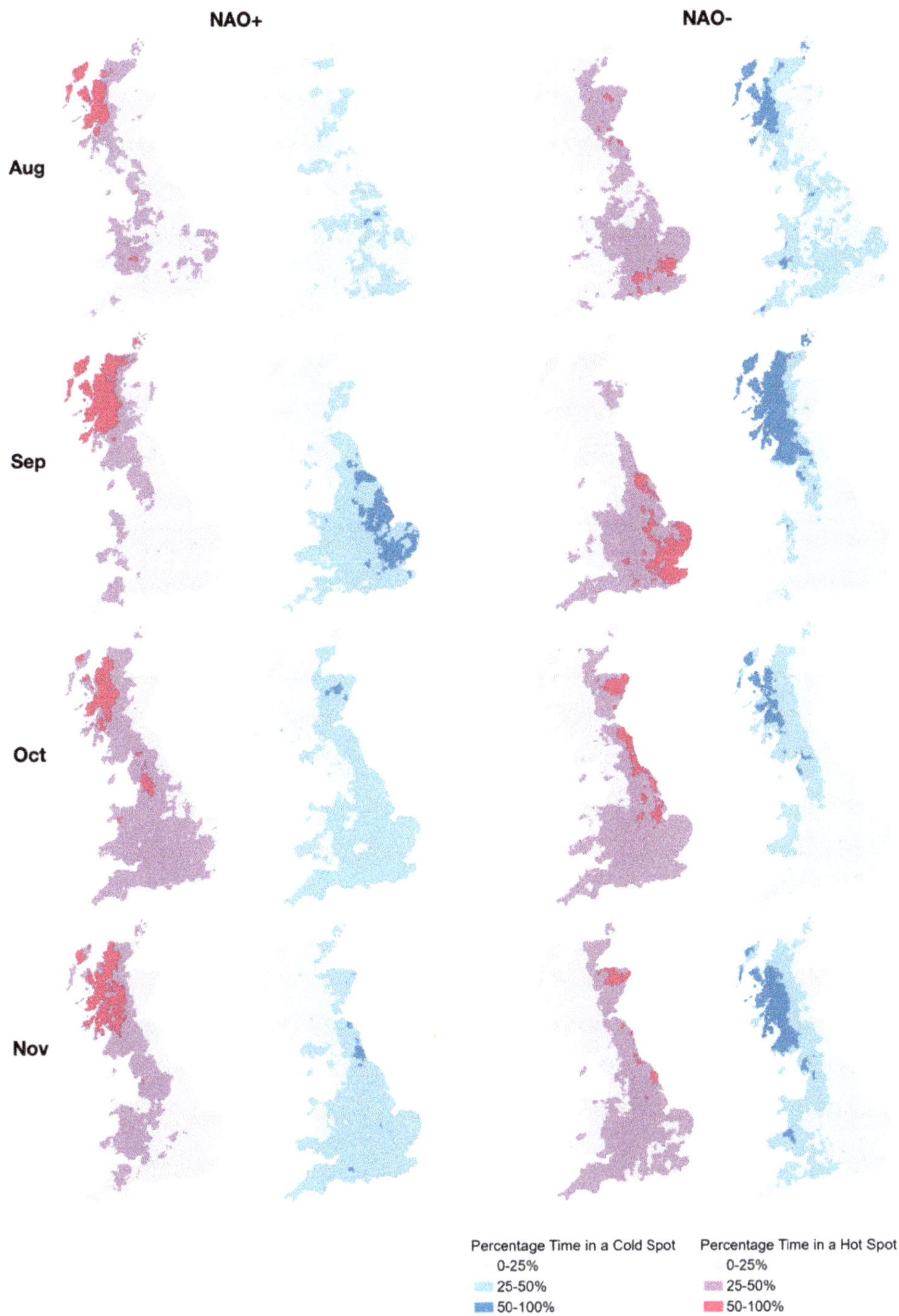

Figure A3. Disaggregated space-time hot spot results (ASON). Values indicate percentage of time in a statistically significant hot/cold spot.

References

1. Wilby, R.L.; O'Hare, G.; Barnsley, N. The North Atlantic Oscillation and British Isles climate variability, 1865–1996. *Weather* **1997**, *52*, 266–276. [CrossRef]
2. Rodwell, M.J.; Rowell, D.P.; Folland, C.K. Oceanic forcing of the wintertime North Atlantic Oscillation and European climate. *Nature* **1999**, *398*, 320–323. [CrossRef]
3. Hurrell, J.W.; Kushnir, Y.; Ottersen, G.; Visbeck, M. An overview of the North Atlantic Oscillation. In *The North Atlantic Oscillation: Climatic Significance and Environmental Impact*; American Geophysical Union: Washington, DC, USA, 2003; Volume 134, pp. 1–35.
4. Woollings, T.; Czuchnicki, C.; Franzke, C. Twentieth century North Atlantic jet variability. *Q. J. R. Meteorol. Soc.* **2013**, *140*, 783–791. [CrossRef]
5. Folwer, H.J.; Kilsby, C.G. Precipitation and the North Atlantic Oscillation: A study of climatic variability in Northern England. *Int. J. Climatol.* **2002**, *22*, 843–866.
6. Afzal, M.; Gagnon, A.S.; Mansell, M.G. Changes in the variability and periodicity of precipitation in Scotland. *Theor. Appl. Climatol.* **2015**, *119*, 135–159. [CrossRef]
7. Rust, W.; Holman, I.; Corstanje, R.; Bloomfield, J.; Cuthbert, M. A conceptual model for climatic teleconnection signal control on groundwater variability in Europe. *Earth Sci. Rev.* **2018**, *177*, 164–174. [CrossRef]
8. Simpson, I.R.; Jones, P.D. Analysis of UK precipitation extremes derived from Met Office gridded data. *Int. J. Climatol.* **2013**, *34*, 2438–2449. [CrossRef]
9. Kosanic, A.; Harrison, S.; Anderson, K.; Kavcic, I. Present and historical climate variability in South West England. *Clim. Chang.* **2014**, *124*, 221–237. [CrossRef]
10. West, H.; Quinn, N.W.; Horswell, M. Regional rainfall response to the North Atlantic Oscillation (NAO) across Great Britain. *Hydrol. Res.* **2019**, *50*, 1549–1563. [CrossRef]
11. Spencer, M.; Essery, R. Scottish snow cover dependence on the North Atlantic Oscillation index. *Hydrol. Res.* **2016**, *47*, 619–629. [CrossRef]
12. Folland, C.K.; Knight, J.; Linderholm, H.W.; Fereday, D.; Ineson, S.; Hurrell, J.W. The Summer North Atlantic Oscillation: Past, Present, and Future. *J. Clim.* **2009**, *22*, 1082–1103. [CrossRef]
13. Hall, R.J.; Hanna, E. North Atlantic circulation indices: Links with summer and winter UK temperature and precipitation and implications for seasonal forecasting. *Int. J. Climatol.* **2018**, *38*, e660–e677. [CrossRef]
14. Phillips, I.D.; McGregor, G.; Wilson, C.J.; Bower, D.; Hannah, D. Regional climate and atmospheric circulation controls on the discharge of two British rivers, 1974?97. *Theor. Appl. Climatol.* **2003**, *76*, 141–164. [CrossRef]
15. Rust, W.; Cuthbert, M.; Bloomfield, J.; Corstanje, R.; Howden, N.; Holman, I. Exploring the role of hydrological pathways in modulating North Atlantic Oscillation. *Hydrol. Earth Syst. Sci.* **2021**, *25*, 2223–2237. [CrossRef]
16. Burt, T.P.; Howden, N.J.K. North Atlantic Oscillation amplifies orographic precipitation and river flow in upland Britain. *Water Resour. Res.* **2013**, *49*, 3504–3515. [CrossRef]
17. Lavers, D.A.; Hannah, D.; Bradley, C. Connecting large-scale atmospheric circulation, river flow and groundwater levels in a chalk catchment in southern England. *J. Hydrol.* **2015**, *523*, 179–189. [CrossRef]
18. Rust, W.; Holman, I.; Bloomfield, J.; Cuthbert, M.; Corstanje, R. Understanding the potential of climate teleconnections to project future groundwater drought. *Hydrol. Earth Syst. Sci.* **2019**, *23*, 3233–3245. [CrossRef]
19. Wilby, R.; Johnson, M. Climate variability and implications for keeping rivers cool in England. *Clim. Risk Manag.* **2020**, *30*, 100259. [CrossRef]
20. UK Hydrological Outlook. UK Hydrological Outlook December 2020. Available online: http://www.hydoutuk.net/files/2816/0743/4122/2020_12_HO_Complete.pdf (accessed on 14 December 2020).
21. Donegan, S.; Murphy, C.; Harrigan, S.; Broderick, C.; Golian, S.; Knight, J.; Matthews, T.; Prudhomme, C.; Quinn, D.F.; Scaife, A.A.; et al. Conditioning Ensemble Streamflow Prediction with the North Atlantic Oscillation improves skill at longer lead times. *Hydrol. Earth Syst. Sci.* In Review. [CrossRef]
22. Comas-Bru, L.; McDermott, F. Impacts of the EA and SCA patterns on European twentieth century NAO−winter climate rela-tionship. *Q. J. R. Meteorol. Soc.* **2014**, *140*, 354–363. [CrossRef]
23. Mellado-Cano, J.; Barriopedro, D.; García-Herrera, R.; Trigo, R.M.; Hernández, A. Examining the North Atlantic Oscillation, East Atlantic Pattern, and Jet Variability since 1685. *J. Clim.* **2019**, *32*, 6285–6298. [CrossRef]
24. Weisheimer, A.; Schaller, N.; O'Reilly, C.; MacLeod, D.A.; Palmer, T. Atmospheric seasonal forecasts of the twentieth century: Multi-decadal variability in predictive skill of the winter North Atlantic Oscillation (NAO) and their potential value for extreme event attribution. *Q. J. R. Meteorol. Soc.* **2017**, *143*, 917–926. [CrossRef] [PubMed]
25. Parker, T.; Woollings, T.; Weisheimer, A.; O'Reilly, C.; Baker, L.; Shaffrey, L. Seasonal Predictability of the Winter North Atlantic Oscillation from a Jet Stream Perspective. *Geophys. Res. Lett.* **2019**, *46*, 10159–10167. [CrossRef]
26. Athanasiadis, P.J.; Yeager, S.; Kwon, Y.-O.; Bellucci, A.; Smith, D.W.; Tibaldi, S. Decadal predictability of North Atlantic blocking and the NAO. *npj Clim. Atmos. Sci.* **2020**, *3*, 1–10. [CrossRef]
27. Smith, D.M.; Scaife, A.A.; Eade, R.; Athanasiadis, P.; Bellucci, A.; Bethke, I.; Bilbao, R.; Borchert, L.F.; Caron, L.-P.; Counillon, F.; et al. North Atlantic climate far more predictable than models imply. *Nature* **2020**, *583*, 796–800. [CrossRef]
28. Pokorná, L.; Huth, R. Climate impacts of the NAO are sensitive to how the NAO is defined. *Theor. Appl. Clim.* **2015**, *119*, 639–652. [CrossRef]

29. National Centre for Atmospheric Research. Hurrell North Atlantic Oscillation Index (PC-Based). Available online: https://climatedataguide.ucar.edu/climate-data/hurrell-north-atlantic-oscillation-NAO$-$index-pc-based (accessed on 30 June 2020).
30. Berton, R.; Driscoll, C.T.; Adamowski, J.F. The near-term prediction of drought and flooding conditions in the northeastern United States based on extreme phases of AMO and NAO. *J. Hydrol.* **2017**, *553*, 130–141. [CrossRef]
31. Tanguy, M.; Fry, M.; Svensson, C.; Hannaford, J. *Historic Gridded Standardized Precipitation Index for the United Kingdom 1862–2015 (Generated using Gamma Distribution with Standard Period 1961–2010) v4*; NERC Environmental Information Data Centre: Wallingford, UK, 2017.
32. Irannezhad, M.; Haghighi, A.T.; Chen, D.; Klöve, B. Variability in dryness and wetness in central Finland and the role of teleconnection patterns. *Theor. Appl. Climatol.* **2014**, *122*, 471–486. [CrossRef]
33. Kingston, D.G.; Stagge, J.H.; Tallaksen, L.M.; Hannah, D.M. European-Scale Drought: Understanding Connections between Atmospheric Circulation and Meteorological Drought Indices. *J. Clim.* **2015**, *28*, 505–516. [CrossRef]
34. Djellouli, F.; Bouanani, A.; Babahamed, K. Climate change: Assessment and monitoring of meteorological and hydrological drought of Wadi El Hamman Basin (NW–Algeria). *J. Fundam. Appl. Sci.* **2016**, *8*, 1037–1053. [CrossRef]
35. Getis, A.; Ord, J.K. The Analysis of Spatial Association by Use of Distance Statistics. *Geogr. Anal.* **2010**, *24*, 189–206. [CrossRef]
36. Esri. Hot Spot Analysis (Getis-Ord Gi*). Available online: https://pro.arcgis.com/en/pro-app/latest/tool-reference/spatial-statistics/hot-spot-analysis.htm (accessed on 4 February 2021).
37. Ord, J.K.; Getis, A. Local Spatial Autocorrelation Statistics: Distributional Issues and an Application. *Geogr. Anal.* **1995**, *27*, 286–306. [CrossRef]
38. Scott, L.; Warmerdam, N. Extended Crime Analysis with ArcGIS Spatial Statistics Tools. Available online: https://www.esri.com/news/arcuser/0405/ss_crimestats1of2.html (accessed on 4 February 2021).
39. Esri. Space Time Cluster Analysis. Available online: https://pro.arcgis.com/en/pro-app/tool-reference/spatial-statistics/space-time-analysis.htm (accessed on 6 August 2020).
40. Rust, W.; Bloomfield, J.P.; Cuthbert, M.O.; Corstanje, R.; Holman, I.P. Non-stationary control of the NAO on European rainfall and its implications for water resource management. *Hydrol. Process.* **2021**, *35*, e14099. [CrossRef]
41. Casanueva, A.; Rodríguez-Puebla, C.; Frías, M.D.; González-Reviriego, N. Variability of extreme precipitation over Europe and its relationships with teleconnection patterns. *Hydrol. Earth Syst. Sci.* **2014**, *18*, 709–725. [CrossRef]
42. Hurrwll, J.W.; Deser, C. North Atlantic climate variability: The role of the North Atlantic Oscillation. *J. Mar. Syst.* **2009**, *78*, 28–41. [CrossRef]

atmosphere

Article

Subseasonal Forecasts of the Northern Queensland Floods of February 2019: Causes and Forecast Evaluation

Wayne Yuan-Huai Tsai, Mong-Ming Lu *, Chung-Hsiung Sui and Yin-Min Cho

Department of Atmospheric Sciences, National Taiwan University, Taipei 10617, Taiwan;
r06229002@ntu.edu.tw (W.Y.-H.T.); sui@as.ntu.edu.tw (C.-H.S.); ymcho@ntu.edu.tw (Y.-M.C.)
* Correspondence: mongminglu@ntu.edu.tw

Abstract: During the austral summer 2018/19, devastating floods occurred over northeast Australia that killed approximately 625,000 head of cattle and inundated over 3000 homes in Townsville. In this paper, the disastrous event was identified as a record-breaking subseasonal peak rainfall event (SPRE). The SPRE was mainly induced by an anomalously strong monsoon depression that was modulated by the convective phases of an MJO and an equatorial Rossby (ER) wave. The ER wave originated from an active equatorial deep convection associated with the El Niño warm sea surface temperatures near the dateline over the central Pacific. Based on the S2S Project Database, we analyzed the extended-range forecast skill of the SPRE from two different perspectives, the monsoon depression represented by an 850-hPa wind shear index and the 15-day accumulated precipitation characterized by the percentile rank (PR) and the ratio to the three-month seasonal (DJF) totals. The results of four S2S models of this study suggest that the monsoon depression can maintain the same level of skill as the short-range (3 days) forecast up to 8–10 days. For precipitation parameters, the conclusions are similar to the monsoon depression. For the 2019 northern Queensland SPRE, the model forecast was, in general, worse than the expectation derived from the hindcast analysis. The clear modulation of the ER wave that enhanced the SPRE monsoon depression circulation and precipitation is suspected as the main cause for the lower forecast skill. The analysis procedure proposed in this study can be applied to analyze the SPREs and their associated large-scale drivers in other regions.

Keywords: S2S prediction; Australian summer monsoon; MJO; subseasonal peak rainfall event; extreme rainfall

Citation: Tsai, W.Y.-H.; Lu, M.-M.; Sui, C.-H.; Cho, Y.-M. Subseasonal Forecasts of the Northern Queensland Floods of February 2019: Causes and Forecast Evaluation. *Atmosphere* **2021**, *12*, 758. https://doi.org/10.3390/atmos12060758

Academic Editor: Ke Fan

Received: 29 April 2021
Accepted: 8 June 2021
Published: 10 June 2021

Publisher's Note: MDPI stays neutral with regard to jurisdictional claims in published maps and institutional affiliations.

1. Introduction

In late January and early February 2019, Queensland was hit by a disastrous rainfall event that caused 18,000 residents to lose power and hundreds of others to evacuate. The floods led to huge losses for farmers, and estimates of 625,000 head of cattle and 48,000 sheep were killed [1–3].

The annual rainfall in Australia shows a relatively high rainfall amount along the northern and eastern coastline. The northern part of Australia receives more than 50% of the rainfall amount of the annual totals in summer, where the water vapor is primarily brought by the Australian summer monsoon [4]. The Australian summer monsoonal region is generally regarded as 115°~150° E, 5°~20° S in many studies [5–7]. The subseasonal variability of monsoonal flow and rainfall are affected by multiple-scale phenomena, such as Madden–Julian oscillation (MJO) [8–10], convectively coupled equatorial waves (CCEWs) [11,12], tropical cyclones (TCs) [13,14], and/or extratropical surges [6]. King et al. [15] pointed out that the extreme rainfall variability is closely related to the mean rainfall variability during austral summer, especially the TCs and east coast low (monsoon depression). They defined the extreme rainfall as the monthly maximum consecutive 5-day precipitation totals. On the other hand, in order to evaluate the usefulness of subseasonal to seasonal (S2S) prediction [16], Tsai et al. [17] defined a subseasonal peak rainfall event

(SPRE) as the maximum successive three-pentad (15-day) precipitation within a three-month time window. The SPRE is an ideal target for assessing the baseline prediction skill of a S2S prediction model before applying model products to local usages [18]. The SPRE by definition is the most significant 2-week rainfall episode in the monthly time scale. Therefore, it is deemed to be the most important S2S precipitation prediction target.

Cowan et al. [1] gave a thorough analysis of the large-scale climate conditions of the 2019 Queensland flood. They found this event was a good example to promote the awareness of the benefit of S2S prediction that showed good skill in forecasting the broad-scale atmospheric conditions north of Australia a week ahead. Motivated by Cowan et al. [1], the purpose of the present paper is twofold. First, it provides additional evidence showing that the 2019 Queensland floods are associated with an SPRE influenced by multiple large-scale factors. Secondly, it designs some informative targets that reflect some essential characteristics of the SPRE and can be used in model prediction verification and comparison. To leverage the dynamic model prediction data from the S2S Project Database [19], the forecast targets will include the influential circulation patterns and subseasonal scale accumulated precipitation.

The paper is organized as follows: Section 2 describes the observational and S2S datasets we use. Section 3 documents the 2019 northern Queensland event and its relationship with monsoon depression, MJO, and CCEWs. Section 4 discusses the prediction performance of S2S models for the 2019 event. The prediction skill is also accessed based on the hindcast data from 1998 to 2018. Section 5 is a summary of findings and discussions.

2. Data and Methods

2.1. Data

The wind field is based on the European Centre for Medium-Range Weather Forecasts Reanalysis Interim (ERA-Interim) [20], which is utilized with a horizontal resolution of $0.75° \times 0.75°$ in longitude and latitude during the period from November 1998 to February 2019. Rainfall data are based on the CPC MORPHing technique (CMORPH) precipitation data with a spatial resolution of $0.25° \times 0.25°$. It includes the raw, satellite only precipitation estimates, corrected, and gauge-satellite blended precipitation products, with calibration against surface gauge observations [21]. Outgoing longwave radiation (OLR) is used as an appropriate proxy for the deep convection, which is based on the interpolated daily OLR version 1.2 from National Oceanic and Atmospheric Administration (NOAA) Climate Data Record (CDR) with the spatial resolution of $1° \times 1°$ [22].

The forecast data are downloaded from the S2S Project Database [19]. Both hindcast and forecast data contributed by 11 forecast centers around the world (Table 1) are available in the database. Total precipitation and wind field are used in this study. In this study, we need to collect the model output with initialization frequency at least once every 5 days. However, at the time when the research was carried out in 2020, only four out of the eleven models, namely the BoM, CMA, ECMWF, and NCEP, provided both hindcast and forecast datasets that satisfy this criterion. Therefore, these four models are presented in this paper for model prediction, verification, and comparison. We have also checked the forecasts produced by JMA, KMA, and UKMO, but the results were not presented due to the sampling limitation.

Table 1. The detailed information about the three S2S models (updated from Vitart et al. [19] according to ECMWF website), where "Time Range" is the forecast lead time, "Ensemble Size" is the number of members in the real-time forecast ensemble, and "Frequency" is the frequency the model is run. "Hindcast" or (reforecast, abbreviated as "Rfc.") are run using the actual forecast model but for the past several years on the same (or nearby) calendar day as the forecast. The "Rfc. Period" is the number of years the reforecasts are run, "Rfc. Frequency" is the frequency the reforecasts are run, and the "Rfc Size" is the number of ensemble members for reforecasts. "Resolution" is longitude and latitude based on the data accessed from the ECMWF website.

Model	Time Range (Days)	Resolution	Ensemble Size	Frequency	Hindcast (Rfc.)	Rfc. Period	Rfc. Frequency	Rfc. Size
BoM	0 to 62	$2.5° \times 2.5°$	33	2/Week	Fixed	1981~2013	6/Month	33
CMA	0 to 60	$1.5° \times 1.5°$	4	Daily	Fixed	1994~2014	Daily	4
ECCC	0 to 32	$1.5° \times 1.5°$	21	Weekly	On the fly	1998~2017	Weekly	4
ECMWF	0 to 46	$1.5° \times 1.5°$	51	2/Week	On the fly	Past 20 yrs.	2/Week	11
HMCR	0 to 62	$1.5° \times 1.5°$	20	Weekly	On the fly	1995~2010	Weekly	10
CNR-ISAC	0 to 32	$1.5° \times 1.5°$	41	Weekly	Fixed	1981~2010	1/Pentad	5
JMA	0.5 to 33.5	$1.5° \times 1.5°$	50	2/Week	Fixed	1981~2012	3/Month	5
KMA	0 to 60	$1.5° \times 1.5°$	4	Daily	On the fly	1991~2010	4/Month	3
CNRM	0 to 61	$1.5° \times 1.5°$	51	Weekly	Fixed	1993~2014	4/Month	15
NCEP	0 to 44	$1.5° \times 1.5°$	16	Daily	Fixed	1999~2010	Daily	4
UKMO	0 to 60	$1.5° \times 1.5°$	4	Daily	On the fly	1993~2016	4/Month	7

2.2. *Identifying MJO and CCEWs*

We utilize the space-time filtering technique developed by Wheeler and Kiladis [23] to identify MJO and CCEWs. The first step is to remove the seasonal cycle by subtracting the long-term mean and the first three harmonics of the daily climatology based on the 1979–2018 period for each grid. To retain the full signal of the waves [24], the anomalies of the symmetric and antisymmetric components of the waves are not decomposed. Then, the Fourier transform in longitude is performed, followed by another transform in time. Fourier coefficients outside the range of the filter are then set to zero, and the filtered data are obtained by performing the inverse transform. The spectral bands used in the space-time filter technique for CCEWs are based on the dispersion relation

$$\omega^2 - k^2 - \frac{k}{\omega} = 2m + 1 \qquad m = 1,\ 2,\ \ldots \tag{1}$$

The wavenumber (k), frequency bands (ω), and equivalent depths (h) for equatorial Rossby wave and MJO are presented in Table 2. The influences of other tropical modes are less significant (figure not shown). Note that the wavenumber of MJO is positive, which means that the wave propagates eastward, and the other is negative. This calculation is provided by Carl Schreck, III (SUNY at Albany, see https://ncics.org/portfolio/monitor/mjo/, last accessed, 9 June 2021), and the function can be applied to the NCAR Command Language (NCL) (see https://www.ncl.ucar.edu/Document/Functions/User_contributed/kf_filter.shtml, last accessed, 9 June 2021).

Table 2. The range of planetary zonal wavenumber, period (days), and equivalent depth (m) chosen for filtering MJO and $n = 1$ equatorial Rossby (ER) wave. "N/A" means the region of filtering does not follow the dispersion curve. The wavenumber-frequency ranges are based on Wheeler and Kiladis [23].

Tropical Modes	Planetary Zonal Wavenumber (k)	Period (ω)	Equivalent Depth (h)
MJO	1 to 5	30 to 96	N/A
Equatorial Rossby (ER) wave	−10 to −1	9.7 to 48	8 to 90

The convection modulation of MJO and individual CCEWs is identified based on the same procedure described in Tsai et al. [17]. We use the rank order of the wave-filtered OLR to flag the dates of the study period as a day of the "convective", "no signal", and "suppressed" phases. The convective status of each wave is determined according to the threshold values of the wave-filtered OLR. The threshold values are defined by the first (Q_{25}) and third (Q_{75}) quartiles of the filtered daily OLR anomalies collected during the boreal winter half year (November-April) from 1998 to 2015 of all grid points in a large domain bounded by the longitudes of 105° E and 135° E and latitudes of 15° S and 15° N. The thresholds for each type of wave are summarized in Table 1 of Tsai et al. [17]. When the filtered OLR anomaly is lower than Q_{25} and the raw OLR value is less than 250 W m^{-2}, the day is flagged as the convective phase, while if it is higher than Q_{75}, the day is flagged as the suppressed phase, otherwise the day is flagged as no signal. For MJO (ER), the Q_{25} and Q_{75} values are −8.84 W m^{-2} (−7.48 W m^{-2}) and 8.69 W m^{-2} (7.33 W m^{-2}), respectively.

3. The 2018/19 Northern Queensland SPRE

3.1. Australia Monsoon Trough

In this subsection, we will discuss the northern Queensland SPRE during the 2018/19 austral summer (December to February) represented by two box areas marked in Figure 1a. The SPRE is identified based on the CMORPH dataset. The time series of the area mean 15-day accumulative precipitation running by pentad from November 2018 to February 2019 is presented in Figure 1b,c. In both areas, the maximum 15-day accumulated rainfall occurred during the successive pentads from P5 to P8 (21 January–9 February). The beginning time of the Box-A SPRE is one pentad earlier than that of the Box-B SPRE. In order to see how extreme the 2018/19 case is compared with the climatological rains, the twenty years (1998–2017) of minimum and maximum 15-day accumulative precipitation are plotted together with the 2018/19 rainfall amount in Figure 1b,c. The gray shaded area in Figure 1b,c marks the range between maximum and minimum 15-day rainfall amount running by pentad. It is evident that the precipitation over the east coast (Box-B) is more abundant than the inland region (Box-A). For Box-A the transition time from dry to wet is around mid-December (P70). For Box-B the transition time appears in two stages with the first in mid-December, which is the same as that of Box-A, and the second in mid-January (P3), where the Box-A shows a similar level of rainfall as in late December without showing any sharp increase. Compared with the historical data, it is clear that the 2018/19 SPRE at Box-B broke the twenty-year pentad rainfall record. It is interesting to see the oscillations on the subseasonal time scale in both areas. In the following, we will discuss the relationship between the subseasonal variations of precipitation and the monsoon trough, MJO, and CCEWs.

The relationship between the monsoon trough and major precipitation areas can be illustrated using the 20-year climatological mean seasonal (DJF) precipitation, sea level pressure (SLP) and 850-hPa wind (Figure 2a). Figure 2a sees Australia covered by a large and elongated low-pressure center characterized by a cyclonic circulation in a clockwise direction. It extends northeastward to the east of New Guinea. The precipitation regions in general correspond to the region with cyclonic flow (negative relative vorticity in southern hemisphere), but the rainfall amount is stronger along the coast, and it decreases southward. The monsoon trough over northern Australia is formed with a strong zonal wind shear that has the westerlies in the north and the easterlies in the south. The westerlies flow from the Indian Ocean through the Timor Sea and the Arafura Sea to the Pacific Ocean, and the easterlies flow from the Pacific Ocean through northern Australia to the Indian Ocean. On the other hand, on the equatorial Pacific side, the monsoon trough straddles the equator with the easterlies to the north of the equator and westerlies blowing from Indonesia and the New Genia islands. The 2018/19 austral summer monsoon trough anomalies presented in Figure 2b clearly show a deepened trough especially in northwestern Australia, with an anomalous low-pressure system and cyclonic flow extended to the Coral Sea and the Western Pacific. The enhanced monsoon low also brings strong westerlies in the north and easterlies in the south. However, the positive rainfall anomaly is confined only along the northern coast of Australia and the Coral Sea; over land area, the rainfall is mainly less than the climatological means.

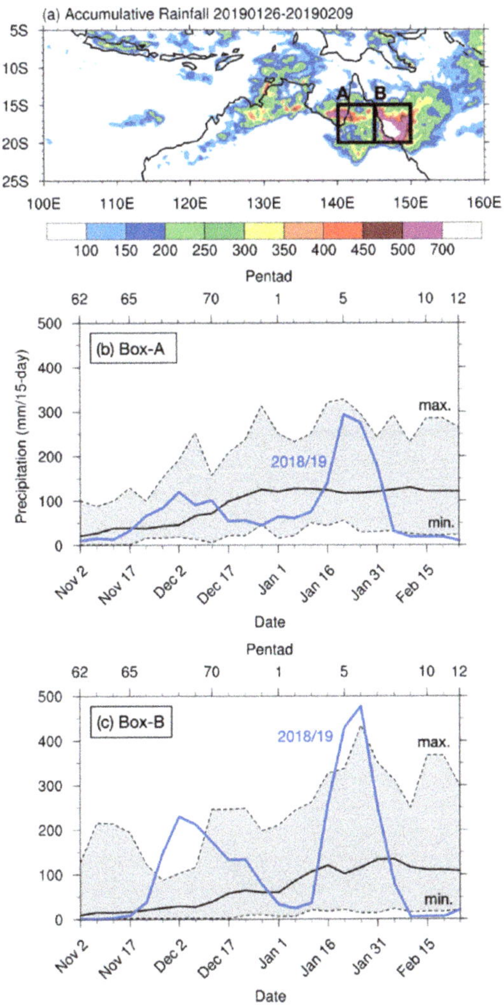

Figure 1. (**a**) The 15-day accumulative rainfall calculated using CMORPH precipitation data from 26 January to 9 February 2019, the northern Queensland flood. The black boxes, Box-A (140°~145° E, 15°~20° S) and Box-B (145°~150° E, 15°~20° S), represent northern Queensland and the marginal sea. (**b**,**c**) The area-mean 15-day accumulative rainfall time series running by pentad is calculated using CMORPH precipitation data over Box-A and Box-B, respectively. The date (the corresponding Julian pentad) marks the first day (pentad) of the 15-day period. The dashed lines represent the maximum and minimum rainfall derived by 1998 to 2017 climatology; the black solid line represents the climatological median. The blue line is the time series in summer 2018/19.

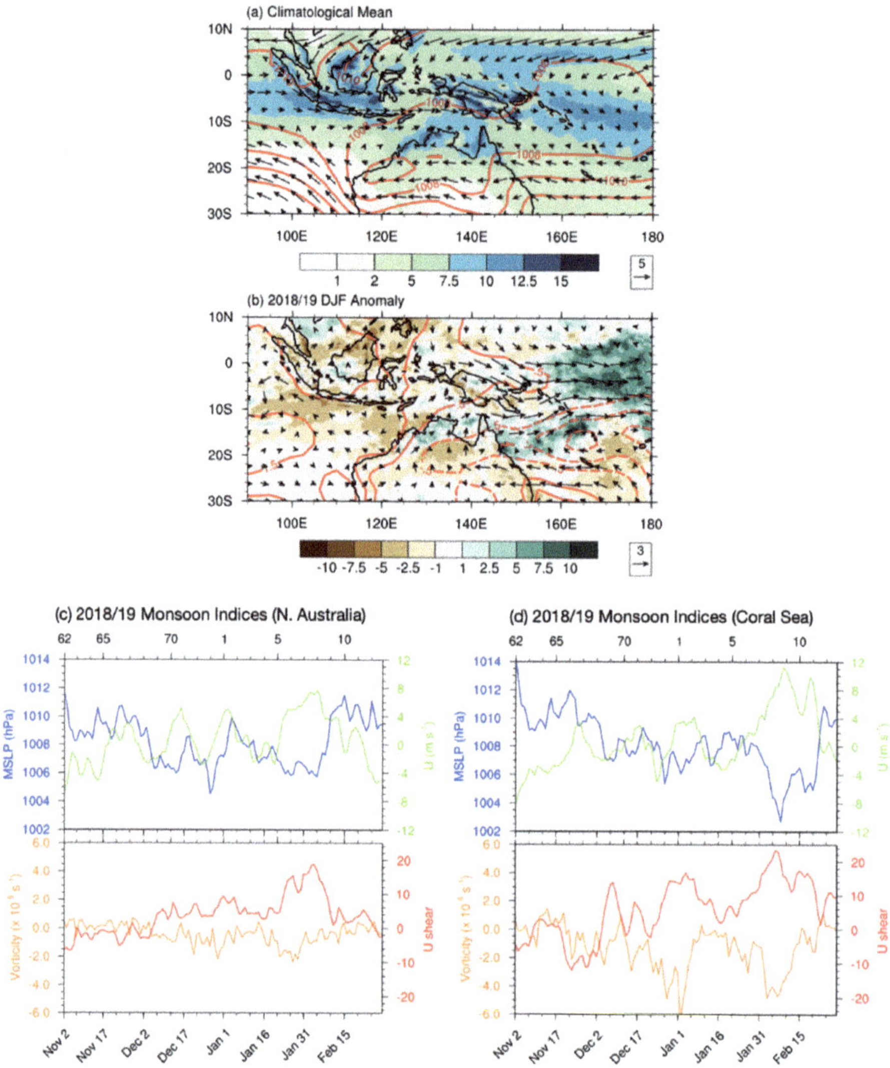

Figure 2. (**a**) Climatological seasonal mean precipitation (shaded, unit: mm day^{-1}), sea level pressure (SLP, contours, level by 2 hPa), and 850-hPa wind (vectors, unit: m s^{-1}), and (**b**) their anomaly fields (SLP contours level by 0.5 hPa) during austral summer (December to February). (**c**) The time series of monsoon indices during austral summer 2018/19 over the Australian summer monsoon region (115°~145° E, 5°~20° S), where the blue, green, orange, and red curves indicate the area-mean SLP, 850-hPa zonal wind, 850-hPa relative vorticity over the monsoon region, and zonal wind shear (U_{850}[5°–15° S, 115°–145° E]–U_{850}[20°–30° S, 115°–145° E]), respectively. (**d**) is the same as (**c**), but for the Coral Sea (145°~165° E, 5°~20° S) and the wind shear is obtained by (U_{850}[5°–15° S, 145°–165° E]–U_{850}[20°–30° S, 145°–165° E]).

The summer monsoon and monsoon depression can be described using the indices such as the area mean SLP, 850-hPa zonal wind and vorticity over the monsoon region (115°~145° E, 5°~20° S) and the Coral Sea (145°~165° E, 5°~20° S), and 850-hPa zonal wind shear (U_{850}[5°–15° S, 115°–145° E]–U_{850}[20°–30° S, 115°–145° E] and U_{850}[5°–15° S, 145°–165° E]–U_{850}[20°–30° S, 145°–165° E], respectively) to quantify the summer monsoon and monsoon depression intensity and variability. This index is a modification of the AUSSM index proposed in Yim et al. [25] based on the climatological monsoon depression in Figure 2a, where the westerlies appear in the north and easterlies in the south. After comparing the difference between the wind shear and the westerly [6,7,26] monsoon indices, we found that the wind shear index has the merit of showing clear seasonal contrast separated by the onset date and the cyclonic structure of the clockwise (cyclonic) monsoon depression over the interior of northern Australia [27]. The subseasonal variations of these indices in summer 2018/19 can be easily identified in Figure 2c,d. A strong westerly surge occurred in late January and lasted until mid-February. Associated with this westerly surge we see strong westerly wind shear. The westerly wind shear sharply dropped during the second week of February, together with a sharp increase in SLP and the anticyclonic vorticity. Comparing Figures 1c and 2c, it is evident that the large-scale monsoonal environment of the SPRE can be characterized by rapid intensification of the low-level westerly flow, the enhanced westerly wind shear and cyclonic circulation, and low surface pressure. The subseasonal enhancement of the monsoon trough over northern Australia created a favorable condition for the precipitation to persist. This is consistent with the findings in Cowan et al. [1].

The subseasonal relationship between the large-scale environment and northern Queensland rains can be clearly seen in the pentad maps from mid-January to mid-February. Figure 3 shows the pentad-mean 850-hPa wind field and precipitation maps from 16 January to 14 February. In two pentads before the SPRE (Figure 3a,b), Queensland received easterlies, implying that the monsoon trough was weak during this period. From 26 January, the monsoonal westerly started to intensify (Figure 3c), and a cyclone formed over the Arafura Sea and northeastern Queensland. The monsoon depression subsequently moved southward and strengthened, which brought abundant rainfall into the continental part of northeastern Queensland. During the peak pentad of the SPRE episode, the center of the monsoon depression was right over northeastern Queensland (Figure 3d), which enhanced the moist westerlies along the north coast and the moist northeasterlies over the Coral Sea that brought the moist air to the east coast of Queensland. This slow-moving monsoon depression appears to be the culprit for the disastrous rainfall event.

3.2. MJO and Equatorial Rossby Wave

Convection activity over northern Australia is frequently influenced by MJO and CCEWs [28]. The MJO circulation pattern reveals a Kelvin–Rossby couplet structure [29], which usually enhances the monsoon depression and leads to rainfall perturbations, especially in phases 5 to 7 [30]. During the occurrence period (26 January–9 February) of the 2018/19 Box-B SPRE, Figure 4a shows that a strong MJO convective phase passed through northern Australia. The MJO phase diagram (Figure 4b) indicates that the MJO remained strong during January and February. At the same time, a westward-propagating ER wave arrived from the South Pacific. Figure 4a shows when the ER moved into the convective zone of MJO in late January, it clearly enhanced the precipitation. Heavy precipitation persisted for more than two weeks before the rain band moved eastward to the South Pacific with the MJO. Note that the westward-propagating ER waves from the Pacific to the Indian Ocean originated from the active deep convection over the tropical central Pacific. The enhanced convection can be identified in Figure 2 as the large area of positive precipitation anomalies over the tropical Pacific to the east of Papua New Guinea. The tropical Pacific during the 2018/19 austral summer was characterized by a transition from a diminishing La Niña in 2018 to the development of a weak El Niño by early 2019 [31]. The central Pacific in the tropics over both northern and southern hemispheres was warmer than normal during January 2019, which enhanced the convection and precipitation near the dateline (160°–180° E).

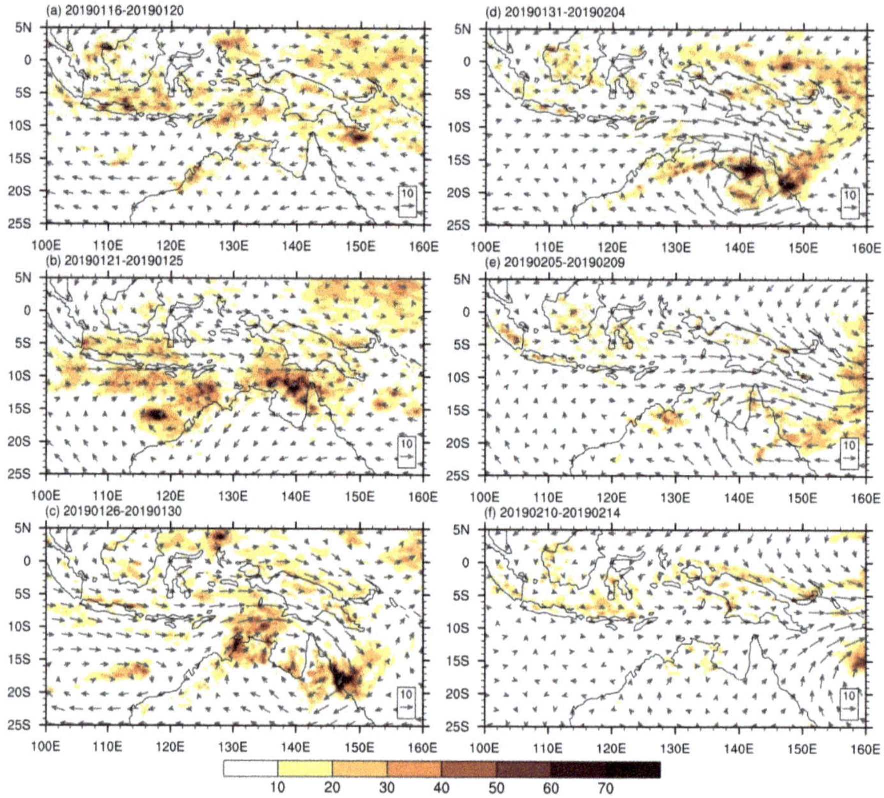

Figure 3. The pentad-mean CMORPH precipitation map (shading, unit: mm day^{-1}), 850-hPa wind field (vectors, unit: m s^{-1}), (**a**–**f**) for the pentads from 16–20 January (Pentad 4) to 10–14 February (Pentad 9) in 2019, which covers the period of 2018/19 SPRE over northeastern Queensland.

The general relationship between the northern Queensland SPREs and the MJOs and ERs is checked based on the frequency that the SPREs and the convective phases of MJO and ER occurred concurrently. The results listed in Table 3 show that for the SPREs in 21 austral summers (DJF) from 1998/99–2018/19, all Box-B SPREs occurred when either the MJO or the ER convective phase passed through, with 57% of the SPREs that occurred when both MJO and ER were in convective phases The years with the largest three SPRE rainfall amounts (2018, 2008, 2006) all occurred concurrently with the convective phases of MJO and ER. Box-A has a higher percentage (76%) of the SPREs that occurred concurrently with the convective phases of MJO and ER than Box-B. However, the Box-B SPRE contributed a higher percentage of the rainfall to the seasonal (DJF) totals compared with the Box-A SPREs. Table 3 shows that for Box-A, the medium of the ratio of an SPRE event to the DJF totals is 31.8%, while for Box-B, the medium is 44.4%. For Box-A, only one year (2018) shows a ratio higher than 50%, while for Box-B seven years show higher than 50% ratios. This suggests that the North Queensland east coast is more sensitive to subseasonal variability. Nevertheless, it is evident in Table 3 that the 2018/19 SPRE is remarkable because for Box-A, the event contributed the largest (50.9%) portion to the seasonal rainfall totals, and for Box-B, the event's total accumulated rainfall amount (477.4 mm) was the largest in 21 years.

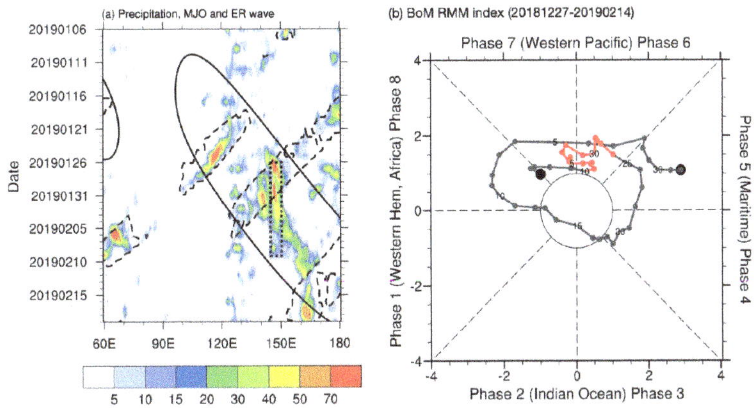

Figure 4. (**a**) The precipitation averaged over 15°~20° S (shading, unit: mm day^{-1}). The solid contours mark out the convective phase of MJO determined by the filtered OLR and the dashed contours mark out the convective phase of the ER waves. The location and period of the 2018/19 SPRE are marked out by the dotted box. Identification procedure of the MJO and ER convective phases is described in Section 2.2. (**b**) The real-time multivariate MJO (RMM) index downloaded from the BoM website (http://www.bom.gov.au/climate/mjo/graphics/rmm.74toRealtime.txt) from 27 December 2018 to 14 February 2019. The period of 2018/19 SPRE is in red.

Table 3. The occurrence pentad (the first pentad), 15-day accumulative rainfall (unit: mm), and the ratio to DJF totals of each SPRE for two box areas over northern Queensland. Checkmarks (Y) are made when convective phase MJO or ER wave occurred during the SPREs.

Year	Box-A (140°~145° E, 15°~20° S)					Box-B (145°~150° E, 15°~20° S)				
	Occurrence Pentad	Rainfall Amount (mm)	Ratio (%)	MJO	ER Wave	Occurrence Pentad	Rainfall Amount (mm)	Ratio (%)	MJO	ER Wave
1998	1	164.5	22.3	Y	Y	2	243.5	32.4	Y	
1999	10	241.1	34.0		Y	10	367.8	40.6		Y
2000	69	252.7	29.8		Y	9	246.5	33.0	Y	Y
2001	9	222.0	32.9	Y	Y	8	161.0	49.3	Y	Y
2002	10	144.4	29.5	Y	Y	10	114.0	44.0	Y	Y
2003	3	180.9	27.2			7	247.8	48.3	Y	Y
2004	71	129.5	24.8	Y	Y	3	197.9	44.4		Y
2005	5	207.7	39.4	Y	Y	4	202.9	51.0	Y	Y
2006	6	296.7	45.1	Y	Y	6	387.9	51.5	Y	Y
2007	8	292.9	30.1	Y	Y	10	315.7	33.5	Y	
2008	5	327.8	30.2	Y		6	434.3	39.8	Y	Y
2009	4	262.8	31.8	Y	Y	4	327.4	47.3	Y	Y
2010	10	285.3	26.4	Y	Y	3	254.2	25.6		Y
2011	5	222.2	32.3	Y	Y	5	223.8	36.1	Y	
2012	3	250.6	48.0	Y	Y	3	261.3	57.7	Y	
2013	6	262.7	32.7		Y	6	223.8	56.0		Y
2014	73	141.4	27.3	Y	Y	8	172.6	39.3		Y
2015	71	173.1	44.5	Y	Y	71	122.8	53.4	Y	Y
2016	1	192.8	35.2	Y	Y	1	191.5	52.0	Y	Y
2017	4	126.2	29.9	Y	Y	10	209.0	43.1	Y	Y
2018	5	294.0	50.9	Y	Y	6	477.4	52.3	Y	Y

To further quantify the MJO and ER modulation on the SPRE rainfall, we calculated the percentage of the time during the 15 days of SPREs that the waves are in convective phases. We use this time percentage as a measure of the temporal modulation of MJO and ER on the SPRE rainfall. We also calculated the ratio of mean rainfall intensity during the convective phase against the mean rainfall intensity during the nonconvective condition (no-signal and suppressed phases) and used the ratio as a measure of intensity modulation of the waves. The results shown in Figure 5 suggest an abnormally strong modulation of MJO and ER on the 2018/19 SPREs over both Box-A and Box-B areas. During the 2018/19 Box-A SPRE, almost all of the 15 days of OLR were in the MJO convective phase, of which the percentage is much higher than the SPRE during other years. The convective phase of ER waves is about 35% which is also above the medium percentage of the 20 years from 1998–2017. On the other hand, Figure 5b shows that during all of the 15 days of the Box-B SPRE the OLR was in convective phase, which is the highest compared with the past 20 years. The percentage of the convective phase of ER also reaches the upper quartile in 20 years, which is clearly above the medium. Therefore, we can conclude that in the 2018/19 case, the temporal modulation of both MJO and ER on the northeastern Queensland SPREs was the strongest since 1998. Regarding the rainfall intensity modulation, we can see in Figure 5c,d that climatologically the rainfall intensity in Box-A is almost two times stronger during the ER convective phase compared with the nonconvective phase, while the difference is smaller in Box-B. On average, the rainfall intensity in two box areas is similar during the ER convective phases, but the variability in Box-A is larger. During the ER nonconvective days, the rainfall intensity in Box-B is slightly larger than that in Box-A. The SPREs in the 2018/19 summer are unusual. Note that the rainfall intensity in Box-B is much larger than in Box-A, although in both areas the intensity is strongest since 1998. The ER modulation is particularly strong. The average rainfall intensity during the ER convective days is almost twice the intensity of the nonconvective ER days.

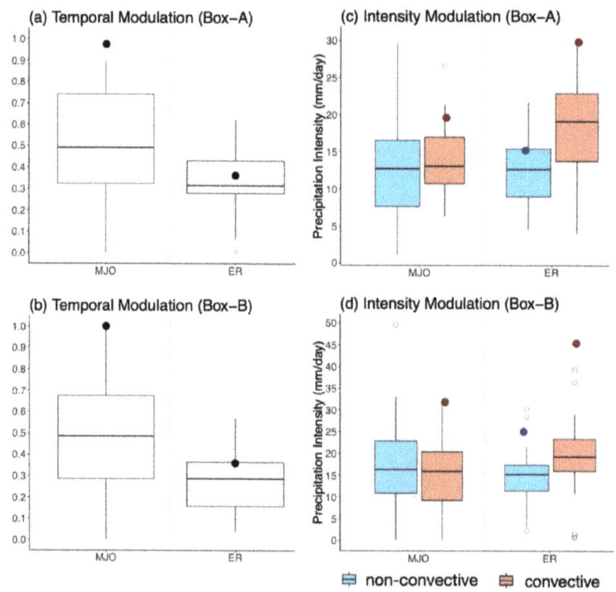

Figure 5. (**a**,**b**) Temporal modulation measured by a percentage in the entire period (15 days) of an SPRE that is associated with the convective phases of MJO or ER wave in (**a**) Box-A and (**b**) Box-B areas. (**c**,**d**) are the intensity modulation measured by a comparison of the mean rainfall intensity during the convective phases of MJO or ER passing through the box areas and the mean rainfall intensity during the nonconvective (suppressed or no-wave days in (**c**) Box-A and (**d**) Box-B. The convectively active CCEWs days are identified when over 50% of the grids achieve the convective threshold for each wave. Boxplots are the statistics using climatological SPREs from 1998 to 2017, and the solid circles are the condition in 2018/19 SPRE. Outliers are estimated using Tukey [32] fences ($[Q_1 - 1.5\text{IQR},\ Q_3 + 1.5\text{IQR}]$) represented in hollow circles.

4. S2S Prediction Evaluation

After seeing the close relationship between monsoon depression and the SPREs, in this section, we will present the assessment of the extended-range forecast quality of the SPREs based on the S2S database. The analysis strategy is, first, to evaluate whether the forecast data can capture the monsoon depression index variability within a 45-day window measured by the zonal wind shear index. Second, we evaluate the rainfall forecast skill with a calibration concept so that all model products can be compared and understood. Two approaches are exercised in this study. One is the percentile rank (PR) of the SPRE rainfall amount based on the percentage distribution of the running 15-day accumulated total rainfall during a season (DJF) obtained from the hindcast database of each model. Another is the percentage contribution of the SPRE rainfall amount to the three-month seasonal (DJF) totals. For the forecast ratio (percentage contribution), the calculation is based on the hindcast database of each model. As the accumulated rain and variability in Box-B is larger than that in Box-A (Figure 1), and the 2018/19 flood is more extreme in Box-B, particularly in Townsville, we will focus on the Box-B SPREs. In the rest of the paper, Box-B is used interchangeably for northern Queensland.

4.1. *Monsoon Depression Index Variability*

Figure 6a shows a composite 45-day time series with the three SPRE pentads from Day(0) to Day(14), the three pentads before the SPREs from Day(-15) to Day(-1), and the three pentads after the SPREs from Day(15) to Day(29). The black curve is the average of the 20-year (1998–2017) 15-day accumulative rainfall running by pentad, and the gray shade shows the range of the 15-day rainfall for each pentad during the period of analysis. The composited monsoon depression index defined by the 850-hPa zonal wind shear (Figure 2c) during the same 45-day time window aligned with the SPREs is presented in Figure 6b. The 45-day mean value is subtracted from the 45-day mean in order to compare the index generated by different models with a specific focus on the relative intensity of the monsoon depression during the SPREs. Figure 6b shows a clear variation pattern associated with SPREs. Monsoon depression shows an intensifying tendency before Day(0) of the SPRE and reaches its peak at the second pentad of the SPRE, then weakens after the SPRE. The coherent relationship between SPRE precipitation and the wind shear monsoon depression index is evident. Therefore, evaluating the relative intensity of the 850-hPa wind shear index during the occurrence period of the SPREs can provide some insight into model forecast performance.

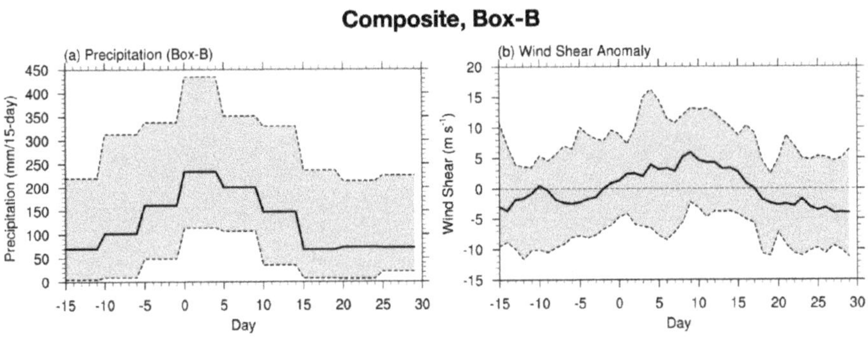

Figure 6. Composited (**a**) 15-day accumulative precipitation (unit: mm) running by pentad averaged over Box-B and (**b**) wind shear anomaly (unit: m s^{-1}) over Australian monsoon region (U_{850}[5°–15° S, 115°–145° E]–U_{850}[20°–30° S, 115°–145° E]), where the anomaly is obtained by subtracting the 45-day-mean in the extended SPRE period (Days 0 to 14 mark the SPRE period). The time range of CMORPH precipitation data ERA-Interim wind is from 1998 to 2017. The solid curves and gray areas indicate the median and the range of the 20-year SPRE cases, respectively.

The 20 years of the daily index anomalies of the SPREs are sorted and grouped into 10 bins, as is shown in the x-axis of Figure 7a–d. Note that here, the index anomalies are defined as the deviation from the 45-day mean value. The forecast data of each model are grouped separately according to the forecast lead times. Then, the cumulative distribution function (CDF) is calculated as the cumulative percentage contributed by each bin during the 15 days of the SPREs (Figure 7a–d). The forecast error presented in Figure 7e is estimated by the integrated difference between the model forecast and ERA-Interim CDF curves. It is clear that the four S2S models of this study can capture the monsoon depression peak tendency reasonably well. The BoM model has the smallest CDF differences in a short-range (1~3 days) forecast, and the ECMWF model has the smallest CDF differences in a medium-range (4~8 days) forecast, and in an extended-range (8~16 days) forecast, the CDF differences of the CMA model are larger than other three models. If the short-range forecast performance is used as a baseline to measure how forecast skill changes with the forecast lead times, Figure 7e shows that ECMWF is the only model that maintains a skill comparable to the short-range up to 10 days. It is worth noting that although the assessment here is measured by the monsoon depression index variability during the SPRE-centered 45 days, the same method can be applied to other studies in different regions.

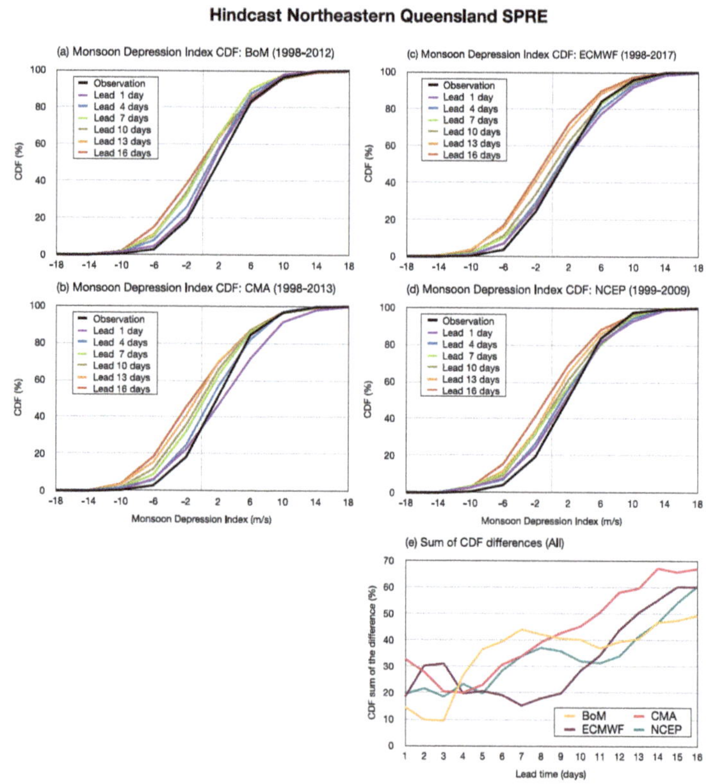

Figure 7. The cumulative distribution function (CDF) of the monsoon depression index anomalies (obtained by subtracting the averages of Days -15 to 29 for each year) represented by 850-hPa zonal wind shear (Figure 2c) in observation (black curves) and (**a**) BoM, (**b**) CMA, (**c**) ECMWF, and (**d**) NCEP S2S models for lead times 1, 4, 7, 10, 13, and 16 day(s) during the 15-day SPREs period in the available years in hindcast datasets. (**e**) illustrates the sum of CDF difference between S2S models and observation varying with lead time. The hindcast years of BoM hindcast model are from 1998 to 2013, CMA from 1998 to 2012, ECMWF from 1998 to 2017, and NCEP from 1998 to 2009.

The verification results for the 2018/19 SPRE are presented in Figure 8. The observational (ERA-Interim) CDF curve suggests that for this case, the monsoon depression index anomalies during almost the entire 15 days are positive. In fact, this feature can be clearly identified in Figure 2c (the bottom figure), where we see within the 45-day period from January 11–February 24, the zonal wind shear monsoon index is particularly strong during the 15 days (January 26~February 9) of the SPRE. The forecast errors or CDF differences in Figure 8e suggest that the CMA model forecast captured the tendency of strong monsoon depression up to the lead time of 16 days. The BoM and ECMWF models also can capture this tendency relatively well, while the NCEP model shows more rapidly growing differences after the 8th day forecast compared with the other model.

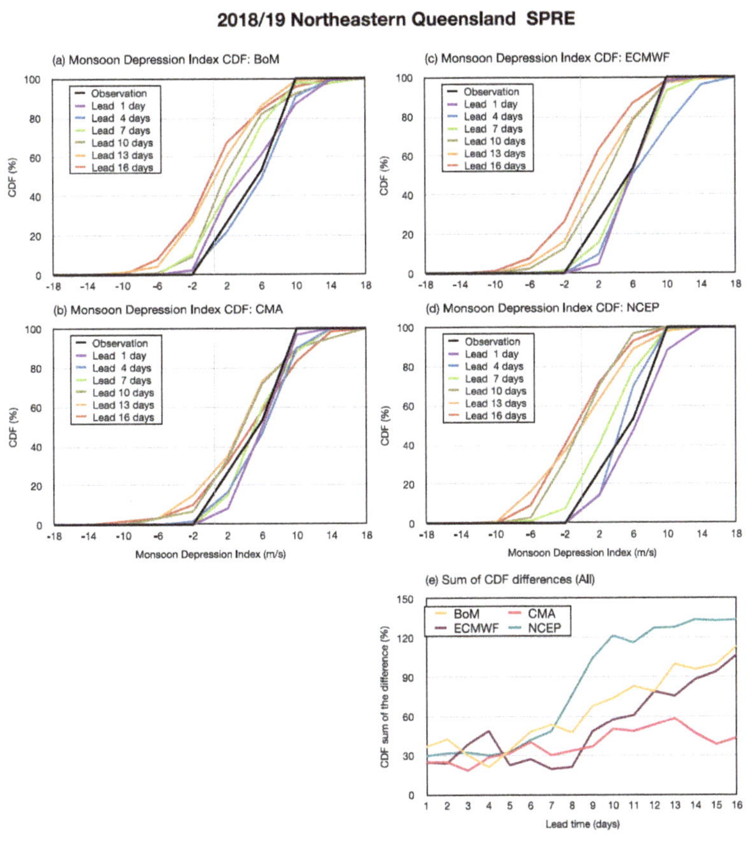

Figure 8. The sane as Figure 7, but for 2018/19 Box-B SPRE.

4.2. SPRE Ranked Rainfall Amount

The rainfall forecast performance is evaluated based on the 15-day accumulated rainfall amount percentile ranks (PRs). The analysis procedure is as follows. First, the 15-day accumulated rainfall amount over Box-B during the three months of from December to February in the 20 years of from 1998 to 2017 is sorted and converted to percentile ranks. The hindcast precipitation data obtained from the S2S Project Database is also converted to PRs. As the ensemble size and forecast frequency vary with models (Table 1), the sample sizes of different models are different. The PR ranges of the SPREs based on CMORPH data are presented in the hollow boxes in Figure 9a–d. As the hindcast years of different models are different, the observational PR ranges formed by the hindcast years are slightly different between models. The blue boxes in Figure 9a–d are the PR distribution based on model hindcast data with different lead times; the red boxes are the PR distribution of the ensemble forecast for the 2018/19 SPRE. The narrow distribution range of the CMA model is due to the small ensemble size (4). When the short-range forecast performance is used as a reference to measure the performance of longer leads, we can see that the NCEP model (Figure 9d) shows the smallest decreasing slope of the difference between the medians of the long-lead members and the short-lead members. The BoM model (Figure 9a) shows that for the 2018/19 SPRE, the forecast up to the lead time of 8 days is better than the historical performance. The CMA model (Figure 9b) shows that the 2018/19 SPRE forecast is exceptionally good. Up to the lead time of 15 days, the PR remains near the top as the observational data shows (Figure 1c). The ECMWF model (Figure 9c) shows that up to 9 days, and the NCEP model (Figure 9d) shows up to 8 days, of the 2018/19 SPRE forecast, the performance is better than the historical statistics. The forecast performance can be summarized using the root mean squared sum of the difference between the forecast and observed PR, which can be interpreted as a kind of root-mean-squared error (RMSE) termed as PR-RMSE. Figure 10 shows that historically, the PR of SPRE-accumulated rainfall is best predicted by the NCEP model, while for the 2018/19 SPRE prediction, CMA prediction is the best.

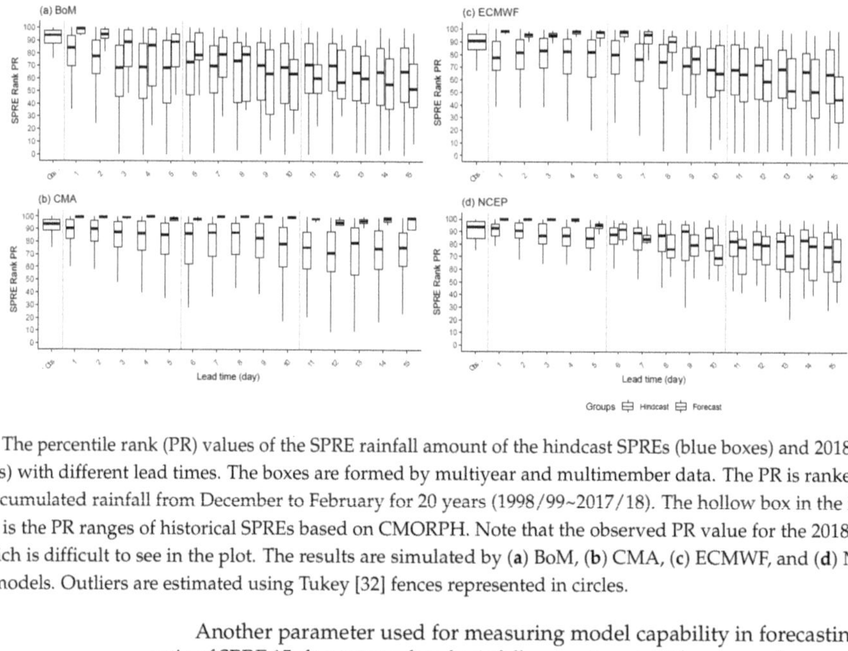

Figure 9. The percentile rank (PR) values of the SPRE rainfall amount of the hindcast SPREs (blue boxes) and 2018/19 SPRE (red boxes) with different lead times. The boxes are formed by multiyear and multimember data. The PR is ranked with all 15-day accumulated rainfall from December to February for 20 years (1998/99~2017/18). The hollow box in the far left for each plot is the PR ranges of historical SPREs based on CMORPH. Note that the observed PR value for the 2018/19 SPRE is 99, which is difficult to see in the plot. The results are simulated by (**a**) BoM, (**b**) CMA, (**c**) ECMWF, and (**d**) NCEP S2S forecast models. Outliers are estimated using Tukey [32] fences represented in circles.

Another parameter used for measuring model capability in forecasting SPRE is the ratio of SPRE 15-day accumulated rainfall amount against the seasonal accumulated rainfall amount. The ratios over Box-A and Box-B are presented in the third and eighth column in Table 3. The square root of the average sum of the differences between the forecast and observation ratios with the forecast lead times from 1 to 15 days, termed as ratio root-mean-squared-errors (ratio-RMSE), for Box-B SPRE is presented in Figure 10. Note that the forecast part of the ratio-RMSE is the multimember forecasts, and the observation part is the CMORPH precipitation. The boxplots are the distribution of the multiyear hindcast database that reflects the expected ratio-RMSE for different models. In order to discern the possible influences of ER and MJO, we selected one year (1999) with ER but without MJO influence and another year (2007) with MJO but without ER influence to compare their ratio-RMSE with the 2019 case, which is a case with strong influences of both ER and MJO. Figure 10 shows that the ratio-RMSE of 1999 is evidently much smaller than the other two years, while the ratio-RMSE of 2019 is the largest. The results suggest the possibility that although MJO is helpful for extended-range rainfall category forecast, the ER influence can overwhelm the predictability associated with MJO influence if the model fails to capture the ER waves. Note that 1999 is a La Niña year, and 2007 is a weak El Niño year. We have compared the ER and MJO modulation on the SPRE intensity during different phases of the ENSO years by plotting the figures similar to Figure 5 and found that the ER modulation is strongest during the La Niña year. The ER wave convection frequently originates at the active SPCZ region. In summary, the ER modulation on average is clearer than MJO, regardless of ENSO. Therefore, the strong ER influence in 2019 seems to be unusual. The strong enhancement by other factors, such as the Warm Air Advection (WAA) described in Callaghan [27], can be important. Further research in this direction is desperately needed to find enough evidence to support the hypothesis.

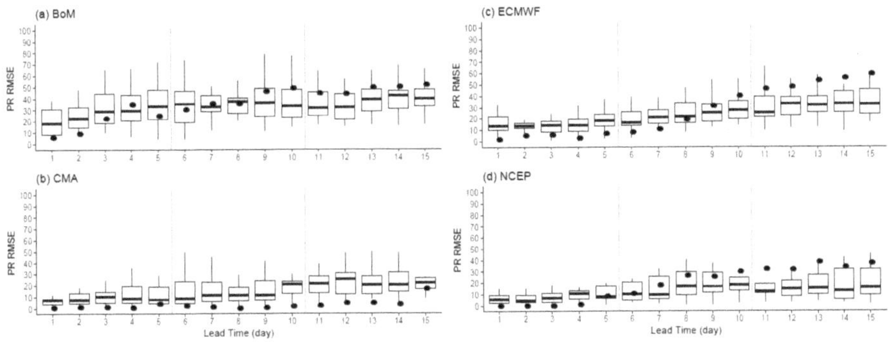

Figure 10. The root-mean-squared error (RMSE) calculated as the difference between the model forecast PR and the observed PR (as in Figure 9), termed as PR-RMSE. The boxes are formed by multiyear and multimember hindcast RMSE, and the black points represent the RMSE of the 2018/19 SPRE. Outliers estimated using Tukey fences [32] are illustrated in white circles.

5. Summary and Discussion

The extreme rainfall event that caused devastating floods in northeastern Queensland in 2019 has been analyzed to understand the relationship between a regional sub-seasonal peak rainfall event (SPRE) and major influential large-scale drivers. Based on the findings of analyzing observational data, we further analyzed the extended-range (8~16 days) forecast skill for northern Queensland SPREs using the S2S database. We found that the 2019 Queensland floods were caused by a strong SPRE that broke the 20-year (1998–2017) record of the SPRE rainfall amount. The strong SPRE was associated with a strong monsoon depression reflected by the 850-hPa wind shear index (Figure 2c) that shows prolonged positive anomalies within a 45-day time window centered at the SPRE (Figure 8). The enhanced monsoon depression anomaly was modulated by the convective phase of the MJO and the convective ER wave originating from the strong convection associated with El Niño over the equatorial Pacific near the dateline. Results of the 20-year climate data analysis show strong modulation of the ER waves on the rainfall intensity of the northern Queensland SPREs, while the MJO modulation is stronger on the number of convective days of the SPRE (Figure 5).

The skill of S2S forecast models on forecasting the northern Queensland SPREs are assessed based on the analysis from two perspectives. The first one is to assess the forecast skill of the monsoon depression anomaly represented by the 850-hPa wind shear. The second is to assess the forecast skill of the precipitation with a calibration concept, which includes calculating the RMSE of the percentile rank (PR) of the SPRE rainfall amount and the RMSE of the ratio (percentage contribution) of the SPRE to the three-month season (DJF). The assessment results of the four S2S models suggest that for the monsoon depression anomaly, the models can maintain a similar skill as the short-range (3 day) forecast up to 8–10 days (Figure 7). On average, the model prediction performance for the 2018 SPRE case was worse than expected, except for the CMA model (Figure 8). The conclusions of the precipitation prediction performance assessment are similar to the ones obtained for the monsoon depression index. In addition, we selected two SPREs in 1999 and 2007 to compare their ratio-RMSEs with the 2019 SPRE. The 1999 SPRE was modulated by ER wave only, the 2007 SPRE was modulated by MJO only, and the 2018 SPRE was modulated by both ER and MJO. It turned out that except for CMA, the other three models show the smallest ratio-RMSE in 1999 and the largest ratio-RMSE in 2018. We suspect that the reason why the CMA model is different from the other three models in a single-year comparison may be due to the small size of its ensemble members, which limits the statistical robustness of the assessment. The model resolution is an important factor to consider when comparing model performance [33]. It is beyond the scope of our current study to carry out detailed analysis of the model physics associated with the predictability of SPREs. More research is needed in this direction for improving our understanding on S2S prediction model capability and predictability.

This study demonstrated a useful analysis procedure that can be used to analyze SPREs and their associated large-scale drivers in other regions. When focusing on the SPREs, the ensemble size and forecast frequency of a model become critical. The modulation of MJO, ER, and ENSO on the SPREs is an extremely important subject for S2S prediction. It is our ongoing work, and the results will be presented in a separate paper.

Author Contributions: Conceptualization, M.-M.L.; methodology, M.-M.L. and W.Y.-H.T.; software, W.Y.-H.T. and Y.-M.C.; formal analysis, W.Y.-H.T. and M.-M.L.; investigation, M.-M.L. and W.Y.-H.T.; resources, M.-M.L. and C.-H.S.; writing—original draft preparation, M.-M.L. and W.Y.-H.T.; writing—review and editing, M.-M.L.; visualization, W.Y.-H.T. and Y.-M.C.; supervision, M.-M.L. and C.-H.S.; project administration, M.-M.L.; funding acquisition, M.-M.L. and C.-H.S. All authors have read and agreed to the published version of the manuscript.

Funding: This research was funded by the Ministry of Science and Technology, Taiwan, Grant MOST 108–2111-M-002-016, MOST 109–2111-M-002-004, MOST 109-2111-M-002-005, and MOST 109-2811-M-002-646-MY2.

Institutional Review Board Statement: Not applicable.

Informed Consent Statement: Not applicable.

Data Availability Statement: CMORPH precipitation data can be downloaded from ftp://ftp.cpc.ncep.noaa.gov/precip/CMORPH_V1.0/ (last accessed on 27 April 2021). Interpolated daily outgoing longwave radiation (OLR) version 1.2 can be downloaded from https://www.ncei.noaa.gov/data/outgoing-longwave-radiation-daily/access/ (last accessed on 27 April 2021). ERA-Interim data can be download from https://www.ecmwf.int/en/forecasts/datasets/reanalysis-datasets/era-interim (last accessed on 31 August 2019). S2S database can be obtained from https://apps.ecmwf.int/datasets/data/s2s-reforecasts-instantaneous-accum-ecmf/ (last accessed: 27 April 2021).

Acknowledgments: Not applicable.

Conflicts of Interest: The authors declare no conflict of interest.

References

1. Cowan, T.; Wheeler, M.C.; Alves, O.; Narsey, S.; de Burgh-Day, C.; Griffiths, M.; Jarvis, C.; Cobon, D.H.; Hawcroft, M.K. Forecasting the extreme rainfall, low temperatures, and strong winds associated with the northern Queensland floods of February 2019. *Weather Clim. Extrem.* **2019**, *26*, 100232. [CrossRef]
2. Gissing, A.; O'Brien, J.; Hussein, S.; Evans, J.; Mortlock, T. Townsville 2019 flood—insights from the field. *Risk Front. Brief. Note* **2019**, *389*, 1–8.
3. Gissing, A. To build or not build: That is the Townsville question. *Risk Front. Brief. Note* **2019**, *388*, 1–5.
4. Zhang, H.; Moise, A. The Australian Summer Monsoon in Current and Future Climate. In *The Monsoons and Climate Change*; de Carvalho, L., Jones, C., Eds.; Springer: Cham, Switzerland, 2016; pp. 67–120. ISBN 9783319216492.
5. Nicholls, N.; McBride, J.L.; Ormerod, R.J. On Predicting the Onset of the Australian Wet Season at Darwin. *Mon. Weather Rev.* **1982**, *110*, 14–17. [CrossRef]

6. Hung, C.-W.; Yanai, M. Factors contributing to the onset of the Australian summer monsoon. *Q. J. R. Meteorol. Soc.* **2004**, *130*, 739–758. [CrossRef]
7. Lisonbee, J.; Ribbe, J.; Wheeler, M. Defining the north Australian monsoon onset: A systematic review. *Prog. Phys. Geogr. Earth Environ.* **2019**, *44*, 1–21. [CrossRef]
8. Madden, R.A.; Julian, P.R. Detection of a 40–50 Day Oscillation in the Zonal Wind in the Tropical Pacific. *J. Atmos. Sci.* **1971**, *28*, 702–708. [CrossRef]
9. Madden, R.A.; Julian, P.R. Description of Global-Scale Circulation Cells in the Tropics with a 40–50 Day Period. *J. Atmos. Sci.* **1972**, *29*, 1109–1123. [CrossRef]
10. Wheeler, M.C.; Hendon, H.H.; Cleland, S.; Meinke, H.; Donald, A. Impacts of the Madden-Julian oscillation on australian rainfall and circulation. *J. Clim.* **2009**, *22*, 1482–1498. [CrossRef]
11. Wheeler, M.C.; Kiladis, G.N.; Webster, P.J. Large-Scale Dynamical Fields Associated with Convectively Coupled Equatorial Waves. *J. Atmos. Sci.* **2000**, *57*, 613–640. [CrossRef]
12. Chen, W.-T.; Hsu, S.-P.; Tsai, Y.-H.; Sui, C.-H. The Influences of Convectively Coupled Kelvin Waves on Multiscale Rainfall Variability over the South China Sea and Maritime Continent in December 2016. *J. Clim.* **2019**, *32*, 6977–6993. [CrossRef]
13. Dare, R.A.; Davidson, N.E.; McBride, J.L. Tropical Cyclone Contribution to Rainfall over Australia. *Mon. Weather Rev.* **2012**, *140*, 3606–3619. [CrossRef]
14. Dare, R.A. Seasonal Tropical Cyclone Rain Volumes over Australia. *J. Clim.* **2013**, *26*, 5958–5964. [CrossRef]
15. King, A.D.; Klingaman, N.P.; Alexander, L.V.; Donat, M.G.; Jourdain, N.C.; Maher, P. Extreme rainfall variability in Australia: Patterns, drivers, and predictability. *J. Clim.* **2014**, *27*, 6035–6050. [CrossRef]
16. Vitart, F.; Robertson, A.W. Introduction: Why Sub-seasonal to Seasonal Prediction (S2S)? In *Sub-Seasonal to Seasonal Prediction*; Robertson, A.W., Vitart, F., Eds.; Elsevier: New Yok, NY, USA, 2019; pp. 3–15.
17. Tsai, W.Y.; Lu, M.-M.; Sui, C.-H.; Lin, P.-H. MJO and CCEW modulation on South China Sea and Maritime Continent boreal winter subseasonal peak precipitation. *Terr. Atmos. Ocean. Sci.* **2020**, *31*, 177–195. [CrossRef]
18. Cowan, T.; Stone, R.; Wheeler, M.C.; Griffiths, M. Improving the seasonal prediction of Northern Australian rainfall onset to help with grazing management decisions. *Clim. Serv.* **2020**, *19*, 100182. [CrossRef]
19. Vitart, F.; Ardilouze, C.; Bonet, A.; Brookshaw, A.; Chen, M.; Codorean, C.; Déqué, M.; Ferranti, L.; Fucile, E.; Fuentes, M.; et al. The Subseasonal to Seasonal (S2S) Prediction Project Database. *Bull. Am. Meteorol. Soc.* **2017**, *98*, 163–173. [CrossRef]
20. Berrisford, P.; Dee, D.P.; Poli, P.; Brugge, R.; Fielding, K.; Fuentes, M.; Kållberg, P.W.; Kobayashi, S.; Uppala, S.; Simmons, A. The ERA-Interim archive Version 2.0. *ECMWF* **2011**, *23*. Available online: https://www.ecmwf.int/node/8174 (accessed on 24 September 2018).
21. Xie, P.; Joyce, R.J.; Wu, S.; Yoo, S.-H.; Yarosh, Y.; Sun, F.; Lin, R. Reprocessed, Bias-Corrected CMORPH Global High-Resolution Precipitation Estimates from 1998. *J. Hydrometeorol.* **2017**, *18*, 1617–1641. [CrossRef]
22. Lee, H.-T. NOAA Climate Data Record (CDR) of Daily Outgoing Longwave Radiation (OLR), Version 1.2. Available online: https://data.nodc.noaa.gov/cgi-bin/iso?id=gov.noaa.ncdc:C00875 (accessed on 24 September 2018).
23. Wheeler, M.C.; Kiladis, G.N. Convectively Coupled Equatorial Waves: Analysis of Clouds and Temperature in the Wavenumber–Frequency Domain. *J. Atmos. Sci.* **1999**, *56*, 374–399. [CrossRef]
24. Kiladis, G.N.; Wheeler, M.C.; Haertel, P.T.; Straub, K.H.; Roundy, P.E. Convectively coupled equatorial waves. *Rev. Geophys.* **2009**, *47*, 1–42. [CrossRef]
25. Yim, S.Y.; Wang, B.; Liu, J.; Wu, Z. A comparison of regional monsoon variability using monsoon indices. *Clim. Dyn.* **2014**, *43*, 1423–1437. [CrossRef]
26. Kajikawa, Y.; Wang, B.; Yang, J. A multi-time scale Australian monsoon index. *Int. J. Climatol.* **2010**, *30*, 1114–1120. [CrossRef]
27. Callaghan, J. Weather systems and extreme rainfall generation in the 2019 north Queensland floods compared with historical north Queensland record floods. *J. South. Hemisph. Earth Syst. Sci.* **2021**, *71*, 123–146. [CrossRef]
28. Wheeler, M.C.; McBride, J.L. Australian-Indonesian monsoon. In *Intraseasonal Variability in the Atmosphere-Ocean Climate System*; Lau, W.K.M., Waliser, D.E., Eds.; Springer: Berlin/Heidelberg, Germany, 2005; pp. 125–173. ISBN 9783540272502.
29. Rui, H.; Wang, B. Development Characteristics and Dynamic Structure of Tropical Intraseasonal Convection Anomalies. *J. Atmos. Sci.* **1990**, *47*, 357–379. [CrossRef]
30. Haertel, P.; Boos, W.R. Global association of the Madden-Julian Oscillation with monsoon lows and depressions. *Geophys. Res. Lett.* **2017**, *44*, 8065–8074. [CrossRef]
31. L'Heureux, M.; Bell, G.D.; Halpert, M.S. ENSO and the tropical Pacific [in "State of the Climate in 2019"]. *Bull. Am. Meteorol. Soc.* **2020**, *101*, S185–S238. [CrossRef]
32. Tukey, J.W. *Exploratory Data Analysis*; Addison-Wesley: Boston, MA, USA, 1977.
33. Hawcroft, M.; Lavender, S.; Copsey, D.; Milton, S.; Rodríguez, J.; Tennant, W.; Webster, S.; Cowan, T. The benefits of ensemble prediction for forecasting an extreme event: The Queensland Floods of February 2019. *Mon. Weather Rev.* **2021**. [CrossRef]

atmosphere

MDPI

Article

The Evolution Characteristics of Daily-Scale Silk Road Pattern and Its Relationship with Summer Temperature in the Yangtze River Valley

Chao Wang [1,*], Ying Wen [1,*], Lijuan Wang [2], Xianbiao Kang [1] and Yunfeng Liu [1]

1 College of Air Traffic Management, Civil Aviation Flight University of China, Guanghan 618307, China; xianbiaokang@cafuc.edu.cn (X.K.); lyf@cafuc.edu.cn (Y.L.)

2 Key Laboratory of Meteorological Disaster, Ministry Education, Joint International Research Laboratory of Climate and Environment Change, Collaborative Innovation Center on Forecast and Evaluation of Meteorological Disasters, Nanjing University of Information Science and Technology, Nanjing 210044, China; 000314@nuist.edu.cn

* Correspondence: wonchaos@cafuc.edu.cn (C.W.); wenying@cafuc.edu.cn (Y.W.)

Abstract: By employing multi-reanalysis daily datasets and station data, this study focuses on the evolution characteristics of the daily-scale Silk Road pattern (SRP) and its effect on summer temperatures in the Yangtze River Valley (YRV). The results manifest that the evolution characteristics of positive- and negative-phase SRP (referred to SRP+ and SRP−) exhibit marked distinctions. The anomaly centers of SRP+ over West Central Asia (WCA) and Mongolia emerge firstly, vanishing simultaneously one week after peak date; however, the Far East (FE) anomaly centers can persist for a longer period. The SRP− starts with the WCA and FE centers, with a rapid decline in the strength of the WCA center and preservation of other anomaly centers after its peak. In the vertical direction, daily-scale SRP mainly concentrates in the mid-to-upper troposphere. Baroclinicity accounts for its early development and barotropic instability process favors the maintenance. Moreover, the SRP+ (SRP−) is inextricably linked to heat wave (cool summer) processes in the YRV. Concretely, before the onset of SRP+ events, an anomalous anticyclone and significant negative vorticities over East Asia related to SRP+ favor the zonal advance between the South Asia high (SAH) and western Pacific subtropical high (WPSH), inducing local descents over YRV area. The sinking adiabatic warming and clear-sky radiation warming can be considered as the possible causes for the YRV heat waves. The adiabatic cooling with the local ascents leads to more total cloud cover (positive precipitation anomalies) and less solar radiation incident to surface of the YRV, inducing the cool summer process during SRP−.

Keywords: Silk Road pattern; evolution characteristics; summer temperature; Yangtze River Valley

Citation: Wang, C.; Wen, Y.; Wang, L.; Kang, X.; Liu, Y. The Evolution Characteristics of Daily-Scale Silk Road Pattern and Its Relationship with Summer Temperature in the Yangtze River Valley. *Atmosphere* **2021**, *12*, 747. https://doi.org/10.3390/atmos12060747

Academic Editors: Ankit Agarwal, Naiming Yuan, Kevin K.W. Cheung, Roopam Shukla and Graziano Coppa

Received: 18 April 2021
Accepted: 7 June 2021
Published: 9 June 2021

1. Introduction

For a long time, meteorologists generally point out that summer weather and climate over East Asia are inextricably linked to the East Asian summer monsoon, Qinghai-Tibet Plateau, and other external forcing factors (i.e., sea surface temperature in the tropical Pacific, Arctic sea ice, solar activity, and soil moisture) [1–3]. Variations of East Asian summer monsoon circulation are particularly complicated owing to the existence and persistence of summer atmospheric teleconnection patterns [4,5]. The well-known Silk Road pattern (SRP) is recognized as one of the dominant modes and most influential patterns during boreal summer over the Eurasian continent [6–9] and is considered as an effective predictor of East Asian summer climate anomalies.

In literature, the SRP features a stationary Rossby wave train pattern trapped along the upper-level westerly jet, geographically fixed over the Eurasian continent along approximately 40° N, which resembles the Eurasian part of a circumglobal teleconnection pattern in mid-latitude circulation of the Northern Hemisphere [10–13]. Previous studies indicated that it can regulate the interaction between the members of the Eurasian circulation systems and markedly affect the summer weather and climate anomalies over East Asia [5,14]. For instance, SRP exerts substantial influence on East Asian summer monsoon circulation [7,15,16]. It is closely linked to the Asian summer monsoon in both subtropical East Asia and India, and it could impact the withdrawal of the South China Sea (SCS) summer monsoon via modulating atmospheric circulation anomalies over the SCS [17]. The SRP contributes to formation of the Bonin high, the western Pacific subtropical high (WPSH), and the Mei-Yu front, affecting the precipitation and temperature around the Asian jet regions [14]. The positive (negative) phases SRP maybe triggers upper-level cyclonic (anticyclonic) anomalies and strong divergence (convergence) in the mid-latitudes over East Asia, with more (less) rainfall in southern China [5]. Moreover, the SRP has a profound effect on the temperature anomalies over the Japan, Europe, and Southeast Asia [18–20]. The combination of SRP and the Pacific–Japan pattern could cause stronger convergence (divergence) and significant anomalies of sea level pressure over Japan, leading to larger surface temperature decrease in the northern part of Japan [18]. Many studies pointed out that the SRP can significantly affect the temperature variation over many parts of Eurasia [11,19,20]. In addition, subsequent studies have indicated that SRP also has a significant effect on climate and weather over the northern China, southern China, and Indian Monsoon regions [7,21–24].

Plenty of works have been devoted to study its evolution characteristics, formation mechanism, and influence on East Asian summer monsoon climate by analyzing monthly fields [13,14]. Recently, various studies have shown that the atmospheric teleconnections can also be identified on daily timescales by analyzing daily fields, revealing their formation mechanism and evolution characteristics [25–28]. Besides, previous studies on the relationship between the SRP and summer temperature mainly focus on the Europe and Japan [18,19]. The middle and lower reaches of Yangtze River Valley in China is significantly affected by East Asian summer monsoon, which is prone to high temperature weather [29]. Moreover, the YRV is one of the most densely populated and economically developed region in China. The heat waves in the YRV have a more important impact on people's production and life. However, the impact of SRP on the summer temperature in the YRV and the role of SRP in temperature anomalies need to be further explored [30]. Therefore, the current work attempts to explore the following issues: (1) what are the evolution characteristics of daily-scale SRP? (2) What are the influence of SRP on summer temperature in the Yangtze River Valley and underlying mechanisms? Addressing these issues will be beneficial to developing a better understanding of the SRP's variation and its effect on the summer climate in China. As we will show, daily analysis unfolds more details of SRP climate effects.

2. Data and Methods

2.1. Data

This primary dataset used in this study is the European Centre for Medium-Range Weather Forecast (ECMWF) Reanalysis data (ERA5), on a horizontal resolution of 0.25° × 0.25° in longitude and latitude, with 37 regular vertical pressure levels. The variables include geopotential height (gpm), wind field (m s^{-1}), vertical p-velocity (Pa s^{-1}), air temperature (K), 2 metre temperature (K), surface solar radiation downwards (kJ m^{-2}), total cloud cover (0–1), and relative vorticity (s^{-1}) [31]. This reanalysis dataset is available online https://cds.climate.copernicus.eu/ (accessed on 7 May 2021). The analyzed time span is selected from 1979 to 2020 for the warm season (June–August).

We also employed the daily precipitation data obtained from China's Ground Precipitation 0.5° × 0.5° Gridded Dataset (version 2.0) [32,33]. This observational dataset is available online http://data.cma.cn (accessed on 1 April 2020).

2.2. Methods

The present study mainly utilizes the methods of empirical orthogonal function (EOF) analysis, composite analysis, and the Student's *t*-test. Only the results satisfying the criteria at the 0.05 confidence level are deemed statistically significant. According to the suggestion of Grumm and Hart [34], a 21-day binomial filter (10 days either side of a specific day) was firstly applied, which can effectively highlight the daily variability of each variable. Then, based on the period of 1979–2020, the daily climatological mean value and standard deviation (σ) of each variable were calculated based on the above smoothed daily fields. Compared with using unsmoothed single-day values, the climatological mean value and standard deviation appear more stable. Such 21-day sliding windows can also highlight the daily variability, which is an effective method for identifying typical synoptic to sub-monthly scale circulation patterns [26,27]. Finally, the normalized anomaly of a variable on a specific day was calculated as follows: the corresponding climatological daily mean was first subtracted to retain the majority of the synoptic signal, and the value was divided by the corresponding climatological daily standard deviation.

In addition, the phase-independent wave-activity flux (WAF) was calculated to describe the wave energy propagation characteristics of the quasi-stationary waves and transient fluctuations, which refers to the two-dimensional formula in the previous studies [35]. The climatological daily mean flow during summer (June–August) from 1979 to 2020 was used as the basic flow of this WAF formula, including zonal and meridional wind fields with zonal nonuniformity. The expression of two-dimensional WAF is:

$$W = \frac{p}{2000\,|\vec{U}|}\left\{ \begin{array}{l} U\left(\Psi'^2_x - \Psi'\Psi'_{xx}\right) + V\left(\Psi'_x\Psi'_y - \Psi'\Psi'_{xy}\right) \\ U\left(\Psi'_x\Psi'_y - \Psi'\Psi'_{xy}\right) + V\left(\Psi'^2_x - \Psi'\Psi'_{yy}\right) \end{array} \right\}$$

where $\vec{U} = (U,\ V)$ denotes the horizontal zonally varying basic flow, U and V represent the zonal and meridional wind component respectively, p signifies the pressure (hPa). Ψ' is the stream function for quasi-geostrophic flow.

3. Identification of Daily-Scale SRP

Teleconnection patterns are usually identified in the summer mean circulation by analyzing monthly-mean fields in most of the previous studies [36,37]. This paper attempts to define the daily-scale SRP by using daily data. To identify the SRP, EOF analysis is performed on 200 hPa daily normalized height anomalies along the Asian jet region of 30–150° E, 30–60° N during the summertime (June–August) of 1979 to 2020. The first EOF mode is the so-called Silk Road pattern, which is wavelike and zonally organized. It explains 21.99% of the total variance (Figure 1). This pattern is very similar to the Silk Road pattern obtained by previous studies based on the monthly-mean fields [7,12,38].

Then, a daily SRP index (SRPI) was defined. Firstly, the three basic points selected to represent three anomaly centers of the SR pattern are (65° E, 40° N) over the West Central Asia (WCA), (100° E, 40° N) over Mongolian (MO) region, and (130° E, 40° N) over the Far East (FE), based on the first mode of EOF analysis (Figure 1). The same basic points were selected in previous studies to analyze the Silk Road pattern by use of monthly-mean data [17,18]. Secondly, SRPI was calculated by use of daily normalized 200 hPa geopotential height anomalies at the three basic points (Equation (1)).

$$\mathrm{SRPI} = \frac{(\mathrm{H_{WCA}} - \mathrm{H_{MO}} + \mathrm{H_{FE}})}{3}, \tag{1}$$

where the $\mathrm{H_{WCA}}$, $\mathrm{H_{MO}}$, and $\mathrm{H_{FE}}$ represent the normalized 200 hPa geopotential height anomalies at the three basic points in the EOF1 mode (the triangles in Figure 1), respectively. The positive phase appears as the SRPI exceeds zero, while the negative phase appears when SRPI is below zero. The peak phase of SRP is defined as the day when the daily SRPI is a local maximum and exceeds +1.0 standard deviation (σ), representing that the geopotential height anomalies of action centers attain maximal value over the Asian jet region.

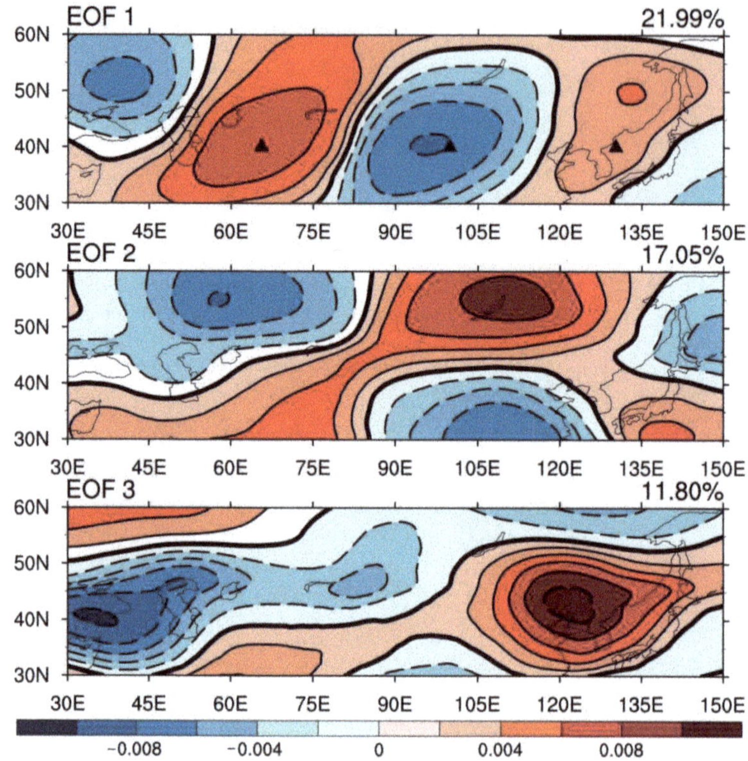

Figure 1. The preceding three EOF modes of daily 200 hPa normalized height anomalies along the Asian jet region (30–150° E, 30–60° N) during the summertime (June–August) of 1979 to 2020. Solid and dashed contours in EOF1 mode denote the positive and negative values, respectively (with interval of 0.002). The triangles represent the three basic points in Equation (1).

4. Evolution Characteristics of Daily-Scale SRP

Composite analysis will be used in this section to investigate the evolution characteristics of the SRP on daily timescales. To ensure that the composite results satisfy the statistical significance, the adequate sample sizes of the positive and negative phases of SR pattern need to be picked out firstly (for convenience, referred to SRP+ and SRP− events respectively in current study).

Based on the daily fields processed by the method in Section 2.2, the more details and some fresh features of evolution characteristics and vertical structure of SRP will be investigated in the following analysis.

Daily-scale SRP+ (SRP−) events were identified, after the SRPI being defined. A typical SRP+ (SRP−) event must meet the following three criteria simultaneously. (1) SRPI is greater than (less than) +1.0 (−1.0) standard deviation (σ) for at least three consecutive days. (2) Height anomalies at 200 hPa show the tripole structure as '+ − +' ('− + −') corresponding to the three basic anomaly centers. (3) Time interval between two SRP events must be more than 10 days. If two or more peak phases occur in less than 10 days, only the first peak is counted to ensure the independence of each peak phase [21].

In total, 40 SRP+ and 40 SRP− events were identified, according to the criteria. The year, start date, peak date, end date, duration, averaged SRPI, H_{WCA}, H_{MO}, and H_{FE} of the identified typical SRP+ and SRP− events are displayed in Tables 1 and 2, respectively. Particularly, on average there is less than one SRP event per summer identified according to the SRPI during 1979–2020 (42 summers). It is possible that though the daily-scale SRP also appears in some summers, its intensity and duration fail to meet the criteria of 'a typical SRP event', which features a daily SRPI value of +1.0 σ or greater persisting for less than three consecutive days. Further considering the complexity of East Asian summer circulation, these identified 40 SRP+ and 40 SRP− events can basically be viewed as the typical cases to investigate the evolution features and the relationship to summer temperature in the Yangtze River Valley.

Table 1. The year, start date, peak date, end date, duration (unit: day), average SRPI, and H_{WCA}, H_{MO}, H_{FE} of the 40 SRP+ events.

Year	Start Date	Peak Date	End Date	Duration	SRPI	H_{WCA}	H_{MO}	H_{FE}
1979	10 Jul	11 Jul	14 Jul	5	1.628	1.797	−1.260	1.828
1980	14 Jul	15 Jul	16 Jul	3	1.180	1.046	−1.353	1.141
1982	3 Aug	7 Aug	11 Aug	9	1.946	1.738	−2.193	1.907
1983	29 Jul	31 Jul	6 Aug	9	2.225	2.440	−1.706	2.475
1983	18 Aug	18 Aug	21 Aug	4	1.685	1.667	−1.811	1.577
1984	24 Jul	28 Jul	30 Jul	7	1.631	1.844	−1.598	1.452
1984	16 Aug	20 Aug	30 Aug	15	2.189	1.902	−2.179	2.486
1985	7 Jul	7 Jul	13 Jul	7	1.292	1.467	−1.378	1.031
1985	24 Jul	28 Jul	31 Jul	8	1.550	1.169	−1.750	1.731
1986	15 Jul	17 Jul	19 Jul	5	1.570	1.461	−1.713	1.536
1986	1 Aug	1 Aug	3 Aug	3	1.852	1.859	−1.694	2.003
1986	16 Aug	17 Aug	19 Aug	4	1.147	0.907	−1.225	1.309
1987	11 Aug	17 Aug	29 Aug	19	1.276	1.202	−0.950	1.676
1988	2 Jul	4 Jul	6 Jul	5	1.529	1.616	−1.817	1.154
1990	19 Jun	22 Jun	26 Jun	8	1.659	1.137	−1.859	1.981
1990	8 Aug	12 Aug	14 Aug	7	1.385	1.404	−1.600	1.151
1991	28 Aug	28 Aug	30 Aug	3	1.683	1.822	−1.746	1.482
1992	20 Jul	22 Jul	23 Jul	4	1.334	1.008	−1.499	1.495
1992	23 Aug	25 Aug	27 Aug	5	1.231	1.216	−1.425	1.052
1994	24 Jun	2 Jul	11 Jul	18	1.539	1.547	−1.174	1.896
1994	4 Aug	5 Aug	7 Aug	4	2.080	2.125	−2.117	1.998
1995	12 Aug	13 Aug	20 Aug	9	1.711	1.245	−1.886	2.002
1997	15 Jul	19 Jul	19 Jul	5	1.392	1.644	−1.379	1.153
1997	17 Aug	19 Aug	20 Aug	4	2.137	2.229	−2.004	2.178
1998	1 Aug	11 Aug	13 Aug	13	1.180	1.462	−0.855	1.223
2001	29 Jun	1 Jul	2 Jul	4	1.451	1.511	−1.333	1.509
2002	9 Aug	10 Aug	11 Aug	3	1.346	1.831	−1.186	1.102
2003	18 Jul	24 Jul	27 Jul	10	1.468	0.977	−1.566	1.861
2003	15 Aug	17 Aug	18 Aug	4	1.407	1.125	−1.359	1.737
2005	25 Jun	27 Jun	28 Jun	4	1.194	1.274	−1.388	0.920
2008	12 Jul	14 Jul	17 Jul	6	1.087	1.295	−1.049	0.917
2008	10 Aug	11 Aug	12 Aug	3	1.268	1.194	−1.066	1.544
2009	14 Aug	14 Aug	17 Aug	4	1.223	1.225	−1.013	1.431
2011	7 Jul	8 Jul	12 Jul	6	1.186	1.309	−1.117	1.132
2012	20 Aug	20 Aug	24 Aug	5	1.606	1.111	−1.656	2.051
2013	29 Jul	1 Aug	9 Aug	12	1.925	1.944	−1.942	1.889
2015	13 Jul	17 Jul	20 Jul	8	1.739	1.363	−1.915	1.939
2017	8 Aug	9 Aug	10 Aug	3	1.201	1.603	−1.008	0.992
2019	22 Jul	23 Jul	24 Jul	3	2.014	2.155	−1.737	2.150
2020	12 Aug	16 Aug	20 Aug	9	1.841	1.797	−1.924	1.758

Table 2. The year, start date, peak date, end date, duration (unit: day), average SRPI, and H_{WCA}, H_{MO}, H_{FE} of the 40 SRP− events.

Year	Start Date	Peak Date	End Date	Duration	SRPI	H_{WCA}	H_{MO}	H_{FE}
1979	6 Jun	12 Jun	16 Jun	11	−1.476	−1.255	1.461	−1.712
1981	12 Jun	13 Jun	19 Jun	8	−1.947	−2.201	1.791	−1.849
1982	22 Jun	24 Jun	24 Jun	3	−1.091	−1.417	1.052	−0.804
1982	5 Jul	8 Jul	9 Jul	5	−1.074	−1.083	1.126	−1.013
1983	6 Jul	7 Jul	11 Jul	6	−1.406	−1.455	1.418	−1.345
1985	10 Jun	12 Jun	18 Jun	9	−1.332	−1.377	1.611	−1.008
1987	9 Jun	21 Jun	25 Jun	17	−1.667	−1.629	1.366	−2.006
1988	15 Jun	15 Jun	19 Jun	5	−1.151	−1.136	1.149	−1.168
1989	11 Jun	20 Jun	22 Jun	12	−1.694	−1.960	1.592	−1.530
1989	3 Jul	4 Jul	6 Jul	4	−1.813	−1.765	1.766	−1.908
1991	19 Jun	22 Jun	28 Jun	10	−1.696	−1.570	1.514	−2.004
1992	6 Jun	10 Jun	20 Jun	15	−2.171	−2.153	2.377	−1.983
1992	10 Aug	11 Aug	13 Aug	4	−1.287	−1.188	0.954	−1.719
1993	2 Jun	7 Jun	9 Jun	8	−1.474	−1.143	1.621	−1.658
1995	22 Jun	26 Jun	28 Jun	7	−1.581	−1.174	1.625	−1.944
1996	15 Jun	17 Jun	19 Jun	5	−1.391	−1.026	1.433	−1.714
1996	5 Jul	7 Jul	8 Jul	4	−1.305	−0.966	1.479	−1.470
1997	11 Jun	12 Jun	15 Jun	5	−1.434	−1.809	1.188	−1.375
1998	7 Jun	15 Jun	18 Jun	12	−1.655	−1.351	2.002	−1.612
1999	10 Jun	10 Jun	13 Jun	4	−1.991	−1.733	2.219	−2.021
2000	3 Jun	10 Jun	14 Jun	12	−2.031	−1.795	1.938	−2.360
2000	18 Jul	24 Jul	26 Jul	9	−1.494	−1.107	1.456	−1.919
2002	12 Jun	16 Jun	20 Jun	9	−1.841	−1.628	1.780	−2.115
2002	7 Jul	9 Jul	16 Jul	10	−1.389	−1.799	1.006	−1.362
2005	10 Jun	12 Jun	20 Jun	11	−1.321	−1.424	1.300	−1.239
2006	5 Jun	6 Jun	11 Jun	7	−2.080	−1.955	1.811	−2.474
2009	18 Jun	23 Jun	27 Jun	10	−2.015	−1.863	2.038	−2.144
2009	23 Aug	24 Aug	25 Aug	3	−1.013	−1.035	0.814	−1.190
2010	21 Jun	23 Jun	24 Jun	4	−1.158	−1.240	1.056	−1.178
2010	5 Jul	6 Jul	7 Jul	3	−1.104	−1.266	0.995	−1.051
2012	8 Jun	11 Jun	14 Jun	7	−1.502	−1.778	1.634	−1.094
2013	17 Jun	17 Jun	20 Jun	4	−1.267	−1.093	1.557	−1.151
2014	7 Jun	11 Jun	13 Jun	7	−1.619	−1.409	1.477	−1.971
2015	25 Jun	27 Jun	28 Jun	4	−1.159	−1.013	0.946	−1.518
2016	23 Jun	25 Jun	26 Jun	4	−1.316	−0.908	1.469	−1.571
2016	16 Aug	17 Aug	18 Aug	3	−1.163	−1.404	1.033	−1.052
2017	10 Jun	12 Jun	17 Jun	8	−1.790	−2.032	1.588	−1.570
2018	19 Jun	22 Jun	29 Jun	11	−1.525	−1.341	1.896	−1.338
2019	20 Jun	23 Jun	25 Jun	6	−1.514	−1.937	1.772	−0.833
2020	28 Jun	29 Jun	1 Jul	4	−1.194	−1.026	1.308	−1.248

As shown in the tables, the typical SRP+ events prevail most frequently during mid-to-late summer (July to August), but SRP− events tend to be more significant during early to mid-summer (June to July). The duration of SRP+ (SRP−) events can be up to 19 (17) days. The following composite analyses are generally based on the identified typical SRP+ and SRP− cases. For convenience, day 0 represents the peak date of SRP events; day (n) denotes the day prior to (negative) and after (positive) the peak date of SRP events.

4.1. The Life Cycle of SRP

To analyze the life cycle of SRP, the composite SRPI indices are displayed in Figure 2. The peak date (day 0) represents the day when the SRPI reaches its maximal value and exceeds +1.0 σ. Positive and negative values of SRPI represent positive and negative phases of SRP respectively, and their absolute values represents the intensity of SRP.

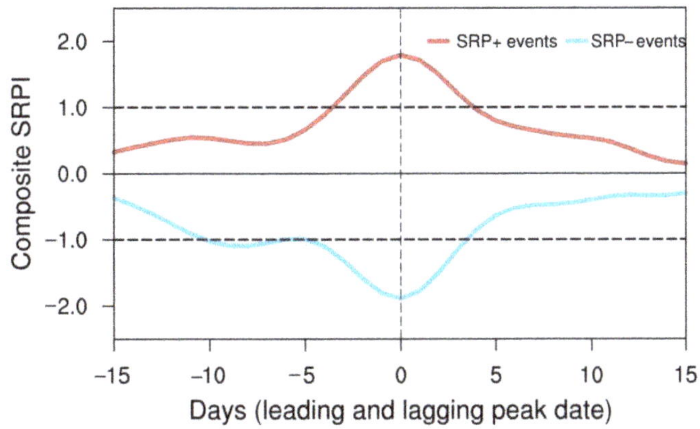

Figure 2. Composite indices of SRP+ (red line) and SRP− (blue line) events from day −15 to day 15. The numbers on the abscissa represent the days leading (negative) and lagging (positive) the peak date of SRP events.

As shown in Figure 2, the amplitude of SRP+ begins to rise rapidly from day −5 onward, experiencing considerable growth in the following 5 days. The SRP+ index exceeds the normal value by +1.0 σ approximately at day −3. By day 0, the SRP+ index attains maximal value, reaching its peak state, with a gradual decline of its intensity after that. The composites of SRP+ and SRP− indices show opposite structure. The SRPI index exceeds the normal value by −1.0 σ approximately at day −10, which is much earlier than that of SRP+. From day −5 onward, the SRI decreases rapidly, and reaches its minimal value on the peak date, and then increases gradually. Moreover, it is also found that the life cycles of both daily-scale SRP+ and SRP− can persist for more than 15 days.

Figure 3 shows the evolutions of the composite geopotential height anomalies field and wave-activity flux (WAF) at 200 hPa during the SRP+ events (day −8 to day 8). Prior to the peak date of SRP+ events, the positive height anomaly center over the West Central Asia and the negative height anomaly center over the Mongolia develop firstly and simultaneously form day −8 onward (Figure 3a), further strengthening in the following 4 days, with pronounced WAF emanating from the West Central Asia center. Meanwhile, the WAF disperses eastward along the Asian jet, stimulating the gradual emergence of the positive height anomaly center over the Far East. These three anomaly centers of SRP+ are consistent with the three basic points of the Silk Road pattern in the first EOF mode (Figure 1). By approximately day 2, the well-organized zonal tripole structure of daily-scale SRP+ completely forms at 200 hPa (Figure 3d). On the peak date, the SRPI index attains its maximal value, the intensity of the positive West Central Asia center and the negative Mongolia center significantly increases, indicating the formation of well-established SRP+, with three significant action centers over the West Central Asia, Mongolia, and Far East (Figure 3e) [7]. During the developing stage, an anomalous strong negative height anomaly center tended to be anchored to the west of the Ural Mountains at high latitudes for several days, and its intensity variation characteristics are similar to that of the negative height anomaly center around the Balkhash Lake, which is consistent with the conclusion of that the SRP is a pronounced circumglobal teleconnection pattern along the summertime Asian jet stream [22,39]. During the decaying period (approximately after day 2), with the dissipation of the upstream WAF, the intensities of West Central Asia and Mongolia centers decrease gradually and disappear one after another, which represents the vanishing of the zonal wave train structure of SRP+ (Figure 3i). In the meanwhile, the positive anomaly center over the Far East maintains its strength for a longer period without decline, even until the SRP turns to its negative phase, which is under the influence of the sustaining eastward wave energy propagation from the Mongolia to the East Asian coast.

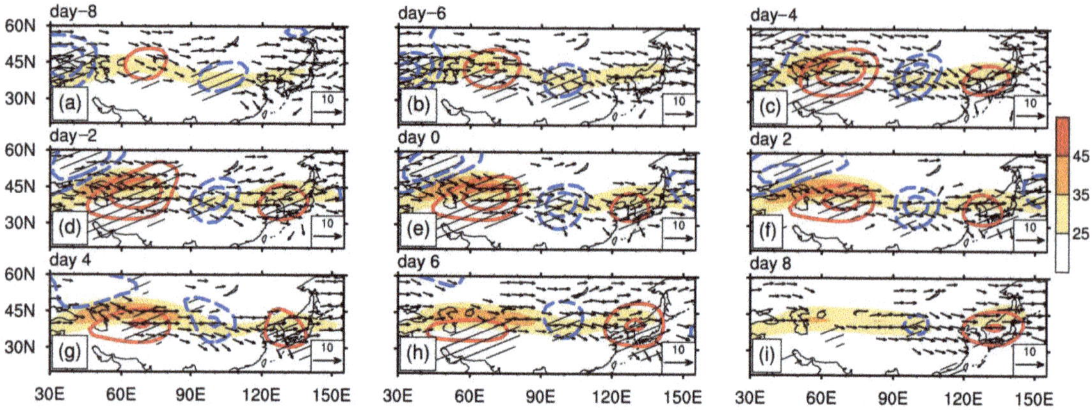

Figure 3. Composite 200 hPa geopotential height anomalies (contours are from −80 to 80 gpm with interval of 40 gpm, red for positive and blue for negative), the zonal wind speed (shading; units: m s^{-1}), and wave-activity flux (vector; units: m^2 s^{-2}) of SRP+ events from day −8 to day 8 (**a–i**) in Table 1. The slash shadings denote that the height anomalies are statistically significant at the 0.05 confidence level. The numbers at the top-left corner above each panel represent the days leading (negative) and lagging (positive) the peak date of the SRP+ events. Wave fluxes smaller than 4 m^2 s^{-2} are not plotted.

This study also examines the composite 200 hPa geopotential height anomalies and WAF of SRP− events. It is evident that both positive and negative phases of SRP have similar and fixed positions of the three anomaly centers; however, the temporal evolution characteristics are different from those of the SRP+. As illustrated in Figure 4, the SRP− events start with a negative geopotential height anomaly center located over the Far East from day −8 onward, experiencing considerable growth during the subsequent 4 days (Figure 4a). In the meanwhile, the negative height anomaly center over the West Central Asia emerges, with a rapid growth in the strength from day −6 to day −4. The positive height anomaly center over Mongolia begins to develop from day −2 onward and then intensifies significantly, indicating the onset of SRP− events. During the peak stage, the well-organized SRP− is characterized by a zonal '− + −' wave train structure, with a maximal amplitude at day 0, with more pronounced WAF emanating from the Caspian Sea regions along this zonal wave train. In the decaying period, the SRP− weakens firstly over the West Central Asia, where the negative anomaly center weaken substantially from day 2 and almost disappeared at day 6, showing a rapid decline compared to other centers. Nevertheless, both the positive anomaly center over Mongolia and the negative anomaly center over the Far East persist for several days with their intensity weakening slightly. In addition, it can be observed that the negative anomaly center over the Far East appears earlier and persists longer than other centers, and the positive anomaly center in Mongolia appears the last but maintain a certain strength for a longer period until the SRP− vanishes.

During the whole life cycle of daily-scale SRP+ and SRP−, the three significant action centers prevail in their fixed positions, with less movement during the development, maintenance, and decaying periods. Thus, the analyses of daily-mean fields show that the Silk Road pattern shows a nature of the quasi-stationary wave train on daily timescales, which is consistent with the previous research conclusions by using monthly-mean data [7,39,40].

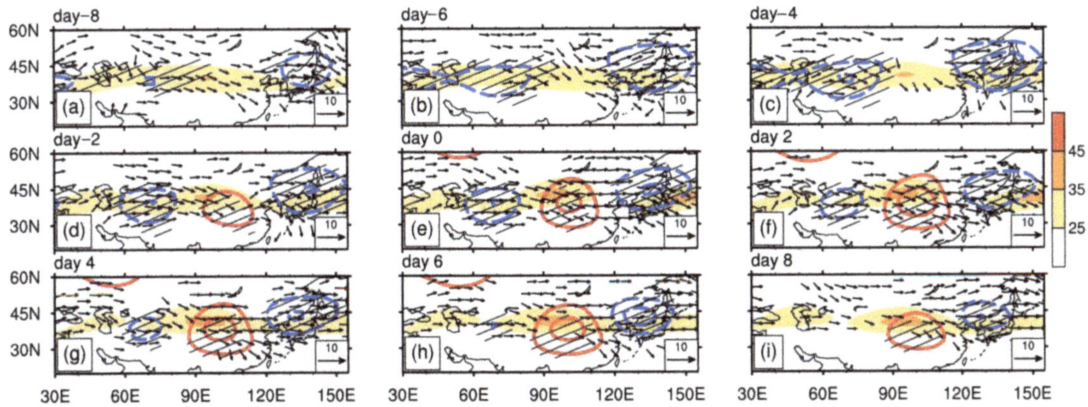

Figure 4. The same as in Figure 3, but for the SRP− events in Table 2.

4.2. Vertical Structure of Daily-Scale SRP

To shed light on the evolution of the SRP, their vertical structures of SRP+ and SRP− are also explored in this section.

As mentioned above, the three anomaly centers of the SRP are all located near 40° N, remaining stable during the life cycle. Therefore, Figures 5 and 6 present the longitude-pressure cross-section of geopotential height anomalies along 42.5° N during the life cycle of SRP+ and SRP− events, respectively. It shows that all the maximum anomaly centers of SR pattern appeared in the mid-to-upper troposphere approximately at 200 hPa. About 8 days prior to the peak date of SRP+ events, the upstream West Central Asia positive anomaly center and the Mongolia negative anomaly center emerged firstly, strengthening remarkably during the subsequent 4 days. In particular, the vertical structure of the West Central Asia center can extend from the 600 hPa to the tropopause (Figure 5a–c). Meanwhile, the downstream Far East positive anomaly center develops rapidly (Figure 5c). By day −4, these three anomaly centers generate the well-established zonal tripole structure before SRP+ gets maximal amplitude. The intensities of the West Central Asia and Mongolia anomaly centers reach the highest at day 0 (Figure 5e), weakening rapidly after the peak date. The positive Far East anomaly center exhibit a tendency of continuous enhancement, which is consistent with the variation characteristics of the SRP+ described in Figure 3.

The variation characteristics of the vertical structure during SRP− events are similar to that of SRP+ events (Figure 6). The maximums of three anomaly centers emerge at the height of 200 hPa, with an equivalent barotropic structure in vertical direction, which were obtained by comparing lower and upper troposphere circulation anomaly fields. The most significant distinction of evolution features between the SRP+ and SRP− lies in the variation characteristics of the Mongolia and Far East anomaly centers. Concretely, prior to the onset of SRP− events, the Mongolia positive anomaly center begins to emerge gradually from day −4, with a vertical structure extending to the height of 400 hPa (Figure 6c). The vertical structure of the Far East negative anomaly center can occupy the whole troposphere, with a stronger strength and longer duration than that of the West Central Asia negative anomaly center, which is consistent with the results obtained in Figure 4.

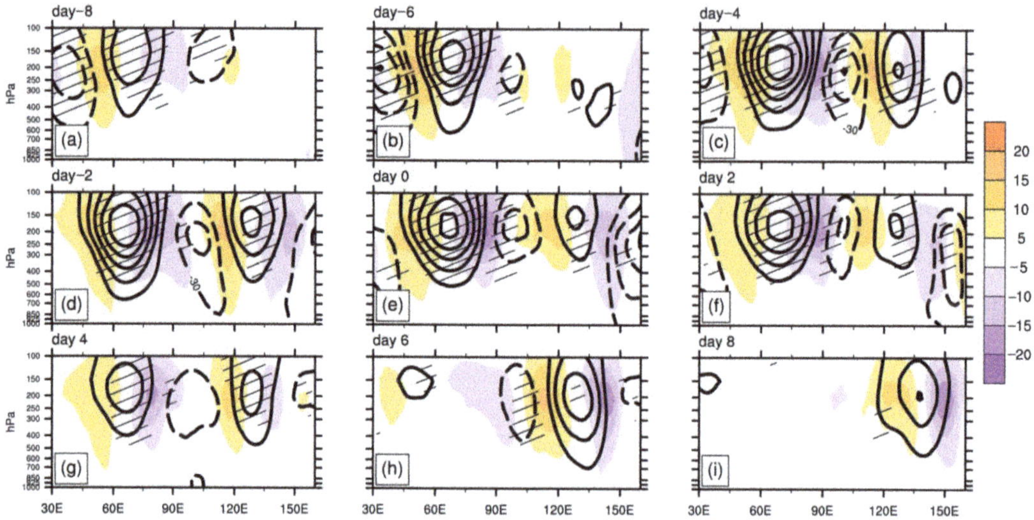

Figure 5. Longitude-pressure cross-section of geopotential height anomalies (contours are from −120 gpm to 120 gpm with the interval of 30 gpm) and meridional wind anomalies (shadings; units: m s^{-1}) along 40° N during the SRP+ events (**a**–**i**: from day −8 to day 8). The slash shadings indicate that the height anomalies are statistically significant at the 0.05 confidence level. The numbers at the top-left corner above each panel represent the days leading (negative) and lagging (positive) the peak date of the SRP+ events.

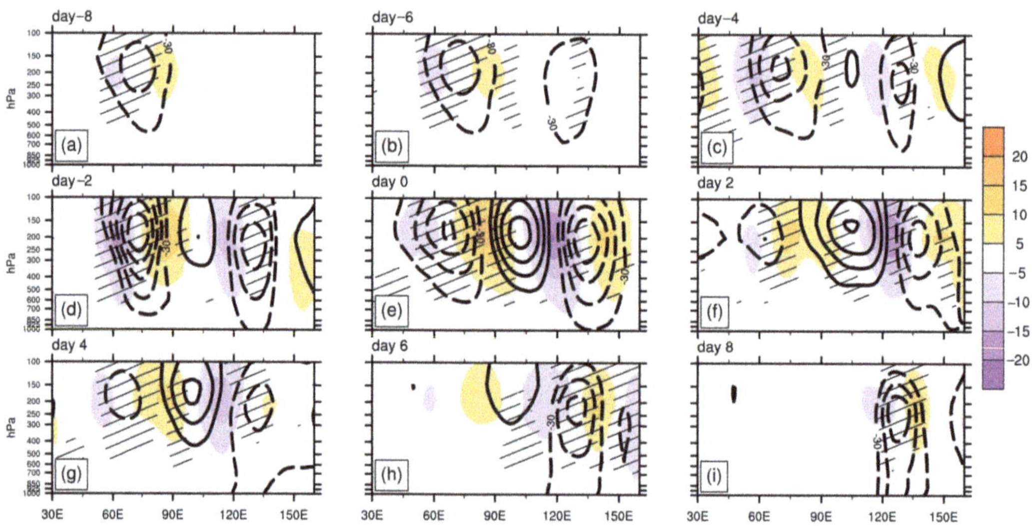

Figure 6. The same as in Figure 5, but for the SRP− events.

Additionally, in the vertical direction, the three anomaly centers of SRP almost exhibit barotropic structures. However, during the development stage, the nascent Mongolia negative anomaly center of SRP+ and the West Central Asia negative anomaly center of SRP− tilt a little westward with height, which shows a baroclinic vertical structure (Figure 5d). As documented in [40,41] that the barotropic instability and baroclinicity could extract the kinetic energy (KE) and available potential energy (APE), respectively, from the basic state, maintaining the development of the Rossby wave train pattern. Thus, the early development of SRP may be attributed to the baroclinicity, and the barotropic instability processes play a significant role in the maintenance of these three anomaly centers.

5. Temperature Anomalies in the YRV Related to SRP

5.1. Influence of SRP on Summer Temperature in China

In this section, the possible influence of the SRP on the summer temperature over China are explored. Based on the samples in Tables 1 and 2, Figure 7 presents the spatial distribution of composite averaged temperature anomalies during the SRP+ and SRP− events.

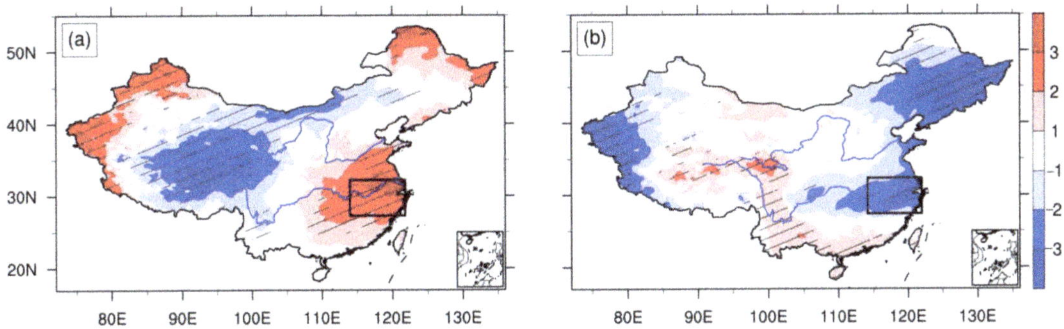

Figure 7. Spatial distribution of the average temperature anomalies ((**a**,**b**); units: °C) during the SRP+ (**a**) and SRP− (**b**) events; The slash shadings indicate that the anomalies are statistically significant at the 0.05 confidence level. Black rectangles in Figure 7a,b denote the YRV regions (114–122° E, 27–32° N) which are selected to calculate the domain-averaged temperature anomalies.

Corresponding to the SRP+ events, positive temperature anomalies dominated most of China, which has three significant positive centers over the middle and lower Yangtze River Valley (YRV), Xinjiang province, and part of Northeast China respectively, with the temperature anomalies having values of 1–3 degrees higher than the climatology mean. In addition, there is a significant center of negative temperature anomaly in Qinghai province and adjacent regions. As for the SRP− events, the distribution of the temperature anomalies has an opposite sign, characterized by negative temperature anomalies over mainland China. Three significant negative centers are located over the YRV, southern region of Xinjiang province, and most of Northeast China, respectively.

It is well known that the Yangtze River Valley, located in central-eastern China, is one of the most densely populated and economically developed regions in China. Thus, high temperatures or extreme heat waves in the YRV have a more important impact on people's production and life. Some studies have shown that the YRV witnessed increased numbers of heat waves in the summer since 1951 [42,43]. Therefore, the following analysis will focus on the SRP-related temperature anomalies in the YRV (114–122° E, 27–32° N) and its possible causes.

Through further analyses, we find that the SRP+ (SRP−) is closely linked to the persistent heat waves and cool summer processes in the YRV. To prove that, the SRP-related heat waves and cool summer process should be defined firstly. The specific criterion and associated anomalies of the temperature and related circulations are provided in the following analysis. According to the previous studies, +1.0 standard deviation (σ) is selected as the threshold to identify the typical extreme weather and climate events [44,45]. In this study, the heat waves (cool summer processes) in the YRV must meet the following criterion: daily normalized domain-averaged temperature anomalies must be greater than (less than) +1.0 σ (−1.0 σ).

As mentioned above, the temperature anomalies in the YRV during the SRP+ and SRP− events exactly show an opposite distribution. Concretely, the SRP+ (SRP−) is closely related to the positive (negative) temperature anomalies signals in the middle and lower reaches of YRV. Therefore, a key region (114–122° E, 27–32° N) is selected to calculate the regional averaged temperature anomalies. As shown in Figure 8, the temperature anomalies increase substantially prior to the peak date of the SRP+ events. Daily normalized domain-averaged temperature anomalies of more than 1 σ can persist from day −1 to day 5, which can keep a certain strength for approximately 7 days, indicating the consecutive summer heat wave cases of YRV. After day 5, the temperature anomalies in the YRV descend significantly. As for the SRP− events, the daily normalized domain-averaged temperature anomalies rapidly descend from day −2, with minimal anomalies less than −1.5 σ at day 3, suggesting a continuous process of cool summer in the Yangtze River Valley (from day 2 to day 5). These results may imply that the daily-scale SRP+ (SRP−) has a profound effect on the summer persistent heat waves (cool summer process) in the YRV.

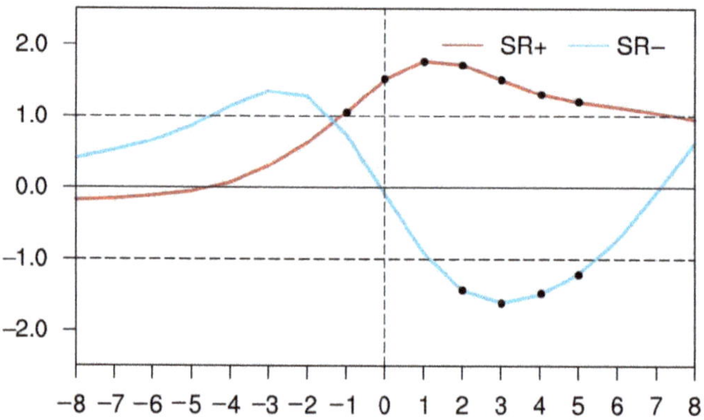

Figure 8. Daily normalized domain-averaged temperature anomalies during the SRP+ and SRP− events from day −8 to day 8 in the YRV (114–122° E, 27–32° N); The solid circles indicate that the anomalies are statistically significant at the 0.05 confidence level.

In order to better explain the important contribution of SRP to the temperature anomaly in the Yangtze River Valley, Figure 9 presents the days of heat wave events and cool summer processes within the YRV in each summer during 1979–2020. Particularly, it has to be stated that the SRP indeed represents only one of the favorable factors responsible for such prolonged extreme temperature anomaly events in the YRV. Accordingly, 171 summer heat wave days can be extracted from the 267 days accumulated by all the identified SRP+ events (Table 1), which account for about 29.03% of the total 589 summer heat wave days in the YRV during 1979–2020. Similarity, during the summers with the SRP− events, 193 cool summer cases can be extracted from the 290 days accumulated by all the identified SRP− events (Table 2), which account for about 30.1% of the total 641 cool summer days in the YRV during 1979–2020. It is also proved that the SRP may play an important role in the summer temperature anomaly of the Yangtze River Valley.

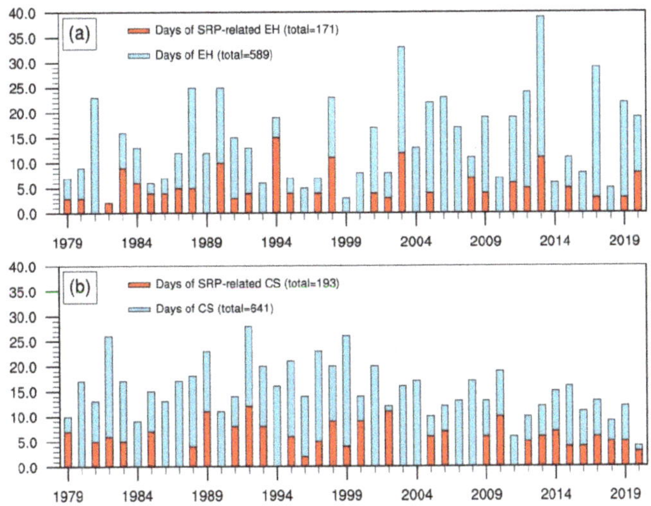

Figure 9. Days of heat wave cases ((**a**), light blue rectangle; red rectangles represent the days of heat wave cases related to SRP+) and cool summer processes ((**b**), blue rectangle; red rectangles represent the days of cool summer cases related to SRP−) within the YRV (114–122° E, 27–32° N) in each summer (June–August) during 1979–2020.

5.2. Possible Causes of SRP-Related Temperature Anomalies in the YRV

The occurrence and maintenance of regional extreme weather and climate events is closely linked to the anomalous regional circulation, and some of the extreme events are the result of the combined anomalies formed by a variety of climatic factors [46,47]. The generation of surface temperature anomalies is usually associated with the specific atmospheric circulation patterns [46]. In the following analysis, this study will further explore the possible causes of temperature anomalies in the middle and lower reaches of YRV from the perspective of Silk Road pattern, revealing the anomalous circulation patterns related to SRP responsible for the YRV temperature anomalies.

As shown in Figure 10, prior to the peak date of typical SRP+ events, an anomalous anticyclone/cyclone/anticyclone wave train ('A/C/A' in Figure 10) aligned in a near-zonal direction along the Asian jet can be identified in the upper troposphere at 200 hPa from day −6 onward, which can also be considered as the embodiment of SRP [14]. It is interesting to note that the anomalous 'A/C/A' wave train is similar to the Silk Road pattern which is mentioned in previous studies [5]. Throughout the maintenance stage of SRP+ events, an anomalous anticyclone dominated over the East Asia to the Northwest Pacific, with significant negative vorticities over the anomalous anticyclone area. Positive (negative) vorticity advection corresponds to the anomalous ascent (descent) motions. To ensure the conservation of potential vorticity, positive (negative) vorticity advection must be accompanied by adiabatic cooling (heating) [48]. Furthermore, another most distinct characteristic of large-scale circulation is the gradual westward extension of the western Pacific subtropical high (WPSH) at 500 hPa and the eastward shift of the South Asia high (SAH) at 200 hPa. The negative vorticities related to the anomalous anticyclonic circulation over East Asia accelerate the zonal advance between the SAH and WPSH, which is consistent with the previous studies [49]. From day −6 onward, they overlap with each other, with the negative vorticities in the overlapping area of the SAH and WPSH. Particularly, the similar zonal approach between the SAH and WPSH has been widely used in different studies, which can be viewed as an effective predictor for the temperature and precipitation anomalies over East Asia [5,26].

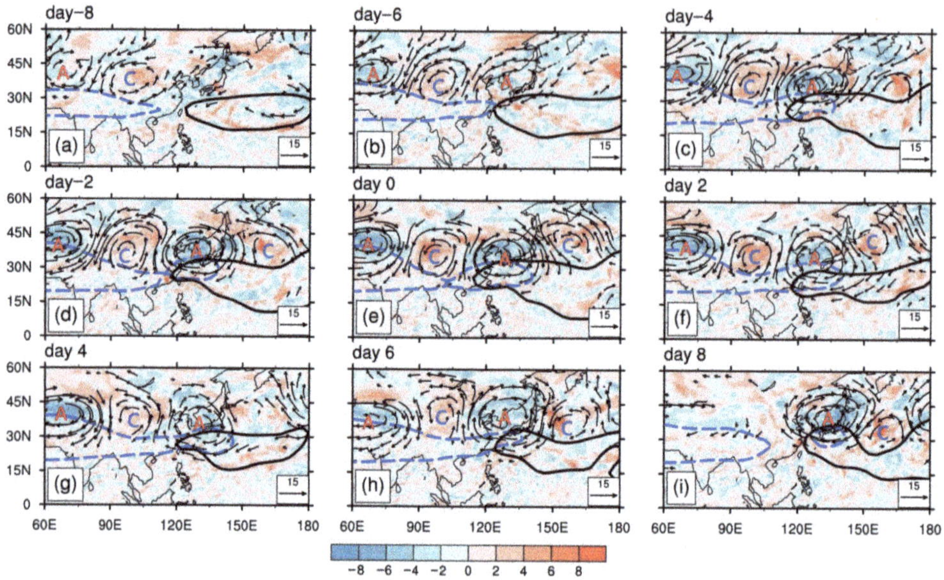

Figure 10. Composite average horizontal wind anomalies at 200 hPa (vector; units: m s^{-1}) and the relative vorticity (shadings; units: s^{-1}) during the SRP+ events from day −8 to day 8 (**a**–**i**). The thick black contours (588 dagpm-contour) and the blue dashed contours (1250 dagpm-contour) denote the activities of WPSH at 500 hPa and SAH at 200 hPa, respectively. The letters 'C' and 'A' represent an anomalous cyclone and anticyclone, respectively.

Figure 11 presents the latitude-pressure cross-section (114–122° E) of vertical *p*-velocity during the SRP+ events (day −8 to day 8). It is shown that both the anomalous anticyclone and overlapping of the SAH and WPSH may favor the anomalous descents above the YRV regions. After their zonal approach, the significant positive vertical *p*-velocity rapidly dominates the YRV areas from day −4, indicating the strong local descents, which can persist throughput the following 8 days. Thus, the persistent sinking adiabatic warming can be regarded as one of the important factors involved in surface warming or heat waves over the YRV area. On the other hand, anomalous descents can reduce the total cloud cover (negative anomalies of total cloud cover in Figure 12c) and increase solar radiation incident to surface of the YRV area (positive anomalies of solar radiation in Figure 12a). Therefore, the summer heat waves in the YRV regions during the SRP+ events may be also determined by the clear-sky warming.

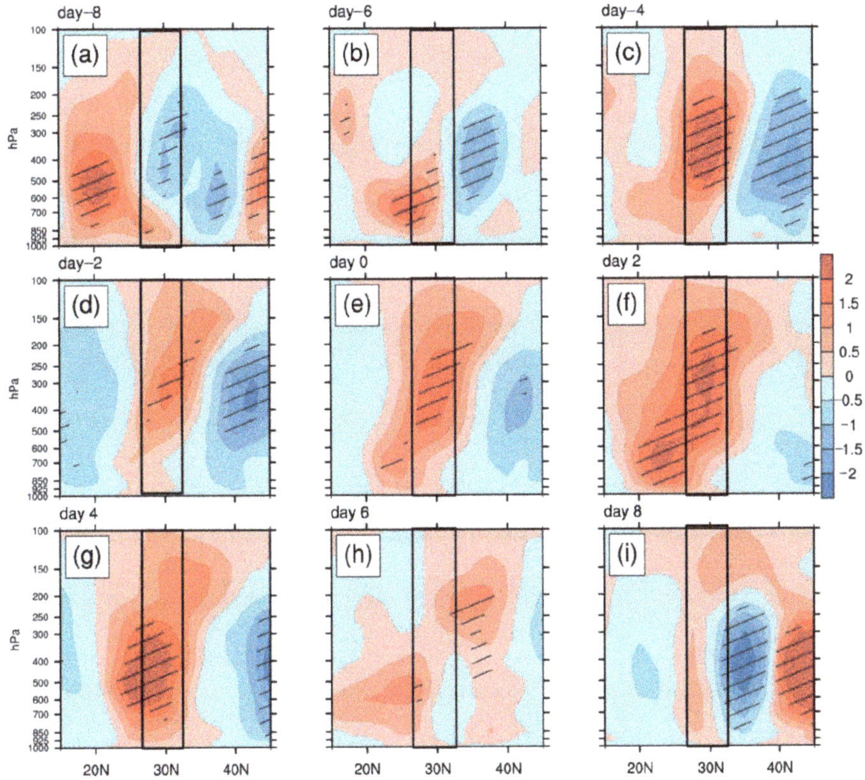

Figure 11. Latitude-pressure cross-section (114–122° E) of vertical *p*-velocity (shading; units: hPa s^{-1}) during the SR+ events from day −8 to day 8 (**a–i**). The slash shadings indicate that the anomalies are statistically significant at the 0.05 confidence level. Black rectangles denote the YRV regions (114–122° E, 27–32° N).

Next, we will explore the possible causes of cool summer processes in the Yangtze River Valley during the typical SRP− events. Two days prior to its onset, a similar anomalous cyclone/anticyclone/cyclone wave train ('C/A/C' in Figure 13) can also be observed along the Asian jet. East Asia to the Northwest Pacific is under the influence of the anomalous cyclonic circulation and positive vorticities. Concurrently, the SAH and WPSH tend to diverge from each other towards the reverse directions and retreat to their normal positions. As illustrated in Figure 13, the positive vorticities between the SAH and WPSH enhanced remarkably around the peak dates, further accelerating the divergence of the SAH and WPSH from day −4 to day −2. By day 0, the eastern boundary of the SAH retreat to the east of 120° E, and the WPSH retreat westward to the western Pacific, which favors the development of the local ascent motions over the YRV area (Figure 14). It is well known the ascent motions may provide the favorable conditions for the YRV summer precipitation, leading to cool summers during the SRP− events. Moreover, the local ascents can also increase the total cloud cover (positive anomalies of total cloud cover in Figure 12d), with less solar radiation incident to the surface of the YRV areas (negative anomalies of solar radiation in Figure 12b), contributing to the development of the negative temperature anomalies in the middle and lower reaches of YRV. Thus, the cool summer processes in the YRV are mainly determined by the adiabatic cooling with the local ascent motions caused by the anomalous cyclonic circulation over the YRV related to the SRP−.

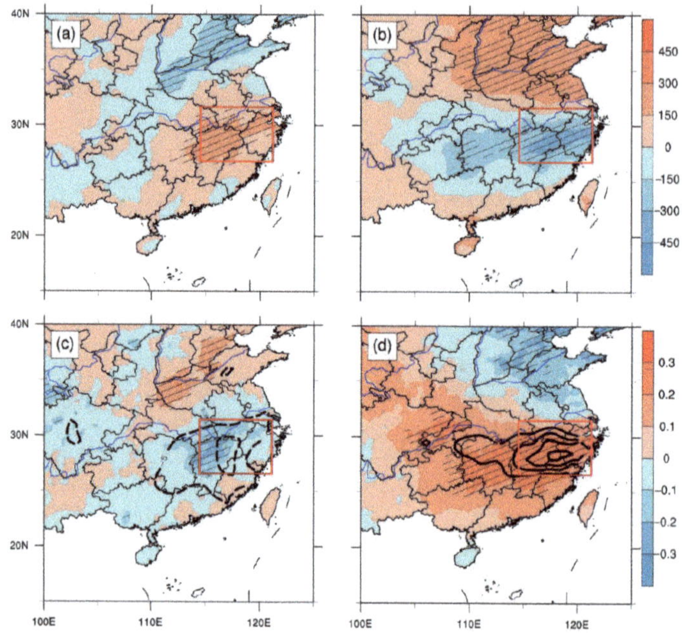

Figure 12. Composite average anomalies of surface solar radiation downwards ((**a**,**b**); unit: kJ m^{-2}; shadings), total cloud cover ((**c**,**d**); unit: 0–1; shadings) and averaged precipitation anomalies ((**c**,**d**); contours are from −20 mm to 20 mm with interval of 5 mm) during the SR+ (**a**,**c**) and SRP− (**b**,**d**) events. The slash shadings indicate that the anomalies of surface solar radiation downwards (**a**,**b**) and total cloud cover (**c**,**d**) are statistically significant at the 0.05 confidence level. Red rectangles denote the YRV (114–122° E, 27–32° N) regions. Only negative and positive precipitation anomalies are plotted in Figure 14c,d, respectively.

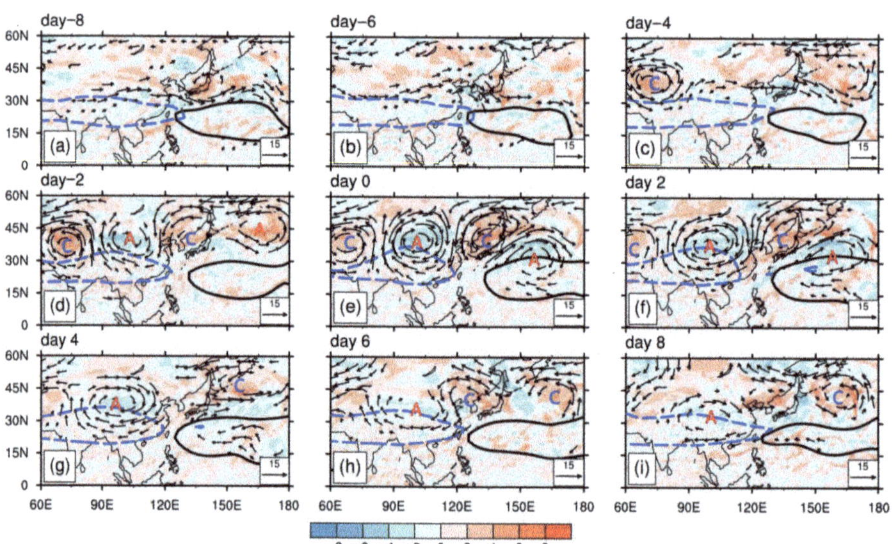

Figure 13. The same as in Figure 10, but for the SRP− events.

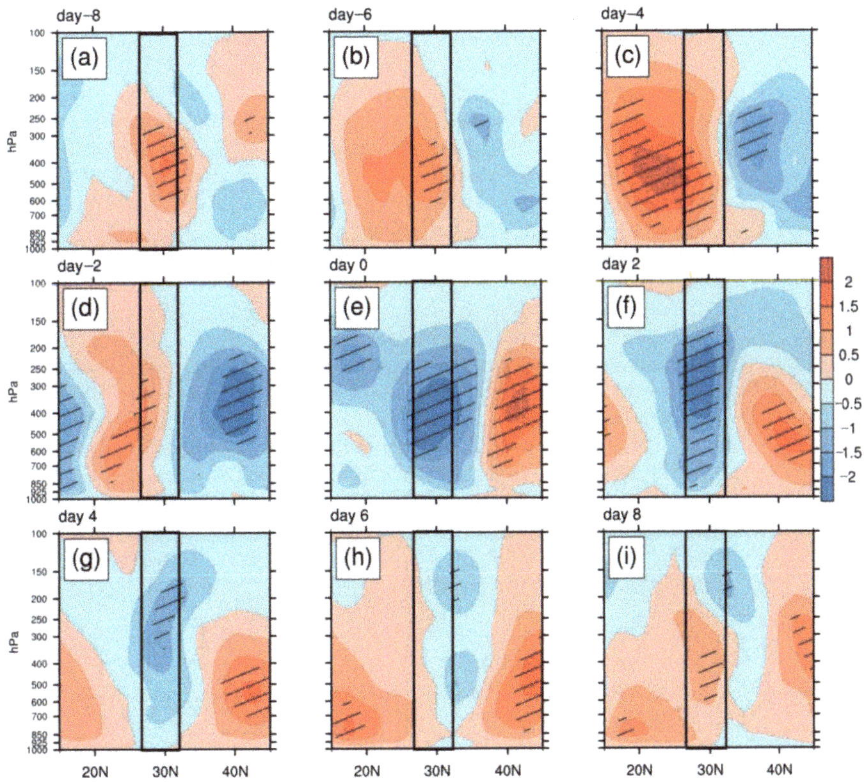

Figure 14. The same as in Figure 11, but for the SRP− events.

6. Conclusions and Discussion

In this study, the evolution characteristics of Silk Road pattern (SRP) and its association with summer precipitation in China are investigated using ERA5 reanalysis daily data. The main conclusions are summarized as follows.

The evolution characteristics of SRP+ and SRP− show marked distinctions, especially on the occurrence, maintenance, and disappearance of their three action centers. Prior to the peak date of SRP+ events, the anomaly centers over West Central Asia and Mongolia firstly emerge, experiencing considerable growth during the subsequent 4 days, with the eastward wave-activity flux (WAF) emanating from the Black Sea regions along the Asian jet. Although the anomaly center over the Far East appears later than other centers (almost from day −4), it can maintain its intensity for a longer period, owing to the convergence of sustaining eastward WAF over the Far East. During the decaying period, the intensities of the West Central Asia and Mongolia centers decrease gradually and vanish one after another. The SRP− events start with a negative anomaly center located over the Far East from day −8 onward, significantly strengthening after that. Meanwhile, the negative height anomaly center over the West Central Asia emerges, with a rapid growth in the strength from day −6 to day −4, and it is under the influence of eastward WAF emanating from the Caspian Sea regions. The positive height anomaly center over Mongolia begins to develop from day −2 onward and then intensifies significantly. During the decaying stage, the intensity of the West Central Asia negative center weakens substantially from day 2 earlier than other centers. Although the positive anomaly center in Mongolia appears last, it can maintain a certain strength for a longer period until the SRP− vanishes. In the vertical direction, SR pattern mainly concentrates in the upper-to-mid troposphere. The baroclinicity contributes to the development of daily-scale SRP, and its maintenance is inextricably linked to the barotropic instability processes.

In addition, as illustrated in the schematic diagram (Figure 15), the SR pattern has a significant effect on the summer temperature anomalies in the Yangtze River Valley (YRV). Concretely, during the SRP+ (SRP−) events, significant positive (negative) temperature anomalies can be observed in the YRV, indicating the heat waves (cool summer) processes. Prior to the peak date of SRP+ (SRP−), an anomalous anticyclone/cyclone/anticyclone (cyclone/anticyclone/cyclone) wave train can be clearly identified along the Asian jet. During SRP+, the anomalous anticyclonic circulation and negative vorticities over East Asia, favor the zonal advance between the SAH and WPSH. The overlapping of these two key systems remains in a favorable position for the violent sinking motions over YRV area. The sinking adiabatic warming can be regarded as one of the important factors involved in surface warming over the YRV area. Furthermore, the anomalous descents can reduce the total cloud cover, providing a favorable condition for the solar radiation incident to surface of the YRV area. Therefore, the summer heat waves in the YRV during the SRP+ events may be also related to clear-sky warming. With respect to SRP− events, owing to the anomalous cyclonic circulation and positive vorticities over East Asia, the SAH and WPSH depart from each other towards the reverse directions, retreating to their normal positions. As a result, the local ascent motions dominate gradually over the YRV, which increase the total cloud cover, with more precipitation signals and less solar radiation incident to surface of the YRV areas, then inducing the negative temperature anomalies (or cool summer processes) in the middle and lower reaches of YRV.

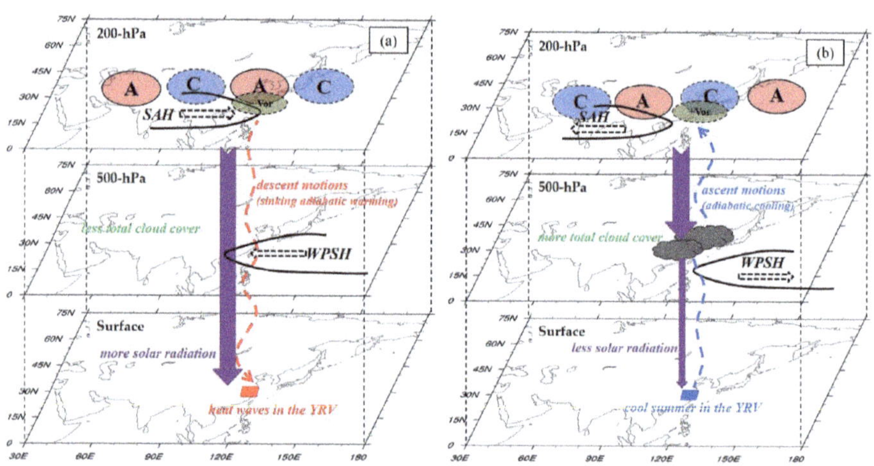

Figure 15. Schematic diagrams for possible causes of the heat waves and cool summer processes in the YRV during the SRP+ (**a**) and SRP− (**b**) events. The letters 'C' (blue shadings) and 'A' (red shadings) represent the anomalous cyclone and anticyclone, respectively. The black solid lines at 200-hPa and 500-hPa denote the boundary of the SAH and WPSH, whose propagation directions are indicated by the dashed blank arrows. The thick and thin purple arrows denote the more and less solar radiation incident to surface of the YRV, respectively. The more total cover is represented by the gray cloud shape. The ascent and descent motions are presented as blue and red dashed lines. The green shadings denote the positive and negative vorticities over the YRV area.

In general, sinking adiabatic warming and clear-sky radiation warming may be the possible causes for the significant heat waves events in the Yangtze River Valley during SRP+. Adiabatic cooling with the local ascents over the YRV area leads to more total cloud cover (precipitation) and less solar radiation incident to surface of the YRV, which are the possible causes of the cool summer process in the YRV during SRP−.

Besides, the SRP seems to exert great influence on the summer temperature anomalies in part of Southeast China, in Qinghai and Xinjiang province (Figure 7); however, the anomalous circulation fields related to the SRP, which can affect the temperature anomalies, need to be further explored. The SRP− is closely associated with the cool summer processes in the YRV. Furthermore, the variation of SRP− leads the variation of cool summer processes for several days. It is possible that the daily SRP index can be considered as a precursor for the persistent temperature anomalies over the YRV.

Author Contributions: Conceptualization, Y.W. and L.W.; methodology, C.W. and L.W.; software, C.W. and Y.L.; validation, Y.W. and C.W.; formal analysis, Y.W.; data curation, L.W. and Y.L.; writing—original draft preparation, C.W. and Y.W.; writing—review and editing, X.K. and Y.L.; visualization, X.K. All authors have read and agreed to the published version of the manuscript.

Funding: This research was jointly funded by the National Natural Science Foundations of China, grant number 41975085, and the Key Laboratory of Flight Techniques and Flight Safety, CAAC, grant number FZ2020ZZ05.

Institutional Review Board Statement: Not applicable.

Informed Consent Statement: Not applicable.

Data Availability Statement: Publicly available datasets were analyzed in this study. The data can be found here: The (ECMWF) Re-Analysis (ERA5) data is available online from https://cds.climate.copernicus.eu/ (accessed on 7 May 2021). The daily precipitation data was available online from http://data.cma.cn (accessed on 1 April 2020).

Acknowledgments: Special thanks to the reviewers for their valuable comments.

Conflicts of Interest: The authors declare no conflict of interest. The funders had no role in the design of the study; in the collection, analyses, or interpretation of data; in the writing of the manuscript, or in the decision to publish the results.

References

1. Wu, G.X.; Liu, Y.M.; Liu, X.; Duan, A.M. How the Heating over the Tibetan Plateau Affects the Asian Climate in Summer. *Chin. J. Atmos. Sci.* **2005**, *29*, 47–56. [CrossRef]
2. Wang, L.J.; Ge, J. Relationship between Low-frequency Oscillations of Atmospheric Heat Source over the Tibetan Plateau and Longitudinal Oscillations of the South Asia High in the Summer. *Chin. J. Atmos. Sci.* **2016**, *40*, 853–863. [CrossRef]
3. Wang, G.L.; Wang, D.G.; Kevin, E.T.; Amir, E.; Miao, Y.; Michael, G.B.; Dana, T.P. The peak structure and future changes of the relationships between extreme precipitation and temperature. *Nat. Clim. Chang.* **2017**, *7*, 268. [CrossRef]
4. Lu, R.Y.; Fu, Y.H. Summer climate variability in East Asian and the western North Pacific and its mechanisms. *Adv. Earth Sci.* **2009**, *24*, 123–131. [CrossRef]
5. Wang, C.; Wang, L.J. Combined Effects of Synoptic-Scale Teleconnection Patterns on Summer Precipitation in Southern China. *Atmosphere* **2018**, *9*, 154. [CrossRef]
6. Zhou, F.L.; Zhang, R.H.; Han, J.P. Relationship between the Circumglobal Teleconnection and Silk Road Pattern over Eurasian continent. *Sci. Bull.* **2019**, *64*, 374–376. [CrossRef]
7. Lu, R.Y.; Oh, J.H.; Kim, B.J. A teleconnection pattern in upper-level meridional wind over the North African and Eurasian continent in summer. *Tellus* **2002**, *54*, 44–55. [CrossRef]
8. Guan, Z.Y.; Yamagata, T. The unusual summer of 1994 in East Asia: IOD teleconnections. *Geophys. Res. Lett.* **2003**, *30*, 235–250. [CrossRef]
9. Liu, Y.; Zhou, W.; Qu, X.; Wu, R.G. An Interdecadal Change of the Boreal Summer Silk Road Pattern around the Late 1990s. *J. Clim.* **2020**, *33*, 7083–7100. [CrossRef]
10. Lin, Z.D.; Lu, R.Y.; Zhou, W. Change in early-summer meridional teleconnection over the western North Pacific and East Asia around the late 1970s. *Int. J. Climatol.* **2010**, *30*, 2195–2204. [CrossRef]
11. Enomoto, T.; Hoskins, B.J.; Matsuda, Y. The formation mechanism of the Bonin high in August. *Q. J. R. Meteorol. Soc.* **2003**, *129*, 157–178. [CrossRef]
12. Hong, X.W.; Xue, S.H.; Lu, R.Y.; Liu, Y.Y. Comparison between the interannual and decadal components of the Silk Road pattern. *Atmos. Ocean Sci. Lett.* **2018**, *11*, 270–274. [CrossRef]
13. Ogasawara, T.; Kawamura, R. Combined Effects of Teleconnection Patterns on Anomalous Summer Weather in Japan. *J. Meteorol. Soc. Jpn.* **2007**, *85*, 11–24. [CrossRef]
14. Hong, X.W.; Lu, R.Y. The meridional displacement of summer Asian jet, Silk Road Pattern, and tropical SST anomalies. *J. Clim.* **2018**, *29*, 3753–3766. [CrossRef]

15. Chen, G.S.; Huang, R.H. Excitation mechanisms of the teleconnection patterns affecting the July precipitation in Northwest China. *J. Clim.* **2012**, *25*, 7834–7851. [CrossRef]
16. Lin, Z.D. Intercomparison of the impacts of four summer teleconnections over Eurasia on East Asian rainfall. *Adv. Atmos. Sci.* **2014**, *31*, 1366–1376. [CrossRef]
17. Hu, P.; Chen, W.; Chen, S.; Liu, Y.Y.; Huang, R.P. Impact of the September Silk Road Pattern on the South China Sea summer monsoon withdrawal. *Int. J. Climatol.* **2020**, *40*, 6361–6368. [CrossRef]
18. Wakabayashi, S.; Kawamura, R. Extraction of major teleconnection patterns possibly associated with the anomalous summer climate in Japan. *J. Meteorol. Soc. Jpn.* **2004**, *82*, 1577–1588. [CrossRef]
19. Sun, X.Q.; Li, S.L.; Hong, X.W.; Lu, R.Y. Simulated Influence of the Atlantic Multidecadal Oscillation on Summer Eurasian Nonuniform Warming since the Mid-1990s. *Adv. Atmos. Sci.* **2019**, *36*, 43–54. [CrossRef]
20. Guan, W.N.; Hu, H.B.; Ren, X.J.; Yang, X.Q. Subseasonal zonal variability of the western Pacific subtropical high in summer: Climate impacts and underlying mechanisms. *Clim. Dyn.* **2019**, *53*, 1577–1588. [CrossRef]
21. Wang, N.; Zhang, Y.C. Evolution of Eurasian teleconnection pattern and its relationship to climate anomalies in China. *Clim. Dyn.* **2015**, *44*, 1017–1028. [CrossRef]
22. Saeed, S.; Lipzig, N.V.; Müller, W.A.; Saeed, F.; Zanchettin, D. Influence of the circumglobal wave-train on European summer precipitation. *Clim. Dyn.* **2014**, *43*, 503–515. [CrossRef]
23. Huang, G.; Liu, Y.; Huang, R.H. The interannual variability of summer rainfall in the arid and semiarid regions of northern China and its association with the Northern Hemisphere circumglobal teleconnection. *Adv. Atmos. Sci.* **2011**, *28*, 257–268. [CrossRef]
24. Huang, R.H.; Liu, Y.; Feng, T. Interdecadal change of summer precipitation over Eastern China around the late-1990s and associated circulation anomalies, internal dynamical causes. *Chin. Sci. Bull.* **2013**, *58*, 1339–1349. [CrossRef]
25. Wang, L.J.; Wang, C.; Guo, D. Evolution mechanism of synoptic-scale EAP teleconnection pattern and its relationship to summer precipitation in China. *Atmos. Res.* **2018**, *214*, 150–162. [CrossRef]
26. Chen, Y.; Zhai, P.M. Synoptic-scale precursors of the East Asia/Pacific teleconnection pattern responsible for persistent extreme precipitation in the Yangtze River Valley. *Q. J. R. Meteorol. Soc.* **2015**, *141*, 1389–1403. [CrossRef]
27. Chen, Y.; Zhai, P.M. Mechanisms for concurrent low-latitude circulation anomalies responsible for persistent extreme precipitation in the Yangtze River Valley. *Clim. Dyn.* **2016**, *47*, 989–1006. [CrossRef]
28. Chen, Y.; Zhai, P.M.; Liao, Z.; Li, L. Persistent precipitation extremes in the Yangtze River Valley prolonged by opportune configuration among atmospheric teleconnections. *Q. J. R. Meteorol. Soc.* **2019**, *145*, 2603–2626. [CrossRef]
29. Li, X.Y.; Lu, R.Y. Subseasonal change in the seesaw pattern of precipitation between the Yangtze River basin and the tropical western North Pacific during summer. *Adv. Atmos. Sci.* **2018**, *35*, 1231–1242. [CrossRef]
30. Gong, Z.Q.; Feng, G.L.; Dogar, M.M.; Huang, G. The possible physical mechanism for the EAP–SR co-action. *Clim. Dyn.* **2017**, *51*, 1499–1516. [CrossRef]
31. Hersbach, H.; Dee, D. ERA5 Reanalysis is in Production. *ECMWF News* **2016**, *147*, 147.
32. National Meteorological Information Center. *Assessment Report of China's Ground Precipitation 0.5° × 0.5° Gridded Dataset (V2.0)*; National Meteorological Information Center: Beijing, China, 2010.
33. Hutchinson, M.F. Interpolation of Rainfall Data with Thin Plate Smoothing Splines-Part I: Two Dimensional Smoothing of Data with Short Range Correlation. *J. Geogr. Inf. Decis. Anal.* **1998**, *2*, 153–167.
34. Grumm, R.H.; Hart, R.H. Standardized anomalies applied to significant cold season weather events: Preliminary findings. *Weather Forecast.* **2001**, *16*, 736–754. [CrossRef]
35. Takaya, K.; Nakamura, H. A Formulation of a Phase-Independent Wave-Activity Flux for Stationary and Migratory Quasi-geostrophic Eddies on a Zonally Varying Basic Flow. *J. Atmos. Sci.* **2001**, *58*, 608–627. [CrossRef]
36. Huang, G. An index measuring the interannual variation of the East Asian Summer Monsoon–the EAP index. *Adv. Atmos. Sci.* **2004**, *21*, 41–52. [CrossRef]
37. Li, X.Y.; Lu, R.Y.; Greatbatch, R.J.; Li, G.; Hong, X.W. Maintenance Mechanism for the Teleconnection Pattern over the High Latitudes of the Eurasian Continent in Summer. *J. Clim.* **2019**, *33*, 1017–1030. [CrossRef]
38. Qu, X.; Huang, G. Impacts of tropical Indian Ocean SST on the meridional displacement of East Asian jet in boreal summer. *Int. J. Climatol.* **2012**, *32*, 2073–2080. [CrossRef]
39. Kosaka, Y.; Nakamura, H.; Watanabe, M.; Kimoto, M. Analysis on the dynamics of a wave-like teleconnection pattern along the summertime Asian jet based on a reanalysis dataset and climate model simulations. *J. Meteorol. Soc. Jpn.* **2009**, *87*, 561–580. [CrossRef]
40. Sato, N.; Takahashi, M. Dynamical processes related to the appearance of quasi-stationary waves on the subtropical jet in the midsummer Northern Hemisphere. *J. Clim.* **2006**, *19*, 1531–1544. [CrossRef]
41. Chen, G.S.; Huang, R.H.; Zhou, L.T. Baroclinic Instability of the Silk Road Pattern Induced by Thermal Damping. *J. Atmos. Sci.* **2013**, *70*, 2875–2893. [CrossRef]
42. Chen, X.L.; Zhou, T.J. Relative contributions of external SST forcing and internal atmospheric variability to July-August heat waves over the Yangtze River valley. *Clim. Dyn.* **2017**, *51*, 4403–4419. [CrossRef]
43. Chen, R.D.; Lu, R.Y. Comparisons of the Circulation Anomalies Associated with Extreme Heat in Different Regions of Eastern China. *J. Clim.* **2015**, *28*, 5844–5930. [CrossRef]

44. Chen, H.S.; Liu, L.; Zhu, Y.J. Possible linkage between winter extreme low temperature events over China and synoptic-scale transient wave activity. *Sci. China Earth Sci.* **2013**, *56*, 1266–1280. [CrossRef]
45. Li, Z.H.; Li, C.Y.; Song, J.; Tan, Y.K.; Li, X. An analysis of the characteristics and causes of extremely high temperature days in the Yangtze–Huaihe River basins in summer 1960–2011. *Clim. Environ. Res.* **2015**, *20*, 511–522. (In Chinese) [CrossRef]
46. Bellprat, O.; Guemas, V.; Doblas-Reyes, F.; Donat, M.G. Towards reliable extreme weather and climate event attribution. *Nat. Commun.* **2019**, *10*, 1–7. [CrossRef] [PubMed]
47. Liu, Y.; Huang, R.H. Linkages between the South and East Asian Monsoon water vapor transport during boreal summer. *J. Clim.* **2019**, *32*, 4509–4524. [CrossRef]
48. Holton, J.R. *An Introduction to Dynamic Meteorology*, 4th ed.; Academic Press: Burlington, VT, USA, 2004; pp. 146–147, ISBN 9780123540157.
49. Li, N.; Xiao, Z.N.; Zhao, L. Analysis on the Mechanism of the 2018 Summer Extreme High Temperature Event in Northeast China. *Clim. Environ. Res.* **2020**, *25*, 469–482. (In Chinese) [CrossRef]

atmosphere MDPI

Article

Refined Characteristics of Moisture Cycling over the Inland River Basin Using the WRF Model and the Finer Box Model: A Case Study of the Heihe River Basin

XiaodUo Pan [1,*], Weiqiang Ma [1,*], Ying Zhang [2] and Hu Li [1,3]

[1] Key Laboratory of Tibetan Environmental Changes and Land Surface Processes, National Tibetan Plateau Data Center, Institute of Tibetan Plateau Research, Chinese Academy of Sciences, No. 16 Lincui Road, Chaoyang District, Beijing 100101, China; lihu@itpcas.ac.cn

[2] Key Laboratory of Remote Sensing of Gansu Province, Northwest Institute of Eco-Environment and Resources, Chinese Academy of Sciences, Lanzhou 730000, China; zhang_y@lzb.ac.cn

[3] The Graduate School (Beijing) of Science and Technology, University of the Chinese Academy of Sciences, Beijing 100049, China

* Correspondence: panxd@itpcas.ac.cn (X.P.); wqma@itpcas.ac.cn (W.M.)

Abstract: The Heihe River Basin (HRB), located on the northeastern edge of the Tibetan Plateau, is the second-largest inland river basin in China, with an area of 140,000 km^2. The HRB is a coupling area of the westerlies, the Qinghai–Tibet Plateau monsoon and the Southeast monsoon circulation system, and is a relatively independent land-surface water-circulating system. The refined characteristics of moisture recycling over the HRB was described by using the Weather Research and Forecasting (WRF) model for a long-term simulation, and the "finer box model" for calculating the net water-vapor flux. The following conclusions were drawn from the results of this study: (1) The water vapor of the HRB was dominantly transported by the wind from the west and from the north, and the west one was much larger than the north one. The net vapor transported by the west wind was positive, and by the north wind was negative. (2) The precipitation over the HRB was triggered mainly by the vapor from the west, which arose from the lower vertical layer to higher one during transporting from west to east. The vapor from the north sank from a higher layer to a lower one, and crossed the south edge of the HRB. (3) The moisture-recycling ratio of evapotranspiration to precipitation over the HRB was much higher than the other regions, which may be due to the strong land–atmosphere interaction in the arid inland river basin.

Keywords: land–atmosphere interaction; water vapor; weather research and forecasting model; precipitation; evapotranspiration; tuning layer

Citation: Pan, X.; Ma, W.; Zhang, Y.; Li, H. Refined Characteristics of Moisture Cycling over the Inland River Basin Using the WRF Model and the Finer Box Model: A Case Study of the Heihe River Basin. *Atmosphere* **2021**, *12*, 399. https://doi.org/10.3390/atmos12030399

Academic Editors: Nicola Scafetta and Ankit Agarwal

Received: 3 February 2021
Accepted: 16 March 2021
Published: 20 March 2021

1. Introduction

Moisture recycling is the contribution of regional moisture to precipitation in a region [1–4]. It has the ability to inhibit extreme precipitation-induced hydrological events [5,6], and reflects the memory of soil moisture and the continuous abnormality of dry and wet conditions [7]. A comprehensive understanding of the regional moisture-recycling process not only is the key to understanding the regional water cycle and surface land–atmosphere interaction [8,9], but also plays the main role in improving future climate-change projections.

Generally, the rate of moisture recycling depends heavily on the spatial scale of the study area and its regional atmospheric circulation. The smaller the spatial scale is, the lower the moisture-recycling rate [1]. The moisture-recycling rate is nearly 1 at the global scale, approximately 40% at the continent scale, and generally less than 20% at a scale of 1000 km^2 [10–12]. However, the contribution of regional evaporation to precipitation is high [13], and far greater than 20% in the inland river basin, which is at the regional scale of 10^6 km^2, due to the strong land–atmosphere interaction. Zhao et al. [14] reported that the contributions of recycled moisture to precipitation in the upper, middle, and lower reaches of the Heihe River Basin (HRB) are approximately 52.4%, 56.5%, and 21.4%, respectively, and that recycled moisture (especially transpiration) plays an important role in regional precipitation redistribution. The inland river basin is located in the hinterland of the continent and far away from the ocean. No sufficient source of water vapor exists, and the runoff generated by rainfall or alpine snow melting in the inland mountainous area cannot flow into the sea, and finally is infiltrated in the local basin. The inland river basin is an ideal hydrometeorological research region that exchanges water and energy with the external world within a clear boundary [15].

However, compared to the relatively simple process of marine inner water cycling, the moisture recycling of the inland river basin is much more complicated because of the spatial heterogeneity and diversity of its complex underlying surface. Generally, the inland river basin originates from a mountainous area and extends to a desert [16]. In the high mountainous upper reaches of the inland river basin, diverse underlying surfaces of glaciers, snow, frozen soil, and alpine forest meadow can be found, and a mountainous "water tower" can be observed overhead, a result of water vapor being intercepted from the outside. In the middle reaches, the Gobi Desert, an oasis, a unique artificial water channel, and farmland can be seen on the surface, with runoff flowing through formed by precipitation from upstream. In the downstream of the inland river basin, the river is injected into the terminal lake or is infiltrated in the desert. These complex underlying surfaces make it difficult to describe the process of precipitation and evapotranspiration in the inland river basin; therefore, the moisture recycling of the basin is much more complex.

Meanwhile, with global climate change, the uncertainty of moisture recycling in the inland river basin has increased. The fifth assessment report of the Intergovernmental Panel on Climate Change (IPCC) noted the lack of doubt regarding the warming of the climate system. Since 1850, the global average temperature increases by 0.85 degrees from 1880 to 2012 [17]. According to the National Oceanic and Atmospheric Administration, the increase in temperatures over the first 15-year period of the 21st century were almost equal to the increases that occurred over the entire last half of the 20th century [18]. The increase in temperature means that increasingly more precipitation reaching the ground will evaporate into the atmosphere. Due to the insufficient precipitation and the fragile ecological environment, the arid inland river basin is vulnerable to global climate change; therefore, moisture recycling and changes mechanism in the arid inland river basin must be carried out to against the background of climate change and to propose regional vulnerability response methods.

Long-term, high-resolution, atmospheric numerical simulation helps to precisely refine the characterization of the moisture-recycling process in the inland river basin. Long-term data help to reduce the bias and to reveal the trend of moisture recycling. Higher resolution improves not only the land-surface representation, but also the ability of regional climate models to simulate important small-scale precipitation processes [19], such as convective phenomena that are generally considered to be sufficiently resolved at a less than approximately 5 km spatial resolution [20].

Generally, the box model is widely used to calculate the regional water and energy mass balance as the mass difference between the input and the output in a certain region at a specific time [21–24]. However, most researchers take the entire research region as a box, which obliterates the jagged complex edge of the natural topography.

In this paper, moisture recycling over the inland Heihe River Basin is calculated based on the Weather Research and Forecasting (WRF) model at a 5 km spatial resolution and an hourly temporal resolution from 2000 to 2015 using the finer box model. This study aims to describe the specific characteristics of water-vapor transport over the inland river basin and calculate the contribution of local evapotranspiration to precipitation over the HRB with a relatively independent land-surface water-circulation system. The WRF model is proved to be credible over the inland river basin [25,26]. The HRB, located on the northeastern edge of the Tibetan Plateau, is the second-largest inland river basin in China, with an area of 140,000 km^2, and is an area coupling the Tibetan Plateau, Mongolia Plateau, and Loess Plateau. This region is mainly dominated by westerly winds, though polar north winds also function as a relatively independent land water-circulating system; hence, it as an ideal research region for examining the atmosphere, hydrology, ecology, and multispheres.

The next section introduces the research region, WRF model configuration, moisture-recycling method, and finer box model. The results are shown in Section 3 and discussed in Section 4, and conclusions are drawn in Section 5.

2. Research Region, WRF Model Configuration, and Methods

2.1. Research Region

The HRB is regarded as an ideal test bed and field laboratory for land-surface or hydrological experiments [27,28], located at the junction of the Qinghai Tibet Plateau, Loess Plateau, and Mongolia Plateau (Figure 1). The HRB originates from glacial meltwater and precipitation in the Qilian Mountains and flows through the provinces of Qinghai, Gansu, and Inner Mongolia. It can be divided into three parts: the upper reaches, with elevations ranging from 2000–5500 m and precipitation ranging from 250~500 mm; the middle reaches, being defined as the reach between Yingluo Gorge and Zhengyi Gorge, with elevations ranging from 1000–2000 m and precipitation ranging from 100~250 mm; and the lower reaches, with elevations of less than 1000 m and precipitation of less than 100 mm. The basin climate is mainly controlled by the westerlies throughout the year [29,30]. From the upper reaches to the lower reaches, glacier, permafrost, oasis, bare, and Gobi Desert areas are distributed. The diverse landscape makes the HRB an ideal watershed for scientific research. Comprehensive experiments, including the Heihe Basin Field Experiment [31], Watershed Allied Telemetry Experimental Research [32], and Heihe Watershed Allied Telemetry Experimental Research [28], have been conducted in the HRB.

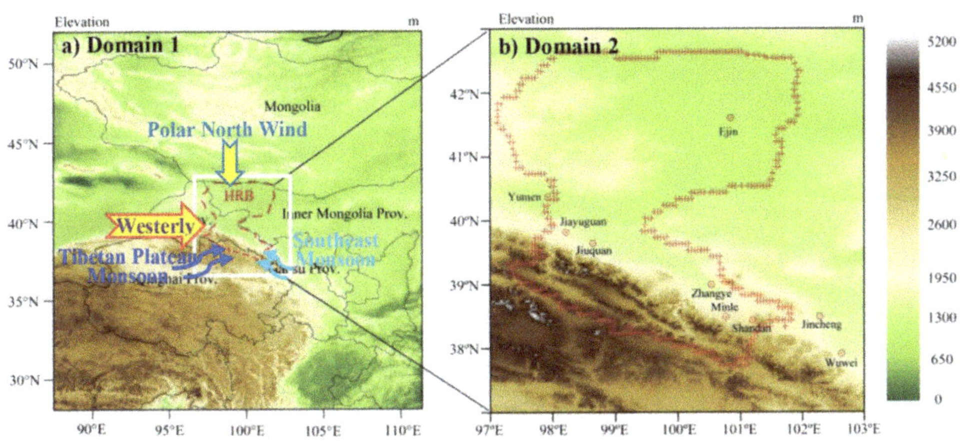

Figure 1. The study region: (**a**) the location of the HRB; (**b**) the research domain.

As shown in Figure 1, the water-vapor transport in the HRB can be controlled by westerly wind, polar north wind, the Tibetan Plateau monsoon and the Southeast monsoon. However, water-vapor transport by the Tibetan Plateau and Southeast monsoons is obstructed by the Qilian Mountains because of their high elevation, so part of it is converted into precipitation in the region south of the HRB, and another part turns eastward [33–35]. Therefore, the water-vapor transport by the Tibetan Plateau and Southeast monsoons is negligible in the HRB.

2.2. WRF Model Configuration

The Advanced Research WRF [36] version 3.5 modeling system was used in this study. The model includes Arakawa C-grid staggering for a horizontal grid, a fully compressible system of equations, terrain-following hydrostatic pressure with vertical grid stretching, and a third-order Runge–Kutta scheme for time-split integration. The WRF physical configuration (Table 1) used in this study consisted of the WRF single-moment 5-class scheme [37] as the microphysics option, the Kain–Fritsch scheme [38] as the cumulus convection parameterization option, the Yonsei University scheme [39] as the planetary boundary layer (PBL) option, a 5-layer thermal diffusion scheme as the land-surface model option, and the Dudhia scheme [40] and the rapid radiative transfer model scheme [41] as the shortwave and longwave radiation options, respectively.

Table 1. Physical configuration of the WRF model used in this experiment.

Physics Processes	Domain 1 (25 km)	Domain 2 (5 km)
Horizontal	60 × 60	130 × 130
Time step	150 s	30 s
Microphysics	single-moment 5-class scheme	single-moment 5-class scheme
Cumulus	Kain–Fritsch scheme	Kain–Fritsch scheme
PBL	YSU scheme	YSU scheme
Shortwave radiation	Dudhia scheme	Dudhia scheme
Longwave radiation	Rapid radiative transfer model	Rapid radiative transfer model
Surface–land	5-layer thermal diffusion	5-layer thermal diffusion
Initial and boundary	NCEP/FNL analysis	Domain 1

This study used two-way nested computational domains with 60 × 60 × 40 and 130 × 130 × 40 grid points and horizontal resolutions of 25 km and 5 km, respectively. Two-way nesting was used to perform the model simulations (Figure 1) in a scheme in which domains at different grid resolutions were run simultaneously and communicated with each other. The coarser domain was forced with the reanalysis data sets of the Final Analysis (FNL) from the National Center for Environmental Prediction (NCEP), and provided initial and boundary conditions for the finer nested domain, which feeds its calculations back to the coarser domain.

2.3. Moisture Recycling Estimation Method

As described in Global Energy and Water Cycle Experiment (GEWEX) [42], the following components of the water cycle exist for a river basin: (1) net inflow of water vapor through the lateral atmospheric boundaries of the basin; (2) net transfer of water from the atmosphere to the surface by excess precipitation over evapotranspiration; and (3) river discharge from the basin to the ocean. Because the HRB is an inland river basin, the water cycle includes only (1) and (2). The water-balance equations for the atmosphere and land are shown as below:

$$d(q_a + q_l)/dt = C_q - N_l - N_q \tag{1}$$

$$dq_a/dt = C_q - (P - E) = \frac{1}{g}\int_0^{p_s} q\,dp \tag{2}$$

$$dq_l/dt = (P - E) - N_l \tag{3}$$

$$C_q = \frac{1}{g}\int_0^{p_s} \nabla\cdot\left(\vec{W}q\right)dp - N_q \tag{4}$$

$$N_q = \beta q_a \tag{5}$$

$$\gamma = \frac{P - C_q - N_q}{P} \tag{6}$$

where q_a and q_l are the total atmospheric and land water masses per unit horizontal area; C_q is the net inflow of atmospheric water to the basin; N_l is the net land runoff, which is the outflow from the area of interest; N_q is the net change in condensed water in the air; P and E are the precipitation and evapotranspiration fluxes at the ground level, respectively; q is the specific humidity; p is the pressure; and p_s is the pressure at the top of the atmosphere. Because the inland river basin is a relatively independent water-circulating system, here, N_l is equal to 0. Generally, N_q is assumed to be negligible; however, warmer temperatures increase the rate of evaporation of water into the atmosphere, in effect increasing the atmosphere's capacity to hold water. With global warming, the water-holding capability improves, which means that the atmospheric water storage increases, and its change N_q cannot be ignored. $\vec{W} = (u, v, w)$ is the wind component at x, y, z directions; g is the gravitational constant; β is the rate of annual change in the total atmospheric water mass; and γ is the regional moisture-recycling ratio.

2.4. Finer Box Model for Calculating the Net Water Vapor Flux

Generally, the water vapor flux is calculated by considering the entire research region as a box; this is known as the box model. This method is quite rough, obliterating the boundary complexity of the research region and making the high-resolution atmospheric simulation data essentially useless. Therefore, in this research, the water vapor of every grid along the boundary was calculated, referred to as the "finer box model", which was designed to improve the accuracy of the net inflow of water vapor through the boundary of the research region based on the high-resolution atmospheric simulation data. These grids are indicated by the red crosses in Figure 1b.

The water vapor flux for each grid is calculated according to Equation (7):

$$q_{(i,j)} = \frac{1}{g}\int_0^{p_s}\left(\vec{W}_{(p,i,j)}q_{(p,i,j)}\cdot\vec{n}_{(i,j)}\right)dp \tag{7}$$

where i and j are the location identification of the boundary grid in the whole research domain, $\vec{n}_{(i,j)}$ is the unit normal vector at the boundary, p is pressure, and $\vec{W}$ is the wind component at x, y, z directions.

The regional net water-vapor flux (Δq) is calculated by Equation (8):

$$\Delta q = \sum_{\substack{0\le i\le M \\ 0\le j\le N}} q_{(i,j)} \tag{8}$$

where M and N are the sizes of the used grids in the adopted unit of space.

3. Results

3.1. Atmospheric Water Storage

As mentioned in Section 2, with global warming, the water-holding capability improves, and the atmospheric water-storage change N_q cannot be ignored. The annual mean atmospheric water mass (Figure 2) was calculated based on the mixing ratios of the water vapor, cloud water, rain water, ice water, and snow water. According to the annual mean atmospheric water mass from 2000 to 2015, though there were minor variations from year to year, the pattern is shown well in the mean that the maximum and minimum of the atmospheric water mass were located in the southeastern and northeastern regions, respectively.

Figure 3 shows that a strong relationship exists between atmospheric water storage, temperature, and precipitation over the HRB. From 2000 to 2015, the change trends of water vapor and temperature were positive and almost synchronous, the change trend of precipitation was also positive, but not so obvious. The reason may be that the higher temperature increased the capacity to contain atmospheric water mass, which is necessary for the occurrence of precipitation [43]. The water vapor atmosphere increased by approximately 0.13 kg/m^2 per year, an important indicator of the net change in condensed water in the air (N_q).

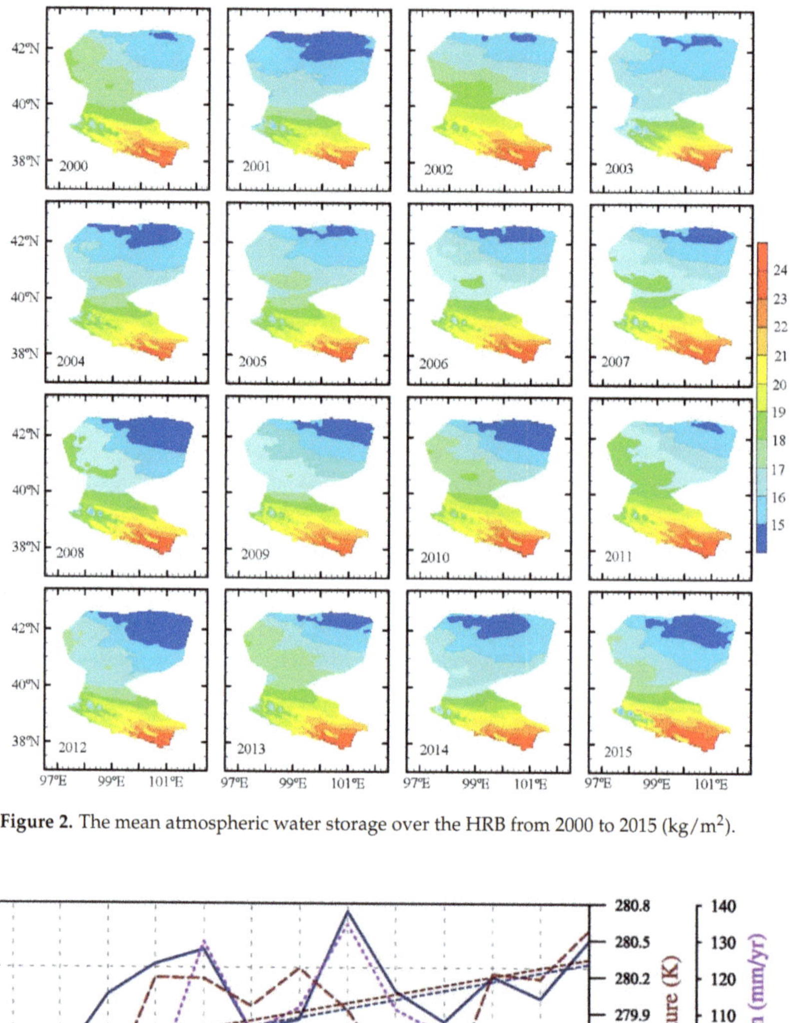

Figure 2. The mean atmospheric water storage over the HRB from 2000 to 2015 (kg/m^2).

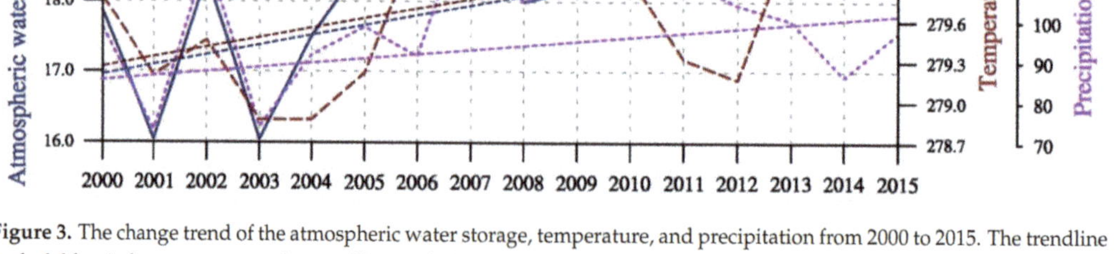

Figure 3. The change trend of the atmospheric water storage, temperature, and precipitation from 2000 to 2015. The trendline in dark blue is for water vapor, the trendline in dark red is for temperature, and the trendline in purple is for precipitation.

3.2. Net Water-Vapor Transport in the X and Y Directions

As analyzed from the NCEP/NCAR reanalysis data set by Wang et al. [33] and Lu et al. [35], the water vapor in the HRB comes mainly from the westerly wind transport via the Black Sea and the Caspian Sea in summer, which is a net income for the water-vapor balance and different from most parts of northwest China; a small part of water vapor in the HRB comes from the polar north wind transport via Mongolia in winter. Meanwhile, according to the net water-vapor transport along the meridional and zonal determined by WRF simulation in this study, it was found that the dominant winds over the HRB were the west wind and the north wind.

The layer-by-layer cumulative analysis shows that the 9th vertical layer (600 $\pm$ 100 hPa) is the tuning layer, the altitude of which can be found in Figure 4. Along the meridional, under the tuning layer, the accumulated water vapor from the west crossing the western boundary as input was much bigger than that crossing the eastern boundary as output; above the tuning layer, the accumulated water vapor from the west crossing the western boundary as input was less than that crossing the boundary as output. However, along the zonal, under the tuning layer, the accumulated water vapor from the north crossing the northern boundary as input was much less than that crossing the southern boundary as output; above the tuning layer, the accumulated water vapor from the north crossing the northern boundary as input was larger than that crossing the southern boundary as output.

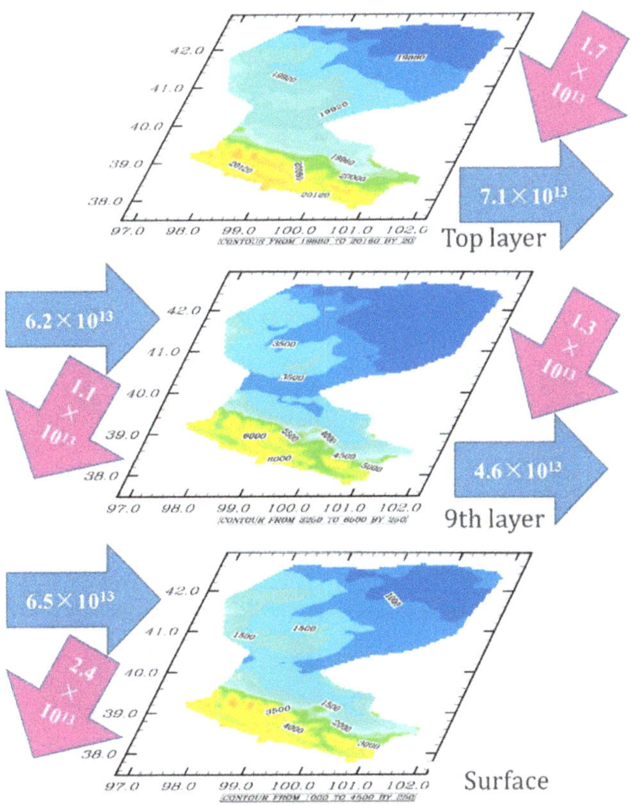

Figure 4. The net water vapor transport in the X and Y directions (kg/year). The blue arrow represents the water vapor transported by west wind, and the pink arrow represents the water vapor transported by north wind.

As shown in Figure 4 and Table 2, the annual cumulative input volumes of water-vapor transport in the upper layers (from the 10th layer to the top) and the lower layers (from the surface to the 9th layer) when crossing the west boundary were 6.2 × 10^{13} kg/year and 6.5 × 10^{13} kg/year, respectively, while the output transport rates when crossing the east boundary were 7.1 × 10^{13} kg/year and 4.6 × 10^{13} kg/year, respectively.

Table 2. The input and output of water vapor by the west wind and north wind over the HRB.

Wind	Layer	Input (kg/year)	Output (kg/year)	Net (kg/year)
West wind	Surface—9th layer	6.5 × 10^{13}	4.6 × 10^{13}	1.9 × 10^{13}
	10th layer—Top	6.2 × 10^{13}	7.1 × 10^{13}	−0.9 × 10^{13}
North wind	Surface—9th layer	1.3 × 10^{13}	2.4 × 10^{13}	−1.1 × 10^{13}
	10th layer—Top	1.7 × 10^{13}	1.1 × 10^{13}	0.6 × 10^{13}

The annual cumulative input volumes of water-vapor transport in the upper layers and lower layers while crossing the north boundary were 1.7 × 10^{13} kg/year and 1.3 × 10^{13} kg/year, respectively, and the output ones while crossing the south boundary were 1.1 × 10^{13} kg/year and 2.4 × 10^{13} kg/year, respectively.

3.3. Precipitation, Evapotranspiration, and Runoff

Figure 5 shows the mean annual spatial distribution of the precipitation, evapotranspiration, runoff, and ratios of evapotranspiration and runoff to precipitation. Much of the precipitation was consumed in evapotranspiration, while some participated in runoff. The figure also indicates that the evapotranspiration was larger than the precipitation in the downstream of the HRB. Parts of the ratios of the evapotranspiration to precipitation in the downstream exceeded 2.0, especially in Juyan Lake, although the runoff over this area was larger than that in other areas of the downstream HRB. The large difference between evapotranspiration and precipitation over this region caused the lake to remain dry for more than 200 days per year. The ratio of runoff to precipitation was much lower than that of evapotranspiration to precipitation.

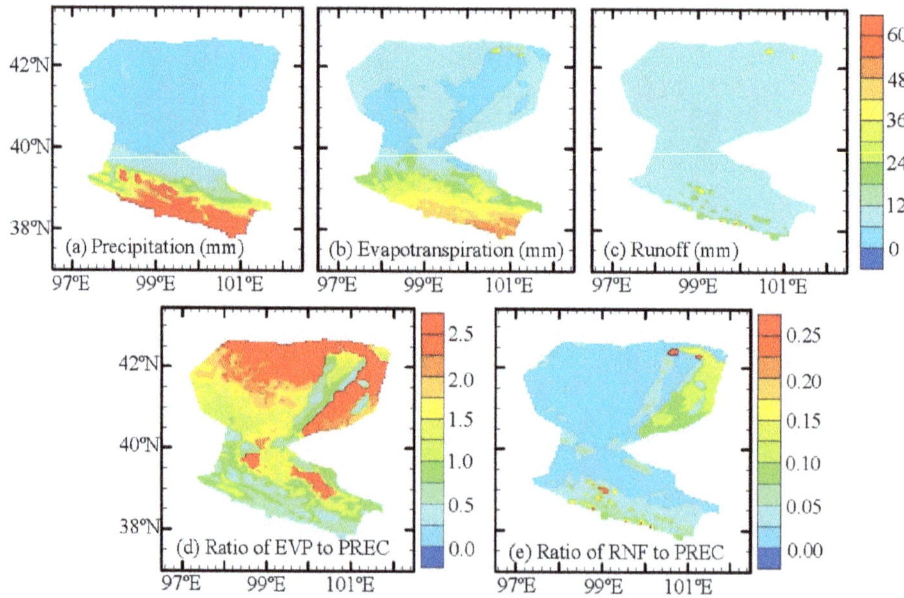

Figure 5. The spatial distribution of the precipitation, evapotranspiration, runoff, and their ratios.

3.4. Moisture Recycling

Based on the WRF simulation and finer box model, the average annual net water vapor was 0.5×10^{13} kg, the annual water vapor storage change was approximately $0.13\ kg/m^2$, the average annual precipitation of the HRB was 103.4 mm, and the area of the HRB was approximately 100,000 km^2. According to Equation (6), i.e., $\gamma = \frac{P-C_q-N_q}{P}$, the moisture-recycling rate of the HRB was approximately 0.52.

4. Discussion

4.1. Water-Vapor Transport in the HRB

Based on different methods and data, scientists have consistently concluded that the water-vapor transmission in the HRB is dominated by both westerly transport from the west to the east and the polar (Siberia) cold-air mass from the north to the south, with the intensity of the westerly wind in summer being greater than that of the northerly wind in winter, and that the water-vapor input mainly occurs in the period from June to September [14,34,35,44,45].

However, different opinions exist on the following issues: (1) whether the net water-vapor transport over the entire HRB is positive or negative; (2) which direction of water-vapor transport is positive, westerly or northerly; and (3) the actual volume of water-vapor transport over the HRB.

Lu et al. [36] not only showed that the net water-vapor transport over the entire HRB was negative (approximately $-48.8\ km^3$), but also indicated that the water vapor transported by the westerlies over the entire HRB was negative in summer, and that transported by northerlies was positive in winter. Wang et al. [44], Jiang et al. [34], and Xu [45], however, reached a completely opposite conclusion, i.e., the net water-vapor transport was positive, while the net water-vapor transport by the westerlies was positive, and that by the northerlies was negative.

The total input volume of water-vapor transport by the westerlies was larger than that output, which means that the net water-vapor transport by the westerlies was positive. In addition, the net water-vapor transport of the lower layers by the westerlies was positive, and that of the upper layers by the westerlies was negative, which may be due to the water transported by the westerlies being forced to uplift by the gradually strengthening monsoon from south, which is conducive to the formation of precipitation in east, and part of uplifting water vapor is exported from the east boundary in the upper layers, as explained in Xu et al. [46]. However, the phenomenon for the polar north wind was the opposite. The total input volume of water-vapor transport by polar north wind was less than that output, which means that the net water-vapor transport by the polar north wind was negative. Meanwhile, the net water-vapor transport of the lower layers by the polar north wind was negative, and that of the upper layers by the polar north wind was positive, which means that the water transported by the polar north wind underwent a divergence process, which is not conductive to the formation of precipitation, and made the HRB winters drier. The water vapor transported across the southern HRB boundary reaches the Qilian Mountains and then falls as snow, which makes the Qilian Mountains abundant in precipitation [47].

Compared to the question of which water-vapor-transport process is positive, disputes on the volume of water vapor transported over the HRB are much greater. In comparison to this study, Wang et al. [44], Jiang et al. [34], and Xu [45] obtained the same results: the net water-vapor transport over the entire HRB was positive, that by westerlies was positive, and that by polar north winds was negative. However, the total volumes of the net transport, that by westerlies and that by polar north winds in these studies were different from those of our study and were different from each other. The volumes of net water-vapor transport were 17.6 billion m^3, 28.8 billion m^3, 22.573 billion m^3, and 4.0 billion m^3 in Wang et al. [44], Jiang et al. [34], Xu [45] and this study, respectively. The net input water vapor was 667.8 billion m^3, 248.4 billion m^3, and 157.0 billion m^3 in Wang et al. [44], Jiang et al. [34], and this study, respectively. The net output water vapor was 650.2 billion m^3, 219.6 billion m^3, and 153.0 billion m^3 in Wang et al. [44], Jiang et al. [34], and this study, respectively. The possible reason for the discrepancies is that the other studies used the box model to calculate the water-vapor transport, which tends to obliterate the jagged complex edge of the natural topography.

4.2. Accuracy of the Moisture Recycling Estimation

In this study, moisture recycling was defined as the contribution of local moisture to precipitation. The accuracy of moisture recycling estimation depends on the following estimation: (1) the accuracy of the water-vapor-transport calculation, (2) the accuracy of the precipitation simulation in the WRF model, and (3) the accuracy of the net change in condensed water in the air. Here, the high resolution of the WRF model in this study was crucial to the moisture-recycling estimation. In addition, compared to the traditional coarse box model, a finer box model was adopted to calculate the net water-vapor flux in detail.

The WRF model is strongly sensitive to the model resolution and underlying surface terrain features in the HRB [48,49]. High spatial resolution not only yields more precise amounts and more reasonable patterns of precipitation (including snow) in complex mountainous regions, but also contributes to more apparent wind-speed patterns [50,51]. Wind speed plays an important role in calculating the water-vapor transport. Additionally, the net change in condensed water in the air (N_q), which was zero in most studies but plays an increasingly important role in the air water cycle, was considered in this study using long-term atmospheric water storage based on a high-resolution simulation in the WRF model.

Currently, the flux volume of water-vapor transport varies from one researcher to another. This study, based on long-term high-resolution simulations in regional climate models and finer box models, was helpful in resolving this issue, and obtained an accurate moisture-recycling rate over the HRB.

4.3. Higher Moisture Recycling over the Inland River Basin

There was a higher moisture recycling and a much stronger land–atmosphere interaction over the inland river basin than the other regions with same spatial scale, which may be due to the inland river basin being located far from the sea. Wu et al. [52] indicated that the Tarim River Basin, the biggest inland river basin in China, had a high local moisture recycling because it is located far away from ocean and next to the Tibetan Plateau, which is similar to the HRB. Keys et al. [53] indicated that the land surface plays a dominant role in mediating variability in moisture-recycling processes in the inland river basins. Keys et al. [54] and Li et al. [55] indicated that the intermountain basin near the Qilian Mountains, as a relatively closed terrain of the region, is beneficial to local moisture recycling.

Meanwhile, the water from the cryosphere belt, which is generally considered as a belt in addition to the vegetation belt, oases belt, and desert belt over the inland river basin, accounts for a large portion of moisture recycling [55,56]. The cryosphere belt saves water for cold seasons and releases in the flood season when the temperature is highest and the evapotranspiration is most strong in a year, which makes the moisture-recycling ratio of the inland river basin higher than that of the other regions at the same spatial scale.

4.4. Tuning Layer and Other Issues

The tuning layer, i.e., the 9th layer in our research, was approximately 600 $\pm$ 100 hPa over the entire HRB. Most atmospheric convergent and divergent activities occur at or near this layer [57]. In this study, the annual cumulative net volume of water-vapor transport in the upper layers (from the 10th layer to the top) was negative, and was positive in the lower layers (from the surface to the 9th layer) while water vapor crossed from west to east. The annual cumulative net volume of water-vapor transport was positive in the upper layers and was negative in the lower layers while water vapor crossed from the north to south. The net volume of vapor transport in the upper layers being positive means that the water-vapor convergence uplifted, and was likely to form precipitation during transport; however, the net volume of vapor transport in the lower layers being positive means that the water-vapor divergence sunk, and was not conductive to forming precipitation.

5. Conclusions

The Heihe River Basin (HRB) is a coupling area of the westerlies and the polar north wind circulation system, with a relatively independent land water-circulating system. The aim of this study was to describe the refined characteristics of the water-vapor transport over the inland river basin, and calculate the contribution of local evapotranspiration to precipitation over the HRB. Based on the long-term WRF simulation and finer box model, the net water-vapor flux as calculated grid by grid, the quantitative change trend of the atmospheric water storage, and the refined characteristics of moisture cycling over the inland Heihe River, the following conclusions were drawn:

(1) The water vapor of the HRB was dominantly transported by the wind from west and from the north, and the west wind was much larger than the north wind. The net vapor transported by the west wind was positive, and by north wind was negative;
(2) The precipitation over the HRB was triggered mainly by the vapor from the west, which arose from the lower vertical layer to the higher one during transport from west to east. The vapor from the north sank from a higher layer to a lower one, and crossed the south edge of the HRB;
(3) The moisture-recycling ratio of evapotranspiration to precipitation over the HRB was much higher than in the other regions, which may be due to the strong land–atmosphere interaction in the arid inland river basin.

Although the climatological pattern of water-vapor transport was the main issue in this study, an issue arose regarding how to calculate the water-vapor budget delicately for the HRB, so it was insufficient that the diurnal/seasonal/yearly variability of water-vapor transport was not analyzed in detail as well; this will be our next step in the future. Furthermore, an extended study of the physical processes based on the combination of models and observations also will be undertaken in the future.

Author Contributions: X.P. conceived, designed and performed the experiments, and wrote the paper; W.M., Y.Z. and H.L. contributed to discussions and revisions. All authors have read and agreed to the published version of the manuscript.

Funding: This work was jointly supported by the Strategic Priority Research Program of the Chinese Academy of Sciences (XDA20060603, XDA19070104) and the National Natural Science Foundation of China (Grant nos. 41471292, 41801271).

Institutional Review Board Statement: Not applicable.

Informed Consent Statement: Not applicable.

Data Availability Statement: The data presented in this study are openly available in National Tibetan Plateau Data Center at doi:10.3972/heihe.019.2013.db. Data citation: PAN Xiaoduo. The atmospheric forcing data in the Heihe River Basin (2000-2018). National Tibetan Plateau Data Center, 2020. doi:10.3972/heihe.019.2013.db.

Acknowledgments: The input data for the WRF model were obtained from the Research Data Archive (RDA), which is maintained by the Computational and Information Systems Laboratory (CISL) at the National Center for Atmospheric Research (NCAR). The original data are available from the RDA (http://rda.ucar.edu/datasets/ds083.2/, accessed on 10 July 2019) in data set number ds083.2. The NCAR Command Language (NCL) (Version 6.1.2, 2013) is available from the UCAR/NCAR/CISL/VETS (http://dx.doi.org/10.5065/D6WD3XH5, Released 7 February 2013), Boulder, Colorado. NCL was used for the data analysis and graphs in this paper. The authors thank the anonymous reviewers and the editor for their very helpful comments.

Conflicts of Interest: The authors declare no conflict of interest. The funders had no role in the design of the study; in the collection, analyses, or interpretation of data; in the writing of the manuscript; or in the decision to publish the results.

References

1. Brubaker, K.L.; Entekhabi, D.; Eagleson, P.S. Estimation of continental precipitation recycling. *J. Clim.* **1993**, *6*, 1077–1089. [CrossRef]
2. Eltahir, E.A.B.; Bras, R.L. Precipitation recycling in the Amazon basin. *Q. J. R. Meteorol. Soc.* **1994**, *120*, 861–880. [CrossRef]
3. Koster, R.; Jouzel, J.; Suozzo, R.; Russell, G.; Broecker, W.; Rind, D.; Eagleson, P. Global sources of local precipitation as determined by the Nasa/Giss GCM. *Geophys. Res. Lett.* **1986**, *13*, 121–124. [CrossRef]
4. Trenberth, K.E. Atmospheric moisture recycling: Role of advection and local evaporation. *J. Clim.* **1999**, *12*, 1368–1381. [CrossRef]
5. Bosilovich, M.G.; Schubert, S.D. Water vapor tracers as diagnostics of the regional hydrologic cycle. *J. Hydrometeorol.* **2002**, *3*, 149–165. [CrossRef]
6. Dirmeyer, P.A.; Brubaker, K.L. Contrasting evaporative moisture sources during the drought of 1988 and the flood of 1993. *J. Geophys. Res. Space Phys.* **1999**, *104*, 19383–19397. [CrossRef]
7. Dirmeyer, P.A.; Brubaker, K.L. Characterization of the global hydrologic cycle from a back-trajectory analysis of atmospheric water vapor. *J. Hydrometeorol.* **2007**, *8*, 20–37. [CrossRef]
8. Hua, L.; Zhong, L.; Ke, Z. Characteristics of the precipitation recycling ratio and its relationship with regional precipitation in China. *Theor. Appl. Clim.* **2015**, *127*, 513–531. [CrossRef]
9. Yao, Y.; Zheng, C.; Tian, Y.; Liu, J.; Zheng, Y. Numerical modeling of regional groundwater flow in the Heihe River Basin, China: Advances and new insights. *Sci. China Earth Sci.* **2015**, *58*, 3–15. [CrossRef]
10. Ciric, D.; Stojanovic, M.; Drumond, A.; Nieto, R.; Gimeno, L. Tracking the origin of moisture over the Danube river basin using a Lagrangian approach. *Atmosphere* **2016**, *7*, 162. [CrossRef]
11. González-Rojí, S.J.; Sáenz, J.; De Argandoña, J.D.; Ibarra-Berastegi, G. Moisture recycling over the Iberian Peninsula: The impact of 3DVAR data assimilation. *Atmosphere* **2020**, *11*, 19. [CrossRef]
12. Sorí, R.; Nieto, R.; Drumond, A.; Gimeno, L. The Niger river basin moisture sources: A Lagrangian analysis. *Atmosphere* **2017**, *8*, 38. [CrossRef]
13. Ma, J.J.; Yu, B.; Gao, X.Q.; Li, J. Change of large scale circulation and its impact on the water vapor over north china. *Plateau Meteorol.* **2008**, *27*, 517–523. (In Chinese) [CrossRef]
14. Zhao, L.; Liu, X.; Wang, N.; Kong, Y.; Song, Y.; He, Z.; Liu, Q.; Wang, L. Contribution of recycled moisture to local precipitation in the inland Heihe River Basin. *Agric. For. Meteorol.* **2019**, *271*, 316–335. [CrossRef]
15. Cheng, G.; Li, X. Integrated research methods in watershed science. *Sci. China Earth Sci.* **2015**, *58*, 1159–1168. [CrossRef]
16. Cheng, G.D.; Zhao, C.Y. An integrated study of ecological and hydrological processes in the inland river basin of the arid regions, China. *Adv. Earth Sci.* **2008**, *23*, 1005–1012. (In Chinese)
17. IPCC. *Climate Change 2013: The Physical Science Basis. Contribution of Working Group I to the Fifth Assessment Report of the Intergovernmental Panel on Climate Change*; Stocker, T.F., Qin, D., Plattner, G.-K., Tignor, M., Allen, S.K., Boschung, J., Nauels, A., Xia, Y., Bex, V., Midgley, P.M., Eds.; Cambridge University Press: Cambridge, UK; New York, NY, USA, 2013; 1535p. [CrossRef]
18. NOAA. National Centers for Environmental Information, State of the Climate: Global Climate Report for January 2016. Available online: https://www.ncdc.noaa.gov/sotc/global/201601 (accessed on 20 December 2020).
19. Olsson, J.; Berg, P.; Kawamura, A. Impact of RCM Spatial Resolution on the Reproduction of Local, Subdaily Precipitation. *J. Hydrometeorol.* **2015**, *16*, 534–547. [CrossRef]
20. Prein, A.F.; Gobiet, A.; Suklitsch, M.; Truhetz, H.; Awan, N.K.; Keuler, K.; Georgievski, G. Added value of convection permitting seasonal simulations. *Clim. Dyn.* **2013**, *41*, 2655–2677. [CrossRef]
21. Jain, R.K.; Cui, Z.; Domen, J. Appendix A—Emission factors for air pollutants related to mining and mineral processing. In *Environmental Impact of Mining and Mineral Processing*; Butterworth-Heinemann: Oxford, UK, 2016; pp. 271–297. [CrossRef]
22. Li, N.; Kinzelbach, W.; Li, W.; Dong, X. Box model and 1D longitudinal model of flow and transport in Bosten Lake, China. *J. Hydrol.* **2015**, *524*, 62–71. [CrossRef]
23. Mareddy, A.R. 5—Impacts on air environment. In *Impacts on Air Environment*; Butterworth-Heinemann: Oxford, UK, 2017; pp. 171–216. [CrossRef]
24. Romero, J.; Navarro-Esbrí, J.; Belman-Flores, J.; Pérez, J.A.R. A simplified black-box model oriented to chilled water temperature control in a variable speed vapour compression system. *Appl. Therm. Eng.* **2011**, *31*, 329–335. [CrossRef]
25. Pan, X.; Li, X.; Cheng, G.; Li, H.; He, X. Development and evaluation of a river-basin-scale high spatio-temporal precipitation data set using the WRF Model: A case study of the Heihe River Basin. *Remote. Sens.* **2015**, *7*, 9230–9252. [CrossRef]
26. Pan, X.; Li, X.; Shi, X.; Han, X.; Luo, L.; Wang, L. Dynamic downscaling of near-surface air temperature at the basin scale using WRF-a case study in the Heihe River Basin, China. *Front. Earth Sci.* **2012**, *6*, 314–323. [CrossRef]
27. Cheng, G.; Li, X.; Zhao, W.; Xu, Z.; Feng, Q.; Xiao, S.; Xiao, H. Integrated study of the water–ecosystem–economy in the Heihe River Basin. *Natl. Sci. Rev.* **2014**, *1*, 413–428. [CrossRef]
28. Li, X.; Cheng, G.; Liu, S.; Xiao, Q.; Ma, M.; Jin, R.; Che, T.; Liu, Q.; Wang, W.; Qi, Y.; et al. Heihe watershed allied telemetry experimental research (HiWATER): Scientific objectives and experimental design. *Bull. Am. Meteorol. Soc.* **2013**, *94*, 1145–1160. [CrossRef]
29. Yang, Y.; Zhang, Z.; Xiang, X. Spatial variation of reference crop evapotranspiration on Tibetan Plateau. *Water Sci. Eng.* **2009**, *2*, 112–120. [CrossRef]

30. Zhao, L.; Yin, L.; Xiao, H.; Cheng, G.; Zhou, M.; Yang, Y.; Li, C.; Zhou, J. Isotopic evidence for the moisture origin and composition of surface runoff in the headwaters of the Heihe River Basin. *Chin. Sci. Bull.* **2011**, *56*, 406–415. [CrossRef]
31. Wang, J.M.; Liu, X.H.; Ma, Y.M. Turbulence structure and transfer characteristics in the surface layer of HEIFE Gobi area. *Acta Meteorol. Sin.* **1992**, *6*, 92–104. (In Chinese)
32. Li, X.; Li, X.; Li, Z.; Ma, M.; Wang, J.; Xiao, Q.; Liu, Q.; Che, T.; Chen, E.; Yan, G.; et al. Watershed allied telemetry experimental research. *J. Geophys. Res. Atmos.* **2009**, *114*. [CrossRef]
33. Wang, K. The westerly fluctuation and water vapor transport over the Qilian-Heihe valley. *Sci. China Ser. D Earth Sci.* **2004**, *47*, 32–38. [CrossRef]
34. Jiang, H.; Wang, K.L.; Cheng, G.D.; Li, H.Y.; Yu, T. Temporal variation and spatial structure of the water vapor transport and budget over the Heihe drainage area. *J. Glaciol. Geocryol.* **2009**, *31*, 311–317. (In Chinese) [CrossRef]
35. Lu, G.H.; Dong, X.; He, H. Characteristics of water vapor transportation and budget over the Heihe drainage basin. *J. Nat. Resour.* **2012**, *27*, 510–521. (In Chinese) [CrossRef]
36. Skamarock, W.C.; Klemp, J.B. A time-split nonhydrostatic atmospheric model for weather research and forecasting applications. *J. Comput. Phys.* **2008**, *227*, 3465–3485. [CrossRef]
37. Hong, S.Y.; Dudhia, J.; Chen, S.H. A revised approach to ice microphysical processes for the bulk parameterization of clouds and precipitation. *Mon. Weather Rev.* **2004**, *132*, 103–120. [CrossRef]
38. Kain, J.S.; Kain, J. The Kain—Fritsch convective parameterization: An update. *J. Appl. Meteorol.* **2004**, *43*, 170–181. [CrossRef]
39. Hong, S.-Y.; Noh, Y.; Dudhia, J. A new vertical diffusion package with an explicit treatment of entrainment processes. *Mon. Weather. Rev.* **2006**, *134*, 2318–2341. [CrossRef]
40. Dudhia, J. Numerical study of convection observed during the Winter Monsoon Experiment using a mesoscale two-dimensional model. *J. Atmos. Sci.* **1989**, *46*, 3077–3107. [CrossRef]
41. Mlawer, E.J.; Taubman, S.J.; Brown, P.D.; Iacono, M.J.; Clough, S.A. Radiative transfer for inhomogeneous atmospheres: RRTM, a validated correlated-k model for the longwave. *J. Geophys. Res. Atmos.* **1997**, *102*, 16663–16682. [CrossRef]
42. CGIP. *Global Energy and Water Cycle Experiment (GEWEX) Continental-Scale International Project: A Review of Progress and Opportunities*; National Academies Press: Washington, DC, USA, 1998; pp. 13–22.
43. Zhang, X.; Xiong, Z.; Yan, X. Modeling precipitation changes in the Heihe River Basin, Northwest China, from 1980 to 2014 with the Regional Integrated Environment Modeling System (RIEMS) nested with ERA-Interim reanalysis data. *Theor. Appl. Clim.* **2018**, *137*, 493–503. [CrossRef]
44. Wang, K.L.; Cheng, G.D.; Jiang, H.; Zhang, L.J. Atmospheric hydrologic cycle over the Qilian-Heihe valley. *Adv. Water Sci.* **2003**, *14*, 91–97. (In Chinese)
45. Xu, B.R.; Zou, S.B.; Du, D.Y.; Xiong, Z.; Lu, Z.X.; Ruan, H.W.; Xiao, H.L. Variation of tropospheric specific humidity upon the middle-lower reaches of the heihe river during 1981–2010. *J. Glaciol. Geocryol.* **2016**, *38*, 57–68. (In Chinese) [CrossRef]
46. Xu, J.J.; Wang, K.L.; Jiang, H.; Li, Z.G.; Sun, J.; Luo, X.P.; Zhu, Q.L. A numerical simulation of the effects of westerly and monsoon on precipitation in the Heihe River Basin. *J. Glaciol. Geocryol.* **2010**, *32*, 489–496. (In Chinese)
47. Ding, Y.J.; Zhang, S.Q. Study on water internal recycle process and mechanism in typical mountain areas of inland basins, Northwest China: Progress and Challenge. *Adv. Earth Sci.* **2018**, *33*, 719–728. (In Chinese) [CrossRef]
48. Pan, X.D.; Li, X. Impact of spatial resolution on WRF model in Heihe river basin. *eScience Technol. Appl.* **2011**, *2*, 126–137. (In Chinese)
49. Pan, X.D.; Li, X.; Ran, Y.H.; Liu, C. Impact of underlying surface information on WRF modeling in Heihe River Basin. *Plateau Meteorol.* **2012**, *31*, 657–667. (In Chinese)
50. Bonekamp, P.N.J.; Collier, E.; Immerzeel, W.W. The impact of spatial resolution, land use, and spinup time on resolving spatial precipitation patterns in the Himalayas. *J. Hydrometeorol.* **2018**, *19*, 1565–1581. [CrossRef]
51. Gimeno, L.; Vázquez, M.; Eiras-Barca, J.; Sorí, R.; Stojanovic, M.; Algarra, I.; Nieto, R.; Ramos, A.M.; Durán-Quesada, A.M.; Dominguez, F. Recent progress on the sources of continental precipitation as revealed by moisture transport analysis. *Earth Sci. Rev.* **2020**, *201*, 103070. [CrossRef]
52. Wu, Y.-P.; Shen, Y.; Larry, P.; Li, B. Possible physical mechanism of water vapor transport over Tarim River Basin. *Ecol. Complex.* **2012**, *9*, 63–70. [CrossRef]
53. Keys, P.W.; Barnes, E.A.; Van Der Ent, R.J.; Gordon, L.J. Variability of moisture recycling using a precipitationshed framework. *Hydrol. Earth Syst. Sci.* **2014**, *18*, 3937–3950. [CrossRef]
54. Keys, P.W.; Porkka, M.; Wang-Erlandsson, L.; Fetzer, I.; Gleeson, T.; Gordon, L.J. Invisible water security: Moisture recycling and water resilience. *Water Secur.* **2019**, *8*, 100046. [CrossRef] [PubMed]
55. Zongxing, L.; Qi, F.; Wang, Q.; Song, Y.; Jianguo, L.; Yongge, L.; Yamin, W. Quantitative evaluation on the influence from cryosphere meltwater on runoff in an inland river basin of China. *Glob. Planet. Chang.* **2016**, *143*, 189–195. [CrossRef]
56. Li, Z.; Gui, J.; Wang, X.; Feng, Q.; Zhao, T.; Ouyang, C.; Guo, X.; Zhang, B.; Shi, Y. Water resources in inland regions of central Asia: Evidence from stable isotope tracing. *J. Hydrol.* **2019**, *570*, 1–16. [CrossRef]
57. De Wekker, S.F.J.; Kossmann, M. Convective boundary layer heights over mountainous terrain—A review of concepts. *Front. Earth Sci.* **2015**, *3*, 1–22. [CrossRef]

atmosphere

Article

Characteristic of the Regional Rainy Season Onset over Vietnam: Tailoring to Agricultural Application

Nachiketa Acharya * and Elva Bennett

International Research Institute for Climate and Society (IRI), The Earth Institute at Columbia University, Palisades, NY 10964, USA; elva@iri.columbia.edu
* Correspondence: nachiketa@iri.columbia.edu

Abstract: Owing to its unique position within multiple monsoon regimes, latitudinal extent, and complex topography, Vietnam is divided into seven agroclimatic zones, each with distinct rainy season characteristics. Variation in the dominant rainfall system across zones affects the rainfall climatology, the primary water resource for regional crops. This study explores the creation of an agronomic rainy season onset based on high-resolution rainfall data for each agroclimatic zone for applications in an agricultural context. Onset information has huge practical importance for both agriculture and the economy. The spatiotemporal characteristics of zonal onset date are analyzed using integrated approaches of spatial and interannual variability, temporal changes, and estimation of predictability using teleconnection with Niño 3.4 sea surface temperature anomalies (SSTA) for 1980 to 2010. Results suggest that northern and southern zones experience regional onset dates in May, while the central zones experience rainy season onset in late August. The regional variability of rainy season onset is lower in the northern and southern zones and higher in the central zones which are latitudinally extended. The interannual variation in rainy season onset date is found to be approximately two weeks across all agroclimatic zones. The significant negative trend in rainy season onset date is found for Central Coast and South Central Coast zones, suggesting that the onset date shifted earlier for the entire period. In the decadal scale, the zonal mean onset date shifted later in the Northwest zone and earlier in the Central Highlands. Out of the seven climate zones, a significant positive correlation is only noticed in the Central Highlands and South zones between zonal mean onset date and Niño 3.4 SSTA for Dec–Jan–Feb, suggesting the potential of seasonal scale predictability of rainy season onset date with respect to preceding El Niño-Southern Oscillation (ENSO) events.

Keywords: zonal rainy season; agronomic onset definition; trend detection; ENSO-teleconnection

Citation: Acharya, N.; Bennett, E. Characteristic of the Regional Rainy Season Onset over Vietnam: Tailoring to Agricultural Application. *Atmosphere* **2021**, *12*, 198. https://doi.org/10.3390/atmos12020198

Academic Editors: Ankit Agarwal and Simone Orlandini

Received: 7 December 2020
Accepted: 28 January 2021
Published: 2 February 2021

1. Introduction

In countries where rainy season rainfall is the main water resource to meet the requirements for agricultural production, reliable determination of the rainy season onset is crucial for agricultural planning. The onset date influences the time of land preparation, sowing and transplanting dates of major crops, mobilization of seed/crop, manpower, and equipment [1]. Accurate knowledge of the onset date reduces the risk of planting and sowing too late or too early [1]. Any irregularity in rainfall during this period affects the prospect of yield production because of delayed transplanting and immature growth of crop owing to water scarcity, heavy infestation by weeds, and outbreaks of diseases and insect pests [2]. Although farmers use a range of non-scientific traditional strategies and criteria (e.g., the observation of the behavior of some birds or insects and flowering of certain trees) to predict onset, these methods are unreliable, especially as global climate change alters long-term climatic patterns. Therefore, a scientific definition of the rainy season onset can aid yield increases among local farmers [3].

Defining the onset of the rainy season is an extremely complex problem and, thus, various definitions have been proposed. These definitions can be broadly classified into two groups: regional to large-scale methods where parameters measure large-scale atmospheric dynamics and local-scale methods using fixed precipitation thresholds selected based on the climatology of rainfall over the region of interest [4]. The local scale definition has direct application for agricultural production which depends on rainfall requirements of specific crops. Therefore, this local scale definition is often referred to as agronomic as it is based on crop-relevant local-scale daily rainfall features rather than regional-scale atmospheric circulation changes [5]. An agronomic definition can be estimated by finding the first rainy day which exceeds a particular rainfall threshold without a potential occurrence of crop-threatening dry spells thereafter to account for false onsets [5].

With spurring growth in agricultural productivity in the 1990s, the sector became a key part of Vietnam's economy, now contributing 18.4% of Vietnam's Gross Domestic Product, and employing 54% of the working population [6]. In recent times, Vietnam became a leading global exporter of several important crops such as rice, coffee, cashew nuts, vegetables, and rubber [7]. Agricultural practices in Vietnam are crucially dependent on rainfall. The annual rainfall over Vietnam is distributed asymmetrically throughout the year, where a major proportion (~70% of annual) of the rainfall occurs during the period from May to September/October [6]. Although most of the gross cultivated area is irrigated, irrigation itself is crucially dependent on monsoon rainfall as it largely depends on the use of surface water. Therefore, the date of onset of the rainy season is of particular importance to the agriculture sector in Vietnam. However, despite the fact that the investigation of rainy season onset dates has a huge practical importance for both agriculture and economy, targeted studies evaluating the timing of the rainy season in Vietnam have been very limited [8–10].

There are previous studies that investigate the climatology, trends, and the predictability of rainy season onset dates in southern Vietnam [8] and the Central Highlands region [9] of Vietnam. Both studies also explored prediction of rainy season onset dates based on statistical modeling using the relationship between local rainy season onset dates and large-scale atmospheric variables (pressure and moist static energy gradients, outgoing long-wave radiation, wind fields and mean sea level pressures) over certain regions. Only one study [10] discussed the climatological pattern of onset dates for all regions of Vietnam but it is limited only to the summer monsoon season while the rainy season of the Central Highlands region is mainly influenced by local rain-producing weather systems [11].

Existing studies are limited either for a particular region of the country or for only the summer monsoon season, while the climate of Vietnam substantially varies from north to south with seven climatic sub-regions which are strongly affected by multiple monsoon systems and local factors (more discussions can be found in the following sections). Thus, a more in-depth study was required to reveal an explicit understanding of the spatiotemporal characteristics of the rainy season onset of Vietnam. For the applications in an agricultural context, this study mainly focuses on the agronomic definition of onset for the primary rainy season during the period from May to September/October which has huge importance for agricultural activities, at different the agroclimatic zones of Vietnam. Nevertheless, this study is motivated to raise the following questions: How is the agronomic rainy season onset distributed over agroclimatic zones of Vietnam? How does rainy season onset vary over the spatiotemporal scale for each zone? Since the El Niño-Southern Oscillation (ENSO) has impact on the rainfall in Vietnam, is there any predictability based on the teleconnection between the onset days for each zone and with ENSO? Therefore, this paper aims to answer these issues by exploring the creation of an agronomic rainy season onset definition for all agroclimatic zones over Vietnam using recently developed high resolution Vietnam-gridded precipitation (VnGP) dataset [12] for the period 1980 to 2010. The novelty of this research lies in the fact that it is the first time that the agronomic rainy season onset for each agroclimatic zone across Vietnam is analyzed by using integrated approaches of inter-annual variability, trend analysis, and estimation of predictability teleconnection. The findings of this study offer a potential understanding of agronomic definition for the local stakeholders and decision-makers for agriculture as well as for the Vietnamese National Hydro-Meteorological Service (VNHMS).

The remainder of this paper is organized as follows: After a short description of the agroclimatic setting of Vietnam in Section 2, the description of the dataset is presented in Section 3. Agronomic rainy season onset definition calculation is shown in Section 4. Spatiotemporal characteristics of onset dates of each zone including interannual variability, temporal changes, together with estimation of teleconnections with El Niño-Southern Oscillation (ENSO) are presented in Section 5. The conclusions of this study follow in Section 6.

2. The Agroclimatic Setting of Vietnam

Vietnam, the easternmost country on the Indochina Peninsula in Southeast Asia, is bordered by China to the north, Laos and Cambodia to the west, the Gulf of Thailand to the south, and the Gulf of Tonkin and the South China Sea to the east. From the Tropic of Cancer (approximately 23.26° N) to the deep tropics (approximately 8.58° N), Vietnam extends 1650-km north to south and has an extensive coastline of 3260 km stretching from the Gulf of Tonkin in the north to the Gulf of Thailand in the south. This long, thin, S-shaped country has a very complex topography with the mountainous North, except for the Red River valley and the coastal plain; the central highlands and coast; and the low-lying South, home to the Mekong River Delta (Figure 1a).

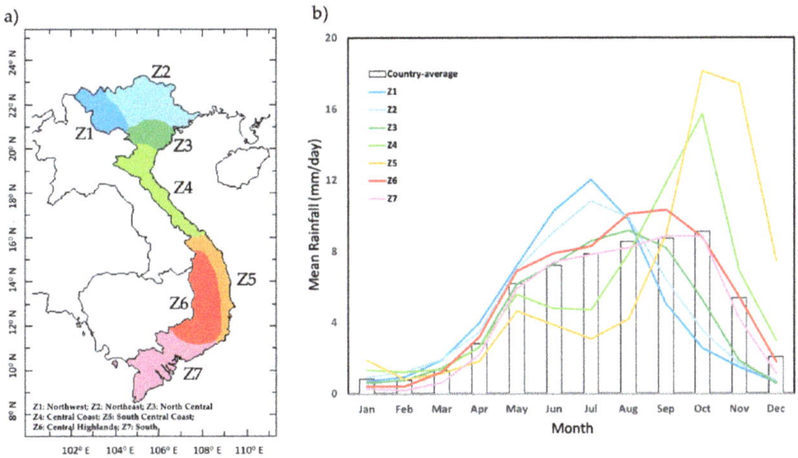

Figure 1. (**a**) The seven agroclimatic zones of Vietnam. (**b**) Monthly mean rainfall for the agroclimatic zones along with country average for 1982–2010.

The total annual rainfall is distributed unevenly throughout the year, where approximately 70 percent of the rainfall occurs during the main rainy season from May to September/October [6]. The northern region (+20° N) rainy season is from May to September with the highest monthly rainfall in August. The central region (11°–20° N) rainy season has an extended bimodal rainy season from May to November with rainfall peaks in July and October. The rainy season tends to be shorter and occur later in the northern part of this elongated region. The central region has varied and unusual monsoon characteristics because of the proximity of the central highlands, eastern coast, and the effects of the adjacent mountains in Cambodia and Laos to the west. The southern region (8–11° N) has a similarly extended rainy season from May to November with near constant high magnitude rainfall from June to October. Therefore, climate regimes in Vietnam are strongly affected by both monsoon systems and local factors [13]. Due to complex topography, an elongated latitudinal extent, and the influence of multiple monsoon systems, like South Asian summer monsoon, East Asian winter monsoon, and western North Pacific monsoon, Vietnam's climate varies significantly from north to south with several sub-climate regions [13]. Vietnam is divided into seven terrestrial sub-regions with distinct climate characteristics based on the duration of the rainy season, the timing of peak rainfall, and the assessment of rainfall station data by VNHMS [10]. These zones are referred to as Northwest (denoted as Z1), Northeast (denoted as Z2), North Central (denoted as Z3), Central Coast (denoted as Z4), South Central Coast (denoted as Z5), Central Highlands (denoted as Z6), and South (denoted as Z7) (Figure 1a).

Almost 35% of the total national land area, including arable land, permanent crop land, and permanent meadow land, is used for agricultural production. Nationally, rice, coffee, tea, and pepper are the predominant crops, but the choice of the crops is specialized by agroecological zone [7]. The Northwest and Northeast zones are mountainous areas with often deficient transportation facilities, poor market access, and limited irrigation systems. The agricultural production in the northern part of Vietnam is mostly consists of industrial crops (tea and rubber) and food subsistence purposes considering that these areas are mountainous with limited irrigation systems. Cash crops (mostly coffee) are mainly produced in the Central Highlands and the South zones. Rice production is concentrated mostly in the two delta regions (Red River Delta in the North Central zone and Mekong River Delta in the South zone). Agricultural production accounts for up to 95% of total water withdrawals in Vietnam while 49% of the total agricultural land is serviced by an irrigation scheme which is primarily designed for rice in the two deltas [7]. Irrigation is crucially dependent on monsoon rainfall because it largely depends on the use of surface water. Any aberration in rainfall during the onset time affects the prospects of a good yield as farmers make many agricultural decisions based on the local monsoon onset, including sowing dates and fertilizer timing. Therefore, it is very important to investigate the rainy season onset characteristic focusing on each sub-climate region.

3. Dataset

The gridded data product named Vietnam-gridded precipitation data (VnGP) is used for all the rainy season onset calculations. The dataset was built from Vietnamese National Hydro-Meteorological Service's 481 rain gauge stations. The Spheremap interpolation method, a modified Shepard's interpolation for spherical coordinates, was used to construct the gridded products from rain gauge stations. These gridded products are generated and archived by the Department of Meteorology and Climate Change, VNU University of Science, Vietnam at daily time scale with spatial resolutions of 0.1° and 0.25°, covering the period from 1980 to 2010. In the present study, the 0.25° version for the entire 31 years (1980 to 2010) is used. The detailed procedure of the generation of VnGP has been discussed in detail by [12] and data are available at http://danida.vnu.edu.vn/cpis/en/content/gridded-precipitation-data-of-vietnam.html.

To investigate the influence of El Niño-Southern Oscillation (ENSO) over zonal rainy season onset, the sea surface temperature anomalies (SSTA) from the second version of the optimum interpolation (OI) analysis (OIv2) produced by the National Oceanic and Atmospheric Administration (NOAA) [14] is used in this study. The OIv2 has a spatial resolution of 1° and a daily timescale, with monthly means from December 1981 to January 2020. The monthly mean SST anomalies from 1981 to 2010 averaged over the Niño 3.4 region of the Pacific Ocean (120° W–170° W, 5° S–5° N), referred to as Nino 3.4 SSTA, are obtained from the Data Library of the International Research Institute for Climate and society at: (https://iridl.ldeo.columbia.edu/SOURCES/.Indices/.nino/.NCEP_OIv2/.NINO34/index.html#info).

4. Methodology

4.1. *Defining Agronomic Rainy Season Onset Based on Local-Scale Climatic Conditions*

In this work, the local rainy season onset for each agroclimatic zone is calculated using the agronomic definition proposed by [5]. As described in Section 1, a local definition like the one used here is of great relevance and can have numerous applications in the development of operational products in agriculture. This agronomic definition is advantageous in regions where dry and rainy seasons are well defined [4], as it is in Vietnam. This agronomic definition, originally developed using historical precipitation patterns from India, allows for the calculation of monsoon timing for any location [5]. According to this definition, rainfall onset date is defined as the first wet day within the first wet spell that is not followed by a dry spell, to avoid the false onset. A wet (dry) day is defined as being greater than (lesser or equal than) to a particular rainfall threshold depending on the local rainfall climatology. Moreover, this definition is based on five basic parameters: the amount of rainfall during the first wet spell, its duration, the duration and intensity of the post-onset dry spells, and the length of the period in which these dry spells are searched. The step by step description of this agronomic definition can be found in [5]. In the present study, the original parameters were adjusted to satisfy agronomic definition parameters based on the climatic features of Vietnam. All parameters for the initial wet spell are based on discussions with local experts at Vietnam National University (VNU) and Vietnamese National Hydro-Meteorological Service (VNHMS). In short, this method defines onset as the first wet day of the first 5-day period with an average daily rainfall equal or larger than the climatological 5-day wet spell for the season and without a 7-day dry spell (no wet days) during the following 20 days to account for false onsets.

As the focus of this paper is the rainy season, which is not limited by monsoon for the central highlands, a wet day is considered any day with at least 1 mm/day to account for sufficient rainfall to penetrate the soil and reach planted seeds. A requirement for five days of 5 mm/day rainfall is translated to 25 mm of rainfall in five days with at least four wet days. The reason behind choosing a higher rainfall threshold to define a wet day rather than the commonly used wet day definition (>0.1 mm/day) is that a low rainfall threshold makes agronomic definition less suitable for agricultural decisions [5].

4.2. Search Window for Onset Definition

Finding the first wet day of the period which meets the subsequent onset date conditions (defined in Section 4.1) is necessary for the calculation of agronomic onset date in a given year; thus, determining the date of the beginning of the search the window in which to look for a day which satisfies the onset definition is a crucial step. In this study, the climatological mean monthly rainfall pattern is used to determine the beginning of the rainy season for each agroclimatic zone. As mentioned in Section 2, the predominant and intersecting monsoon systems over the different regions of Vietnam influence the onset of the rainy season across the agroclimatic zones. The East Asian Monsoon is the primary monsoonal influence on the Northern zones of Vietnam, while the South Asian monsoon dominates the southern zones. The central coastal zones and the central highlands are located between these two systems and, thus, are dominated by local rainfall patterns. The variation in the dominant rainfall system across zones affects the monthly rainfall climatology and contributes to the regional primary crop; for example, coffee in the central highlands is suited to more dispersed rainfall throughout the year with a later onset of the rainy season.

The monthly mean rainfall for each agroclimatic zone along with country average is shown in Figure 1b. It is seen that the main rainy season for northern and southern zones is during May to September, while the peak in rainfall is from July to August. For central zones it shows a decrease in rain relative to the rest of the country in the summer months and a later rainy season from August to December with a peak in rainfall in October. Further, the figure shows that the mean monthly rainfall for the whole country of Vietnam has an extended rainy season of moderate rainfall which slowly increases from May to October, and a period of low rainfall from December to April. For the 1980–2010 time period, the majority of agroclimatic zones, excluding the northwestern mountains and the central coast, experience rainy season onset after mid-April, and thus the search period begins around this time for these zones (Z2, Z3, Z6, and Z7). The Northwest (Z1) has a consistently earlier rainy season onset, while the Central Coast and South Central Coast zones (Z4 and Z5) are found to have a significantly delayed rainy season onset, and thus the search window shifts forward and backward, respectively, for these zones. Additionally, Summer monsoon rainfall in Vietnam has a wet-dry phase shift of 2–10 days [10] and, thus, the dry spell period for the false start parameter is seven consecutive days within the twenty-day period following the first wet day. This requirement is used to avoid false-start onsets. The search window for the rainy season onset date for each zone is 60-days to capture interannual rainy season onset date variation.

The onset date coverage is determined by the percentage of years that satisfy the agronomic onset definition criteria within the search period for each grid point for the entire 31-year study period (Figure 2). Years that do not satisfy the agronomic definition and, thus, are not included, may have had an anomalously early or late onset or unusual rainfall patterns. High onset date coverage suggests the suitability of the agronomic definition to each of the agroclimatic zones. The initial date for the search window was determined using this method of testing possible dates and assessing suitability based on coverage. The earliest date which had high coverage was deemed suitable to capture the onset of the rainy season while representing its variability of the 31-year time period. The Northwest zone (Z1) is found to have the earliest zonal rainy season onset date and the beginning day of the search window for the rainy season is set to April 15th (Figure 2a) to fully capture the interannual variability of the rainy season onset. Other beginning days of the search window were tested and found that the April 25th date (Figure 2b) captures the largest spread of onset date for the majority of zones (Z2, Z3, Z6, and Z7). Lastly, due to latitudinal extent and unique topography, the Central Coast and South Central Coast zones (Z4 and Z5) are found to have a much more variable rainy season and a later onset date. The beginning day of the search window for these zones is determined to be August 5th (Figure 2c).

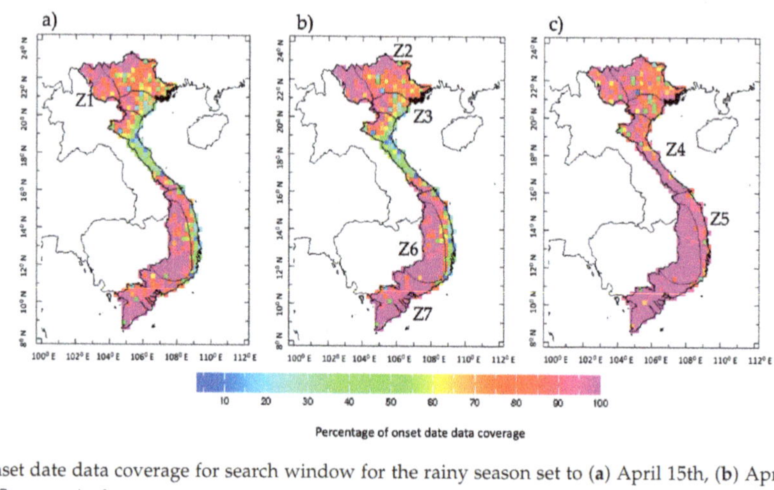

Figure 2. Onset date data coverage for search window for the rainy season set to (**a**) April 15th, (**b**) April 25th, and (**c**) August 5th. Coverage is shown as a percentage, where 100% is a grid point which satisfies the agronomic definition for all years of the 1980–2010 period.

5. Results and Discussion

5.1. Spatiotemporal Characteristics of Regional Onset Date

Following the above-mentioned procedure, for each agroclimatic zone, the onset dates are estimated for each grid point independently for each year during the time period (1980–2010). The zonal mean of the onset dates for each agroclimatic zone are estimated by taking the mean of the onset dates across the time period (31 years) for each grid point within that agroclimatic zone, and, then, averaging the mean onset date across all grid points. It is found that the zonal mean of rainy season onset for the Northeast (Z2), North Central (Z3), and South (Z7) zones is in mid to late May (May 20th, May 21st, and May 14th, respectively). However, for the Northwest (Z1) and Central Highlands (Z6), the mean onset days are in early May (May 9th and May 11th, respectively). These onset dates are closely related to the summer monsoon system (as described in Section 5.1). Although these results are in agreement with earlier studies [8,10], they are not exactly the same as a different definition is used to determine the onset dates. Lastly, the Central Coast and South Central Coast zones (Z4, Z5) have a noticeably late rainy season onset in late August (August 24th and August 29th, respectively). This is likely due to their unique topography, between the Cambodian mountain range to the west and the sea to the east, and position on the border of interacting monsoon systems.

The zonal spatial variability of the onset dates, or the variation of the mean onset date between grid points across each agroclimatic zone, is calculated to assess how coherent the onset dates are across each agroclimatic zone. The spatial variability is calculated by taking the standard deviation of all mean onset dates (mean of the onset dates across the time period for each grid point) across all grid points in a given agroclimatic zone. It is expected that the mean onset dates occur at a similar time scale across the agroclimatic zone (i.e., low zonal spatial variability of onset dates); however, it is found that the mean onset dates vary from 4 to 8 days within each zone. Moreover, spatial variability is lower in more compact zones with a single dominant monsoon regime (Z1, Z2, Z3, Z6, and Z7) at around 4–6 days and higher for latitudinally extended zones on the border of monsoon regimes (Z4 and Z5) at 7–8 days.

The zonal interannual variability of onset date, or temporal variability, is also estimated to illustrate the year-to-year variation of onset dates in a given agroclimatic zone. This is calculated by taking the standard deviation of the onset dates across the time period for each grid point in a given zone, and, finally, averaging the standard deviation of onset date across all grid points. Interannual temporal variability of rainy season onset was found to be approximately two weeks across all agroclimatic zones. Specifically, the majority of zones (Z1, Z2, Z3, Z6, and Z7) have an interannual variability between 11 to 13 days, while the Central Coast and South Central Coast zones (Z4 and Z5) have a slightly higher interannual variability of more than 14 days. This is notable because an interannual variation of approximately two weeks presents a challenge to the prediction of rainy season at a seasonal timescale. Table 1 summarizes the mean rainy season onset day, interannual temporal variability, and spatial variability for each climatic zone.

Table 1. Mean and variability (spatial and interannual) of zonal onset date for each agroclimatic zone.

Zone	Mean Onset Date	Spatial Variability (Days)	Interannual Variability (Days)
1	May 9	4.5	13.5
2	May 20	6.4	13.4
3	May 21	5.4	13.2
4	Aug 24	8.1	14.1
5	Aug 29	7.1	14.5
6	May 11	5.4	11.5
7	May 14	5.2	13.3

5.2. *Temporal Changes of Regional Onset Date*

To study the temporal changes in the onset dates for each agroclimatic zone across the time period, the Mann–Kendall (MK) test [15], a rank-based non-parametric test, is performed on the regionally averaged onset dates across all grid points (average onset dates over all grid points for each year) at the 90% statistical significance level. The MK test is widely used for trend analysis because of its robustness for non-normally distributed data and low sensitivity to an abrupt change. However, the Mann–Kendall test gives the direction (a positive or negative value of MK's test statistic indicates that the rainy season onset date is trending later or earlier respectively) but not the magnitude of the significant trends. Therefore, in addition, the Sen's slope estimator [16], a non-parametric test to identify the slope of the trend, is calculated to estimate the magnitude of change. The MK's test statistic and magnitude of Sen's slope for the entire study period (1980–2010) and each decade (1980s, 1990s, and 2000s, respectively) for each agroclimatic zone is shown in Figure 3. It is found that for the entire 31 years, a significant negative trend in rainy season onset date is only observed for the Central Coast and South Central Coast zones (Z4 and Z5) out of all seven zones. The negative trend for both the zones represent that the onset dates shifted earlier across the entire period. The Sen's slope for the South Central Coast (Z5) has a greater magnitude than that of the Central Coast (Z4), 0.44 compared to 0.34, which shows that the annual zonal mean rainy season onset date in the South Central Coast (Z5) shifted earlier more rapidly than in the Central Coast (Z4). Mostly there is no significant trend found for individual decades for each zone except for the Northwest and Central Highlands (Z1 and Z6) in the 1990s. For the second decade (1991–2000), a significant positive and negative trend with Sen's slope 1.647 and −2.011 are observed for the Northwest and Central Highlands (Z1 and Z6), respectively. This means that the zonal mean onset date shifted later in the northwest and earlier in the Central Highlands in this decade.

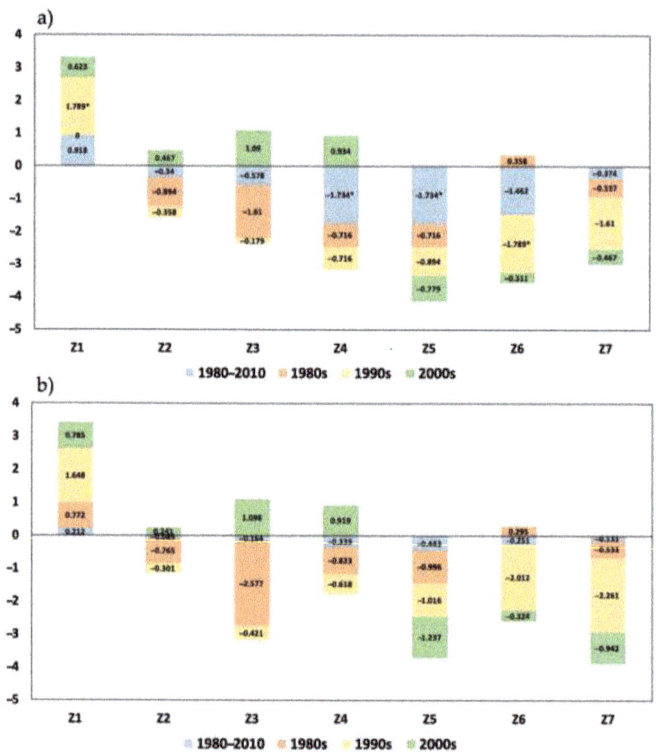

Figure 3. (**a**) Mann–Kendall's test statistic and (**b**) Sen's slope of the trend of rainy season onset for the for the entire period (1980–2010) and each decade (1980s, 1990s, and 2000s) for Northwest (Z1), Northeast (Z2), North Central (Z3), Central Coast (Z4), South Central Coast (Z5), Central Highlands (Z6), and South (Z7) agroclimatic zones. Asterisk indicates that the Mann–Kendall's test statistic values is statistically significant at 90% level.

For a visual representation, the yearly zonal mean onset dates with a linear trend line for the entire period (1980–2010) and each decade is presented for each agroclimatic zone in Figure 4.

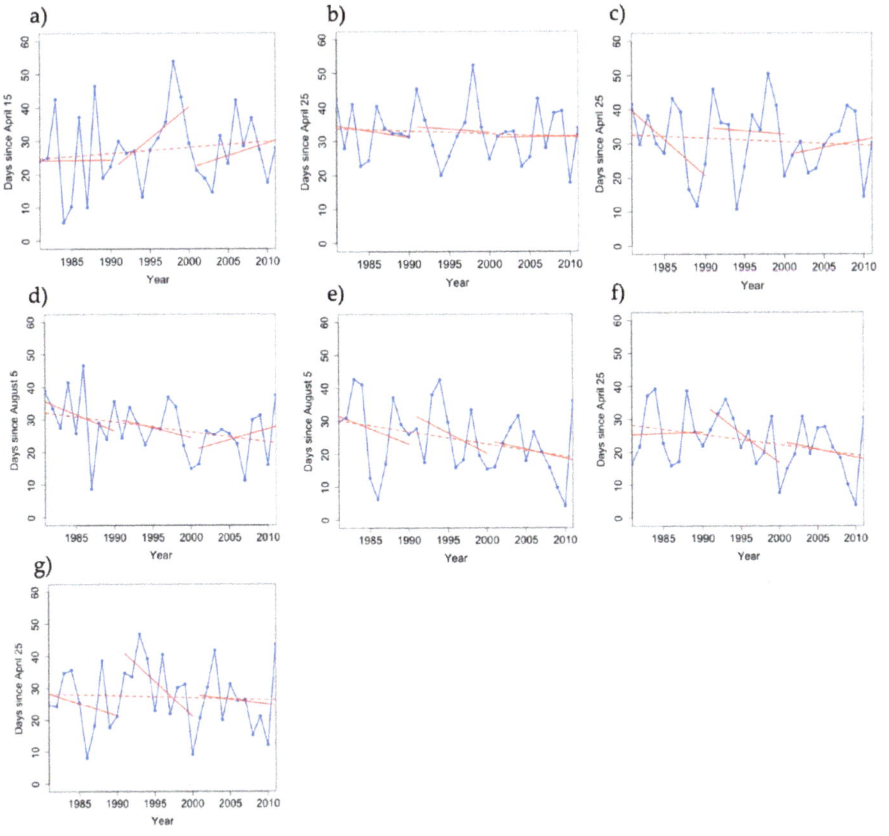

Figure 4. Yearly zonal mean onset dates (blue line) with linear trend line for the entire period (1980–2010) (dashed red line) and each decade (red line) for (**a**) Northwest (Z1), (**b**) Northeast (Z2), (**c**) North Central (Z3), (**d**) Central Coast (Z4), (**e**) South Central Coast (Z5), (**f**) Central Highlands (Z6), and (**g**) South (Z7) agroclimatic zones.

5.3. Teleconnection between El Niño-Southern Oscillation (ENSO) and Regional Onset Date

The influence of the Pacific Ocean climate drivers, namely El Niño-Southern Oscillation (ENSO), on the rainfall in Vietnam has been studied extensively [17,18]. Since the ENSO has an impact on rainfall in Vietnam, it may also impact the timing of the rainy season. In this section, we explore the teleconnections between the tropical Pacific SST (as represented by the Niño 3.4 SSTA) anomalies and the rainy season onset for each agroclimatic zone to investigate potential sources of seasonal predictability. The Pearson's correlation coefficient (r), a measure of linear relationship, is calculated between zonal onset date for each zone and Niño 3.4 SSTA for the seasonal mean of December to February (DJF) period considering the fact that ENSO amplitude is nearing its peak around this time of the year [19]. The student's t test is performed to determine statistical significance of the correlation coefficient at the 95% level. The scatter plot between zonal onset dates and Niño 3.4 SSTA along with the Pearson's correlation coefficient is shown in Figure 5. Out of the seven agroclimatic zones, a statistically significant positive correlation is only noticed in the Central Highlands and the Southern zones (Z6, Z7) between zonal mean onset date and Niño 3.4 SSTA for DJF, while there is no correlation with ENSO found for the rest of the region. This suggests that the El Niño (La Niña) events result in a delayed (early) rainy season onset date in these two agroclimatic zones. This finding suggests the potential of seasonal prediction of rainy season onset date in these zones with respect to preceding

ENSO events. As both the zones are significantly important for agricultural production in Vietnam (the South zone is the location of the Mekong River Delta) in terms of coffee and rice cultivation, seasonal prediction of rainy season onset would be very useful to many Vietnamese farmers in these agroclimatic zones.

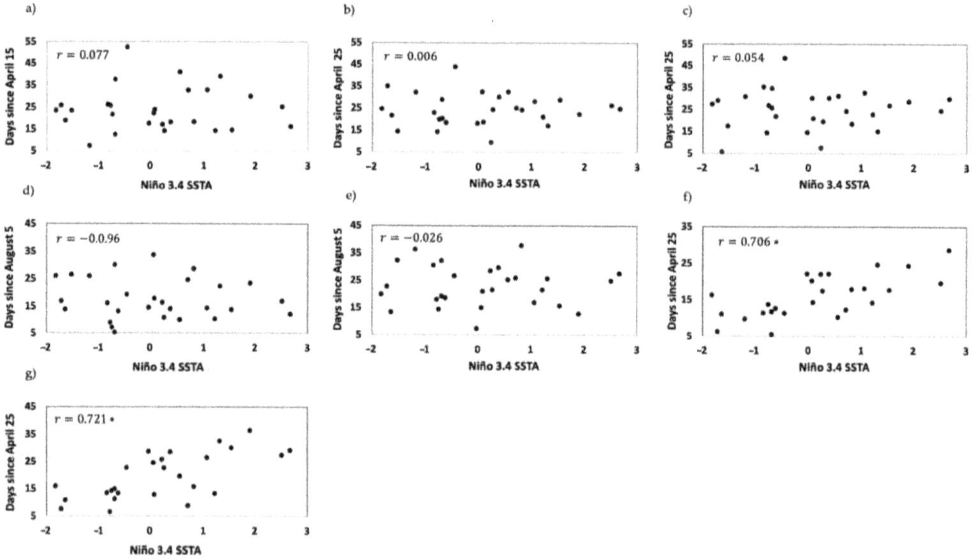

Figure 5. Scatter diagram between Niño 3.4 seas surface temperature anomaly (SSTA) (x-axis) and zonal mean onset dates (y-axis) for (**a**) Northwest (Z1), (**b**) Northeast (Z2), (**c**) North Central (Z3), (**d**) Central Coast (Z4), (**e**) South Central Coast (Z5), (**f**) Central Highlands (Z6), and (**g**) South (Z7) agroclimatic zones during 1980–2010. The Pearson's correlation coefficient (r) is also displayed for each zone. Asterisk indicates that the r values are statistically significant at 95% level.

6. Concluding Remarks

The primary goal of this study is to define a rainy season onset tailored to an agricultural application over the seven primary agroclimatic zones of Vietnam and assess its spatiotemporal characteristics by using integrated approaches of spatial and interannual variability, temporal changes, and estimation of predictability using teleconnection with El Niño-Southern Oscillation (ENSO) for 1980 to 2010. The novelty of this research is that a tailored agronomic definition based on high-resolution rainfall data is used for the first time over Vietnam at a regional climate scale. The results of this study would be highly relevant to local farmers and decision-makers in Vietnam where the productivity and profitability of the agricultural sector is highly dependent on rainfall and the timing of the rainy season.

In summary, the major findings of the present study are:

- Rainy season onset date varies across the agroclimatic zone. For the Northern and Southern zones, the zonal mean of rainy season onset dates is primarily in May which associated with summer monsoon system. Due to their unique topography and position on the border of interacting monsoon systems, Central Coast and South Central coast have rainy season onset in late August.
- The zonal spatial variability of onset dates, defined as the standard deviation zonal mean onset dates across all grid points in a given agroclimatic zone, varies from 4 to 8 days within each agroclimatic zone. However, the variability is lower in a single dominant monsoon regime (northern and southern zones) and higher in latitudinally extended zones on the border of monsoon regimes (Central Coast and South Central Coast).

- The interannual variation in rainy season onset date is approximately two weeks (11 to 14 days) across all agroclimatic zones of Vietnam. This is result suggest that the prediction of rainy season onset is highly challenging work due to such high temporal variation.
- The trend analysis for the zonal onset dates of each agroclimatic zone has been analyzed by using the Mann–Kendall (MK) Test to gather with Sen's Slope Estimator for the entire 31-year period and each decade. The rainy season onset date is significantly trended earlier in the Central Coast and South Central Coast zones. For 1990s (second decade) the zonal mean onset date shifted later in the northwest and earlier in the central highlands at 90% significance level.
- The teleconnections between Niño 3.4 SSTA for Dec–Jan–Feb season and onset date in the Central Highlands and South zones are found (positive) to be statistically significant at 95% level. This finding suggests the potential of seasonal prediction of rainy season onset date in these zones with respect to preceding ENSO events.

Considering the fact that the seasonal prediction of rainy season onset at each climate zones would be very beneficial to the agriculture sector; the future scope of the present work includes the development of an operational forecast system using state-of-art general circulation models outputs.

Author Contributions: Conceptualization, N.A.; methodology, N.A. and E.B.; data curation, N.A. and E.B.; formal analysis, E.B.; resources, N.A.; writing—original draft preparation, N.A. and E.B.; writing—review and editing, N.A. and E.B.; visualization, N.A. and E.B.; supervision, N.A.; funding acquisition, N.A. All authors have read and agreed to the published version of the manuscript.

Funding: This research work and the APC was funded by the Columbia World Project, ACToday, Columbia University in the City of New York (https://iri.columbia.edu/actoday/). E.B. acknowledges the support from the Climate and Society MA program at Columbia University in the City of New York.

Institutional Review Board Statement: Not applicable.

Informed Consent Statement: Not applicable.

Data Availability Statement: Publicly available datasets were analyzed in this study. The website to access the data can be found in Section 2 of this paper.

Acknowledgments: This work is undertaken as part of the Columbia World Project, ACToday, Columbia University in the City of New York. We acknowledge the support of VNU, NOAA, and IRI personnel in creating, updating, and maintaining the dataset used in this study. We are also grateful for the effort of Rémi Cousin for making the onset calculation function in IRI's data library and for the thoughtful discussion and suggestions of scientists at the Vietnamese National Hydro-Meteorological Service (VNHMS) and the Department of Meteorology and Climate Change, VNU University of Science, Vietnam to improve the manuscript. Sincere thanks are due to the four anonymous reviewers for their constructive suggestions to enhance the quality of the manuscript.

Conflicts of Interest: The authors declare no conflict of interest.

References

1. Laux, P.; Kunstmann, H.; Bárdossy, A. Predicting the regional onset of the rainy season in West Africa. *Int. J. Clim.* **2008**, *28*, 329–342. [CrossRef]
2. Robertson, A.W.; Acharya, N.; Goddard, L.; Pattanaik, D.R.; Sahai, A.K.; Singh, K.K.; Ghosh, K.; Agarwal, A.; Buizer, J.L. Subseasonal forecasts of the 2018 Indian summer monsoon over Bihar. *J. Geophys. Res. Atmos.* **2019**, *121*, 13861–13875. [CrossRef]
3. Ati, O.F.; Stigter, C.J.; Oladipo, E.O. A comparison of methods to determine the onset of the growing season in northern Nigeria. *Int. J. Clim.* **2002**, *22*, 731–742. [CrossRef]
4. Bombardier, R.J.; Moron, V.; Goodnight, J.S. Detection, variability, and predictability of monsoon onset and withdrawal dates: A review. *Int. J. Climatol* **2020**, *40*, 641–667. [CrossRef]
5. Moron, V.; Robertson, A.W. Interannual variability of Indian summer monsoon rainfall onset date at local scale. *Int. J. Clim.* **2014**, *34*, 1050–1061. [CrossRef]
6. Tran, V.S.; Boyd, W.B.; Slavich, P.; Mai Van, T. Agriculture and climate change: Perceptions of provincial officials in Vietnam. *J. Basic Appl. Sci.* **2015**, *11*, 487–500. [CrossRef]

7. Nguyen, T.T.N.; Roehrig, F.; Grosjean, G.; Tran, D.N.; Vu, T.M. *Climate Smart Agriculture in Vietnam;* CSA Country Profiles for Asia Series; International Center for Tropical Agriculture (CIAT); The Food and Agriculture Organization: Hanoi, Vietnam, 2017; p. 28. Available online: https://hdl.handle.net/10568/96227 (accessed on 7 December 2017).
8. Thanh, P.X.; Fontaine, B.; Philippon, N. Onset of the summer monsoon over the southern Vietnam and its predictability. *Theor. Appl. Clim.* **2010**, *99*, 105–113. [CrossRef]
9. Pham-Thanh, H.; van der Linden, R.; Ngo-Duc, T.; Nguyen-Dang, Q.; Fink, A.H.; Phan-Van, T. Predictability of the rainy season onset date in Central Highlands of Vietnam. *Int. J. Clim.* **2019**, *40*, 3072–3086. [CrossRef]
10. Nguyen-Le, D.; Matsumoto, J.; Ngo-Duc, T. Climatological onset date of summer monsoon in Vietnam. *Int. J. Clim.* **2014**, *34*, 3237–3250. [CrossRef]
11. Chen, T.-C.; Tsay, J.-D.; Yen, M.-C.; Matsumoto, J. Interannual variation of the late fall rainfall in Central Vietnam. *J. Clim.* **2012**, *25*, 392–413. [CrossRef]
12. Nguyen-Xuan, T.; Ngo-Duc, T.; Kamimera, H.; Trinh-Tuan, L.; Matsumoto, J.; Inoue, T.; Phan-Van, T. The Vietnam gridded precipitation (VnGP) dataset: Construction and validation. *Sci. Online Lett. Atmos. SOLA* **2016**, *12*, 291–296. [CrossRef]
13. Nguyen, D.M.; Nguyen, V.T. Definition of new summer monsoon index for Vietnam region. *Vietnam. J. Sci. Technol. Eng.* **2018**, *60*, 90–96. [CrossRef]
14. Reynolds, R.W.; Rayner, N.A.; Smith, T.M.; Stokes, D.C.; Wang, W. An improved in situ and satellite SST analysis for climate. *J. Clim.* **2002**, *15*, 1609–1625. [CrossRef]
15. Kendall, M.G. *Rank Correlation Measures;* Charles Griffin: London, UK, 1975; pp. 25–55.
16. Sen, P.K. Estimates of the regression coefficient based on Kendall's tau. *J. Amer. Stat. Assoc.* **1968**, *63*, 1379–1389. [CrossRef]
17. Gobin, A.; Nguyen, H.T.; Pham, V.Q.; Pham, H.T.T. Heavy rainfall patterns in Vietnam and their relation with ENSO cycles. *Int. J. Clim.* **2016**, *36*, 1686–1699. [CrossRef]
18. Duc, H.N.; Bang, H.Q.; Quang, N.X. Influence of the Pacific and Indian Ocean climate drivers on the rainfall in Vietnam. *Int. J. Clim.* **2018**, *38*, 5717–5732. [CrossRef]
19. Santoso, A.; McPhaden, M.J.; Cai, W. The defining characteristics of ENSO extremes and the Strong 2015/2016 El Niño. *Rev. Geophys.* **2017**, *55*, 1079–1129. [CrossRef]

atmosphere

MDPI

Article

Trends and Variabilities of Thunderstorm Days over Bangladesh on the ENSO and IOD Timescales

Md Wahiduzzaman [1,*], Abu Reza Md. Towfiqul Islam [2], Jing-Jia Luo [1,*], Shamsuddin Shahid [3], Md. Jalal Uddin [4], Sayed Majadin Shimul [2] and Md Abdus Sattar [5]

1 Key Laboratory of Meteorological Disaster of Ministry of Education, Collaborative Innovation Center on Forecast and Evaluation of Meteorological Disasters, Institute for Climate and Application Research (ICAR), Nanjing University of Information Science and Technology, Nanjing 210000, China

2 Department of Disaster Management, Begum Rokeya University, Rangpur 5400, Bangladesh; towfiq_dm@brur.ac.bd (A.R.M.T.I.); sayedmajadins@gmail.com (S.M.S.)

3 Department of Water & Environmental Engineering, School of Civil Engineering, Universiti Teknologi Malaysia (UTM), 81310 Johor, Malaysia; sshahid@utm.my

4 Collaborative Innovation Center on Forecast and Evaluation of Meteorological Disasters, School of Atmospheric Physics, Nanjing University of Information Science and Technology, Nanjing 210000, China; dmjalal90@nuist.edu.cn

5 Department of Disaster Risk Management, Patuakhali Science and Technology University, Patuakhali 8602, Bangladesh; abdus.sattar@pstu.ac.bd

* Correspondence: wahid.zaman@nuist.edu.cn (M.W.); jjluo@nuist.edu.cn (J.-J.L.)

Received: 1 October 2020; Accepted: 28 October 2020; Published: 30 October 2020

Abstract: Thunderstorms (TS) are one of the most devastating atmospheric phenomena, which causes massive damage and adverse losses in various sectors, including agriculture and infrastructure. This study investigates the spatiotemporal variabilities of TS days over Bangladesh and their connection with El Niño Southern Oscillation (ENSO) and Indian Ocean Dipole (IOD). The TS, ENSO and IOD years' data for 42 years (1975–2016) are used. The trend in TS days at the spatiotemporal scale is calculated using Mann Kendall and Spearman's rho test. Results suggest that the trend in TS days is positive for all months except December and January. The significant trends are found for May and June, particularly in the northern and northeastern regions of Bangladesh. In the decadal scale, most of the regions show a significant upward trend in TS days. Results from the Weibull probability distribution model show the highest TS days in the northeastern region. The connection between TS days and ENSO/IOD indicates a decrease in TS activities in Bangladesh during the El Niño and positive IOD years.

Keywords: thunderstorm; variability; trend; ENSO; IOD; Bangladesh

1. Introduction

Thunderstorms (TS) are a severe hazard in Bangladesh that cause immense death and adverse loss in agriculture, infrastructure and livestock during pre-monsoon and monsoon months. In the most severe situation, it can also create very destructive tornadoes [1], due to hot and humid air in the lower atmosphere from the southeastern direction and the opposite cold and dry air from the northwestern direction. TS predominantly occur in Bangladesh during the premonsoon and monsoon seasons, with the maximum frequency in May. The country experienced TS strikes an average of nine days in May before 1981, but later, it increased to 12 days. According to the Bangladesh Meteorological Department (BMD), a total number of 1476 people have died in Bangladesh by TS since 2010. The Comprehensive Disaster Management Programme under the Ministry of Disaster Management and Relief reported 180 deaths by TS only in 2016. The Government of Bangladesh declared TS as natural disasters on 17 May 2016 by considering their significant impacts on human life and economy [2].

El Niño Southern Oscillation (ENSO) and Indian Ocean Dipole (IOD) are associated with monthly or seasonal climate anomalies at many places around the globe [3,4]. They might have an effect on TS activity over Bangladesh due to its tropical climate. Therefore, further understanding the effects of ENSO on TS variations over Bangladesh is necessary for disaster prevention and mitigation. Some studies have performed globally to investigate the effects of ENSO on TS activity [5–10]. Manohar et al. [5] found the influences of El Niño on thunderstorm occurrences during the Indian monsoon season. Allen and Karoly [7] showed a significant influence of ENSO on the spatial distribution of thunderstorms in Australia. Yuan and Di [11] found a decrease in thunderstorms in Eastern China during ENSO episodes. Pinto [10] found an increasing tendency in thunderstorm activity in Southern Brazil during the ENSO warm phase.

A substantial number of studies have been conducted on thunderstorms worldwide in recent years, e.g., Kunkel et al. [12] in the USA, Mir et al. [13] in Pakistan, Kunz et al. [14] in Germany, Pinto et al. [9,10] in Brazil, Enno et al. [15] in Europe, Allen and Karoly [7] in Australia, Zheng et al. [16] in China, Singh and Bhardwaj [17] in India, Saha and Quadir [2] in Bangladesh and Araghi et al. [18] in Iran. Most of the earlier studies focused on the synoptic, dynamic and physical aspects of TS events, as well as modelling or predicting TS occurrence. However, the spatiotemporal variabilities of TS days linked to ENSO and IOD have been less investigated in the prevailing literature.

A few studies have also been conducted to explore the TS events in Bangladesh [2,19,20]. These prior studies mostly investigated time and space variations and the origin and frequency of TS over Bangladesh. However, these earlier works in Bangladesh were very limited to a specific site or a short period like the premonsoon season only.The spatial variabilities in the occurrence of TS days for different timescales and the relation of TS days with ENSO and IOD are not clear yet for Bangladesh. The present study would contribute to filling this knowledge gap by investigating the spatiotemporal variabilities in monthly, seasonal, annual and decadal TS days and exploring the link between TS days and ENSO/IOD over Bangladesh.

2. Data and Methods

2.1. Study Area

Bangladesh is a South Asian country located between 20.57° N to 26.63° N and 88.02° E to 92.68° E with a tropical and subtropical monsoon climate [21]. Bangladesh often faces severe natural disasters during the premonsoon, monsoon and postmonsoon seasons. Geographically, the Indian States of West Bengal, Assam, Meghalaya and Tripura border Bangladesh in the west, north and east, respectively. Myanmar forms the southern part of the eastern frontier. The Bay of Bengal is on the southern side. The topography of Bangladesh is extreme lowlands, with most of the land below 10 m above the mean sea level. The Brahmaputra, Ganges (Padma) and the Meghna influence the main river system of Bangladesh. There are four seasons in Bangladesh: pre–monsoon (March–May), monsoon (June–September), post–monsoon (October and November) and winter (December–February). Southwest and northeast monsoons have a major influence on the country's climate, resulting in marked seasonal rainfalls and temperatures. In this study, we selected 29 weather stations (Figure 1).

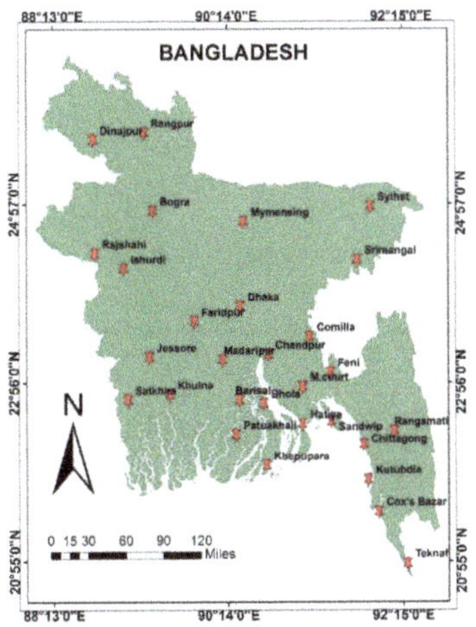

Figure 1. Meteorological stations over Bangladesh used in this study.

2.2. *Data Source and Quality Control*

Daily thunderstorm data are collected from the BMD for the period of 42 years (1975–2016). A storm with thunder and lightning formed by the rapid upward movement of warm moist air is considered as a thunderstorm. The mechanism of thunderstorm formation is different for different seasons in Bangladesh. Therefore, thunderstorm data are analyzed separately for each season in this paper. ENSO and IOD data are collected from the National Oceanic and Atmospheric Administration (NOAA) for the same period, which are represented in Table 1. These years are selected based on the definition of the NOAA. Meteorological data often contain inhomogeneity, which results in erroneous and false analysis and prediction. Data collection techniques, data processing methods, relocation of the stations, lack of proper equipment and the drift of equipment are the causes of data inhomogeneity [22]. In this study, the Standard Normal Homogeneity Test (SNHT) was used to assess the homogeneity in the collected TS data. Data for all the stations found homogeneity at a significance level of 95% or more.

Table 1. Identified years under various climate modes during 1975–2016.

Climate Modes	Years
El Niño	1976, 1977, 1986, 1987, 1988, 1997, 1998, 2006, 2007, 2014
La Niña	1975, 1983, 1984, 1985, 1989, 1996, 1999, 2000, 2001, 2005, 2008, 2010, 2012, 2013
Normal	1978, 1991, 1992, 1994, 1995, 2002, 2003, 2004, 2005, 2009, 2015, 2016
IOD-positive	1982, 1983, 1994, 1997, 2006, 2012,2015
IOD-negative	1975, 1981, 1989, 1992, 1996, 1998, 2010, 2014, 2016

2.3. *Methods*

2.3.1. Mann–Kendall (MK) Trend Test

The trend in a data series is most popularly detected using the Mann–Kendall (MK) test [23]. Wilks et al. [24] identified the following equations from the original version of the MK test:

$$s_t = \sum_{c=1}^{n-1} \sum_{d}^{n} sign(x_d - x_c) \tag{1}$$

where

$$\text{sign}(x_d - x_c) = \begin{cases} +1 & \text{if } (x_d - x_c) > 0 \\ 0 & \text{if } (x_d - x_c) = 0 \\ -1 & \text{if } (x_d - x_c) < 0 \end{cases} \tag{2}$$

The trend significance is estimated using Z statistics,

$$Z = \begin{cases} \frac{S-1}{\sqrt{\text{Var (S)}}} & \text{if } S > 0 \\ 0 & \text{if } S = 0 \\ \frac{S-1}{\sqrt{\text{Var(S)}}} & \text{if } S < 0 \end{cases} \tag{3}$$

where

$$\text{Var(S)} = \frac{n(n-1)(2n+5) - \sum_{i=1}^{m} t_i(t_i - 1)(2t_i + 5)}{18} \tag{4}$$

where n represents the data size, m represents the tied groups with repeated values that are indicated by j and t_j, is the number of repeated values in and t_i represents the number of data in the tied group, i.

The existence of serial correlation and a seasonality pattern in a time series data can have a major effect on the results of the MK test. Autocorrelation estimation is one of the simplest methods to check for the existence of seasonality patterns or serial correlations in any time series. Plotting autocorrelation coefficients versus lags are called a correlogram, which is usually used for detecting seasonality patterns and serial correlations in a time series. The autocorrelation coefficient for lag can be calculated as:

$$r_k = \frac{\sum_{i=1}^{n-k}[(x_i - x_-)(x_{i+k} - x_+)]}{\sqrt{\left[\sum_{i=1}^{n-k}(x_i - x_+)^2\right]}\sqrt{\left[\sum_{i=k+1}^{n}(x_i - x_+)^2\right]}} \tag{5}$$

The subscripts "−" and "+" in Equation (6) indicate the sample averages over the first and last $n - k$ values in the time series, respectively. To judge, if the time series is serially correlated, the significance of the lag–1 autocorrelation coefficient at a significance level of $\alpha = 0.10$ of the two–tailed t–test is assessed using Equation (6).

$$\frac{-1 - 1.645\sqrt{n-2}}{n-1} \leq r_k \leq \frac{-1 + 1.645\sqrt{n-2}}{n-1} \tag{6}$$

If the time series has a positive (negative) lag–1 autocorrelation coefficient, then the variance estimation will be less (more) than the actual value and based on Equation (7); this will increase or decrease the MK Z–value erroneously. When the lag–1 autocorrelation coefficient is significant, or, in other words, when there is serial correlation in a time series, then the modified version of the MK test should be used as follows:

$$\text{var}(S') = \frac{1}{18}[n(n-1)(2n+5)]\left[\frac{n}{n_e}\right] \tag{7}$$

$$\frac{n}{n_e} = 1 + \left(\frac{2}{n^3 - 3n^2 + 2n}\right)\sum_{f=1}^{n-1}(n-f)(n-f-1)(n-f-2)\,\rho_e(f) \tag{8}$$

$$\rho(f) = 2\sin\left[\frac{\pi}{6}\,\rho_e\,(f)\right] \tag{9}$$

2.3.2. Weibull Probability Distribution Model

Wallodi Weibull invented the Weibull distribution for parameter estimation of the frequency distribution of data. Weibull distribution is commonly used in the probability density function. The detailed information regarding Weibull probability distribution can be found in Islam et al. [20].

2.3.3. Spearman's Rho Test

The Spearman's rho (SR) test is a technique with uniform power for linear and nonlinear trends [20]. It is commonly used to verify the absence of trends. The null hypothesis (H_0) of the test is that all the data in the time series are independent and identically distributed, while the alternative hypothesis (H_1) is that increasing or decreasing trends exist. Positive values of the standardized test statistic, SR_Z, indicate upward trends, while the negative values of SR_Z indicate downward trends in the time series. The SR is calculated by the following Equation (10),

$$R = 1 - \frac{6\sum d^2}{n^3 - n} \tag{10}$$

where R denotes the spearman's rank correlation, d denotes the difference in the rank and n is the total number of data.

3. Results

3.1. Monthly and Seasonal Variation of TS Days

The monthly and seasonal variabilities of TS days over Bangladesh are presented in Figures 2 and 3, respectively. The highest average TS days are observed in May, while the lowest in December (Figure 2). The least TS activity is found during the postmonsoon and winter seasons because of low temperatures and lower moisture. The premonsoon and the monsoon seasons have comparatively higher TS days than that of the winter season (Figure 3). The average monthly value of TS days in the monsoon and pre–monsoon seasons is 7.96 and 7.95, respectively, while the average TS days in the postmonsoon and cold winter are four and one, respectively. The seasonal variation of TS days during 1975–2016 shows a sharp increasing trend in TS days during the monsoon, with R^2 = 0.41, and almost no change during the premonsoon season.

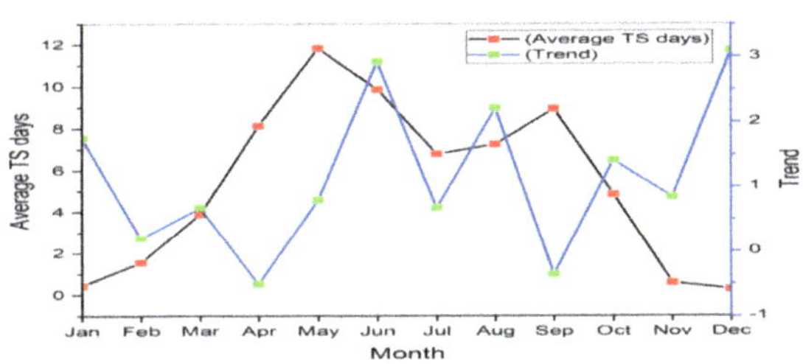

Figure 2. Average monthly thunderstorm (TS) days in Bangladesh during 1975–2016.

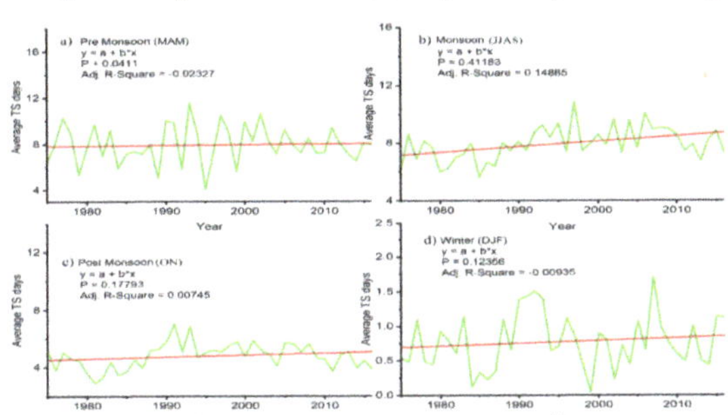

Figure 3. Seasonal variation days (**a**) premonsoon (March–May (MAM)); (**b**) monsoon (June–September (JJAS)); (**c**) postmonsoon (October–November (ON)); (**d**) winter (December–February(DJF)) of TS days in Bangladesh during 1975–2016.

3.2. *Annual Variation of TS Days*

More than 60% of stations revealed an upward trend of TS days in most parts of Bangladesh, except in the southwest and the northeast. Atmospheric disturbance, unstable temperature and uneven rainfall variability may be the possible reasons for TS day variability in Bangladesh. The annual average TS days in Bangladesh over the study period are shown in Figure 4. The total number of annual TS days in Bangladesh is 65. The descriptive statistics for the total number of annual TS days are given in Table A1. The TS days in the country vary from zero in Taknaf to 7.5 days per decade in Sylhet (Table A1).

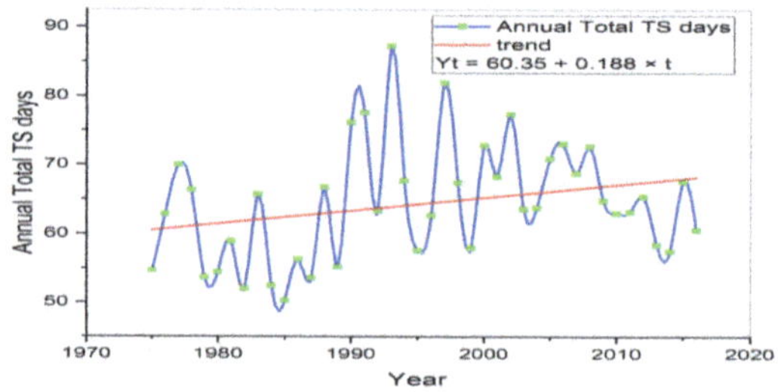

Figure 4. Same as Figure 3, but for annual variation.

3.3. *Spatial Distribution of TS Days*

Monthly Spatial Distribution of TS Days

Spatial distributions of average TS days in Bangladesh for different months are shown in Figure 5. It can be observed that TS mainly happen in the eastern and south–central regions of the country. However, it dominates in the eastern, central and southeastern parts during the premonsoon and monsoon months (March–August). Faridpur experiences the highest number of TS days, with 2.95 and six in January and February, respectively. The maximum TS days during March and April are found in the northeast, north and south–central parts. Sylhet located in the northeastern part of Bangladesh experiences the highest TS days (average 9.85 days) during these months. The maximum amount of TS days, 22.7, 19.7 and 15.8, are noticed at Sylhet, Mymensingh and Rangpur stations, respectively, in May. The highest number of TS days in June are noticed at Sylhet (21.5) and Jessore (15.3), while the highest TS days are observed at Sylhet, Srimangal and Mymensingh (17.6, 14 and 13.4, respectively, in July and 19.8, 16.5 and 13.2, respectively, in August). The TS days are found more in the south and east of Bangladesh during October. The number of TS days was nearly zero in most of the country during November and December. Only a few TS days (less than two days on average) are found in the coastal region during these two months. Figure 6 shows the spatial distributions of the average TS days in Bangladesh for the ENSO (and Figure 7, the IOD) years for different seasons. It can be observed that the highest TS mainly happen in the eastern regions of the country for all seasons, except winter for both the ENSO and IOD years. Among all four seasons, the monsoon season contributes the highest.

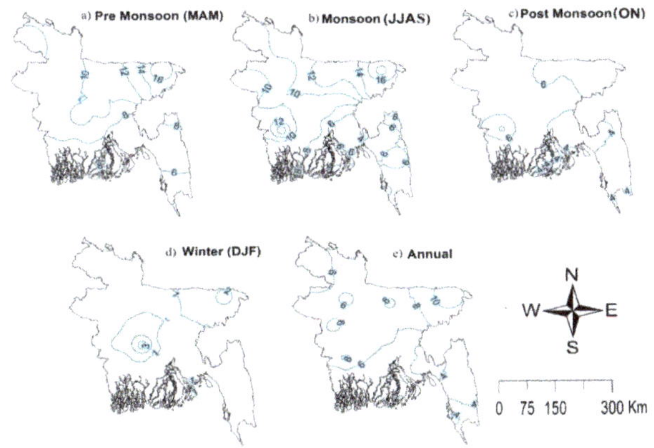

Figure 5. The average number of TS days (**a**) premonsoon (March–May (MAM)); (**b**) monsoon (June–September (JJAS)); (**c**) postmonsoon (October–November (ON)); (**d**) winter (December–February(DJF)); (**e**) annual during 1975–2016 in Bangladesh.

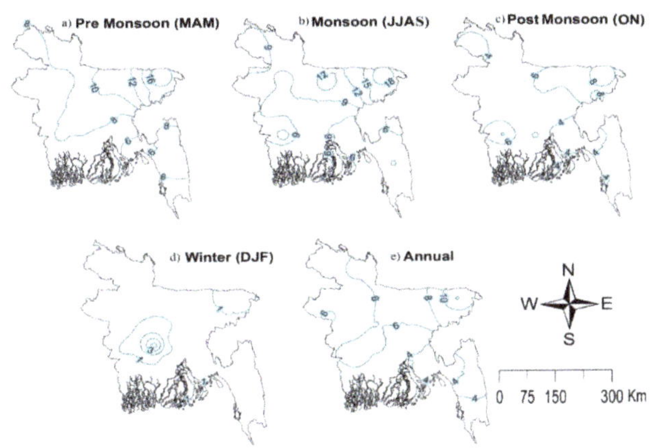

Figure 6. Spatial distribution of TS days (**a**) premonsoon (March–May (MAM)); (**b**) monsoon (June–September (JJAS)); (**c**) postmonsoon (October–November (ON)); (**d**) winter (December–February(DJF)); (**e**) annual in the ENSO years.

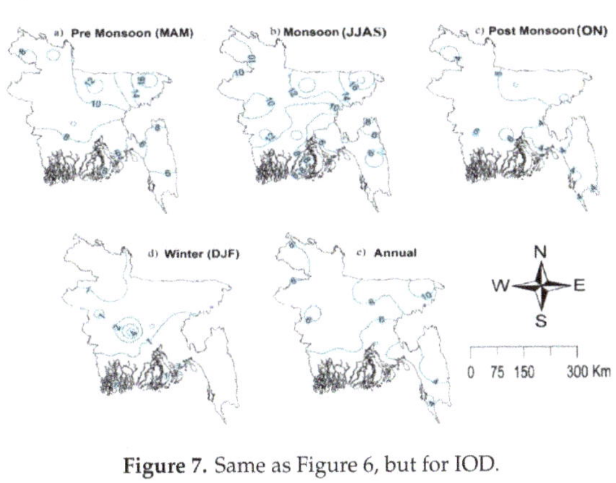

Figure 7. Same as Figure 6, but for IOD.

3.4. *Spatiotemporal Pattern of TS Days*

Spatial Patterns in the Monthly Trends of TS Days

Most parts of the country, except the central part, show a negative insignificant trend in TS days in December, November, January and February (Figure 8). A positive insignificant trend, indicating a slight ascension in TS days, is noticed in the northern part (Rangpur and Dinajpur). The TS days in March are noticed to increase significantly at the Faridpur, Rangamati and Hatiya Stations, while a mix of insignificant trends ($p > 0.05$) are noticed in the northeast and northwest parts. The TS days in April and May in most parts of the country are observed to increase significantly. It is also found to increase significantly in June over a large area in the northeastern and some eastern regions. A similar trend is observed in July and August in the northern region and some parts of southern stations, including Patuakhali, Khepupara and M. Court. In October, a significant positive trend is detected only in M. Court Station and a negative trend in Faridpur Station.

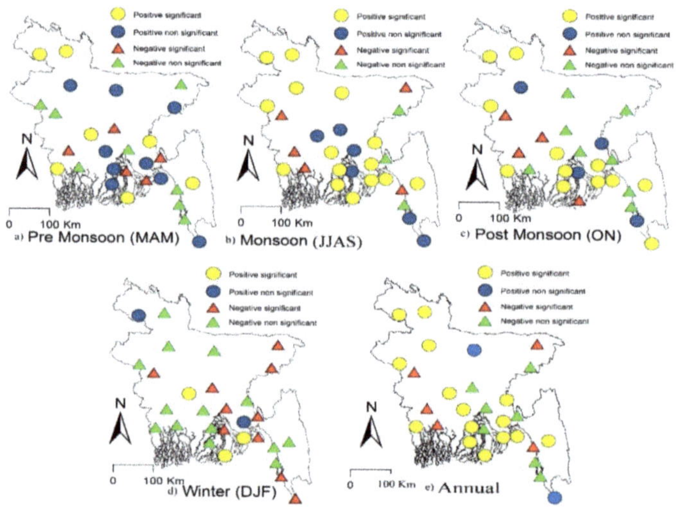

Figure 8. Distributions of the trends in monthly TS days (**a**) premonsoon (March–May (MAM)); (**b**) monsoon (June–September (JJAS)); (**c**) postmonsoon (October–November (ON)); (**d**) winter (December–February(DJF)); (**e**) annual in Bangladesh during 1975–2016. Positive significant (insignificant) is shown as yellow (blue) color and negative significant (insignificant) is shown as red (green) color.

3.5. *Probability of TS Days over Bangladesh*

The Weibull distribution model with a plotting position method was applied to estimate the probable maximum number of TS days for 5, 10, 15 and 20 years in Bangladesh. The results are shown in Table A2 and Figure 9. The results showed the maximum number of TS days in the northeastern region, whereas the lowest was in the southwestern parts of the country close to the coastal area. The results indicate that northeast Bangladesh as most prone to TS.

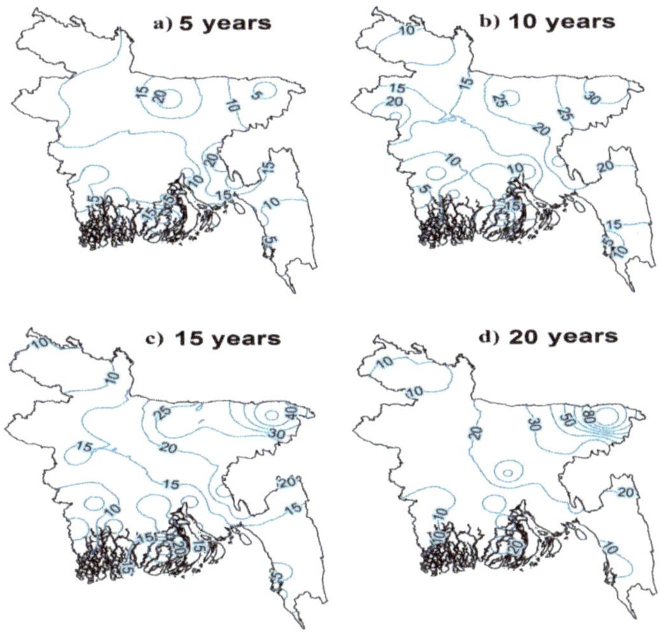

Figure 9. Probability of maximum annual TS days for different periods (**a**) 5 years; (**b**) 10 years; (**c**) 15 years; (**d**) 20 years in Bangladesh estimated using the data from 1975 to2016.

3.6. The Relationship between ENSO/IOD and TS Days

The annual averages and anomalies of TS days in the El Niño, La Niña and neutral years are shown in Figures 10 and 11, respectively. A smaller number of average premonsoon TS days compared to neutral years is observed for the La Niña years, while a higher number of average premonsoon TS days compared to neutral years is noticed in the El Niño years. The anomalies of TS days (Figure 11) revealed a large positive anomaly in the pre–monsoon season in the La Niña years and a large negative anomaly in the monsoon season in the El Niño years.

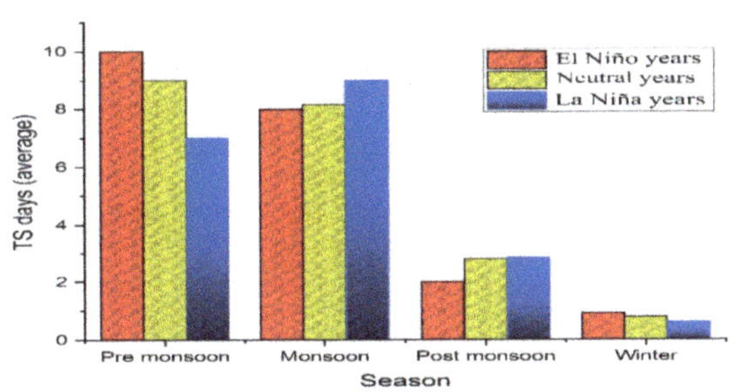

Figure 10. Seasonal (premonsoon, monsoon, postmonsoon and winter) variation of annual average TS days during the El Niño (indicated as red), neutral (yellow) and La Niña (blue) years.

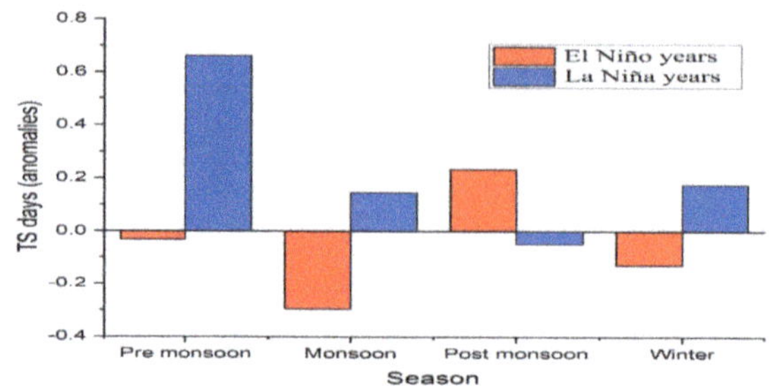

Figure 11. Same as Figure 10, but for composite anomalies of the average TS days during the El Niño (indicated as red color) and La Niña (blue) years.

The annual averages and anomalies of TS days in the positive, negative and neutral IOD years are shown in Figures 12 and 13, respectively. The average premonsoon TS days are found less in the IOD–negative years compared to the IOD–neutral years, while they were not found to vary for the IOD–positive and IOD–neutral years. The influence of IOD on the average monsoon TS days is found much less. Almost no variation in the average monsoon TS days was noticed for IOD–positive and negative years. The anomalies of the average TS days in IOD–positive and negative years (Figure 13) show a large increase in premonsoon TS days in negative IOD years, while a decrease in monsoon TS days during IOD–positive years.

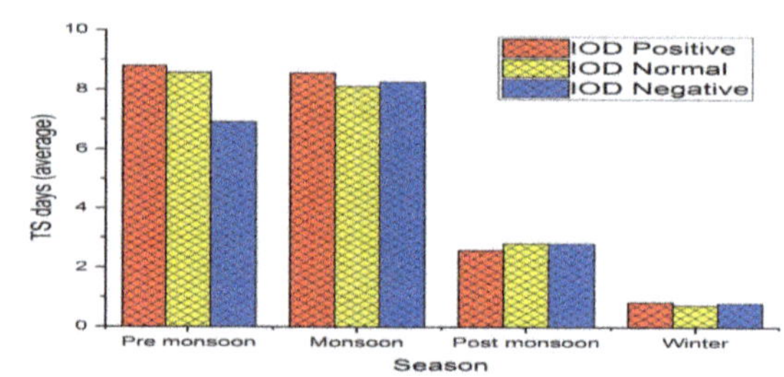

Figure 12. Same as Figure 10, but for IOD.

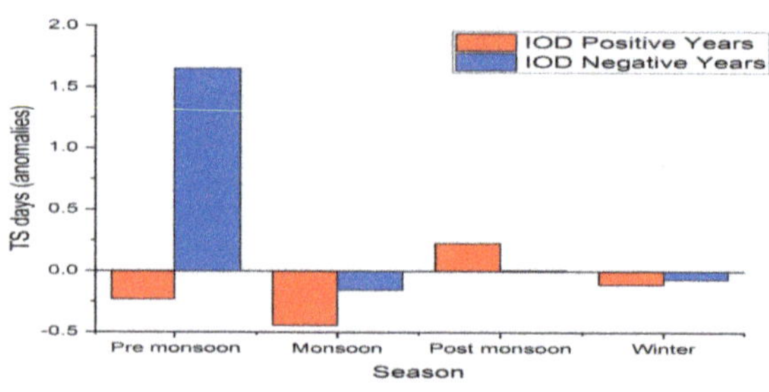

Figure 13. Same as Figure 11, but for IOD.

4. Discussion and Conclusions

The mean annual TS days at the studied 29 stations in Bangladesh vary between 2.8 and 11.8. The annual monthly mean value for TS days for the studied stations is approximately 7.1. The heating of the lower atmosphere moist air during the summer makes the air lighter and move upward through convection and form clouds. A thunderstorm occurs when any atmospheric activity causes a rapid upward movement of air. Generally, moist air from the Bay of Bengal acts as the trigger of a thunderstorm. An interaction of moist air from the Bay of Bengal with the hills in the north causes a rapid uplifting of air and triggering of thunderstorms [25]. The topography and wind regimes made Bangladesh and Northeast India bordering Bangladesh highly favorable for the occurrence of thunderstorms [26]. Therefore, the region experiences the maximum thunderstorm activities compared to other parts of South Asia. A higher number of thunderstorms in Bangladesh are noticed in the northeast region due to the proximity to uplifted land.

Our study shows that the annual average TS days in Bangladesh are increasing by about 1.9 days/decade, which is consistent with an earlier study by Saha et al. [2] and Islam et al. [20]. These outcomes support the findings of the MK test of this study. However, it contradicts the findings in most of India, where a declination of both premonsoon and monsoon thunderstorms was reported by Bhardwaj and Shingh [27]. However, Singh et al. [25] also reported increasing TS days in some parts of West Bengal on the western border of Bangladesh. Bhardwaj and Shingh [27] also reported an increasing trend in thunderstorm–related rainfall activity in Northeast India, despite a decrease in the rest of India. The present study reports an increase in the annual number of TS days mainly due to an increase in TS days in monsoon season. The higher surface temperature caused an increase in the convective available potential energy (CAPE) in the region [28,29]. However, an increasing number of TS days in Bangladesh may be due to an increased number of thunderstorms triggering moist air circulation from the Bay of Bengal. Stronger and more continuous winds from the Bay of Bengal in recent years due to the increase of sea surface temperature have been reported. Glazer et al. [29] evaluated the changes in thunderstorms in Bangladesh due to climate change and reported an increase in severe TS days in most parts of the country.

The monthly spatial distribution of TS shows that TS days are high in the south–central region and very little in the northeastern and northern regions during November–February. This is due to the low sea surface temperature and northeast wind direction from the Bay of Bengal and vapor flux availability in the regions. In March to October, most TS occur in the northeast part of Bangladesh. Besides, high TS days are detected in the north, central, southwest and south of Bangladesh. The higher number of TS days in May and June is noticed in the northeast region, near to Cherapunji, where the cloud formation is the maximum and the mountain ranges produce a large amount of vapor flux and rainfall. Most TS occurs in May, when the average TS days are more than 11 days. The spatial distribution of the annual TS days obtained in this study was found to be consistent with earlier studies [2,20]. For instance, Bangladesh showed an increase in the monthly, seasonal annual and decadal TS days during the last four decades. It has been shown that the increment of TS days is strongly correlated with the strengthening of the sunspot and East Asian summer monsoon, which is the key source of moisture and dynamic force conducive for the premonsoon season climate across the Bay of Bengal.

The monthly, seasonal, decadal and annual trend patterns of TS days show an insignificant decreasing trend at most of the stations in November to February due to the little amount of TS days. By contrast, in May, a positive trend is detected in the southern, northeastern and northern regions. A high positive trend is also found in June. The seasonal and annual trend patterns are similar to the previous observations of Saha et al. [2]. The southeastern air masses when it passes the equator and turns into a southwestern monsoon due to Ferrell's law, which carries a huge amount of water vapor from the Bay of Bengal. This warm and cold monsoon air produces a higher number of TS in Bangladesh in this period [2,19]. The results of the TS days are validated with the earlier literature observed around the world in the last four decades. Most of the stations show an increasing trend in TS days over Bangladesh. This is probably due to the rising CAPE and moisture contents in the Bay of Bengal [15].

The trend of TS days was mostly positive in all months except December and January. The trends were significant in May and June. The linear regression and Weibull distribution model showed a higher number of annual total TS days at almost all stations, except Teknaf and Sawndip. The highest number of TS occurred in May, whereas the lowest in December. This was expected, as the higher surface temperatures and soil moistures in May make the period favorable for the formation of thunderstorms, while the low surface temperatures and less soil moistures make December the least favorable for the formation of thunderstorms.

The magnitude of the average TS days anomaly for the monsoon and premonsoon seasons are found positive for the La Niña years and negative for the El Niño years. This indicates a decrease in TS activities in Bangladesh in the El Niño years. Similar results are found in India and the nearby regions. Kulkarni et al. [6] evaluated the association of TS days over India with the ENSO and showed a reduction of TS days during the El Niño episodes. Yuan and Di [11] found a decrease in thunderstorms in Eastern China during the ENSO episodes. Kulkarni et al. [6] explored the teleconnections between TS days and ENSO for the period 1998–2013 and reported a decrease in TS days during the ENSO years. The anomalies of the average TS days in IOD–positive and negative years also show a large increase in premonsoon TS days in negative IOD years, while a decrease in monsoon TS days during IOD–positive years. This indicates a negative influence of the IOD on TS activities in Bangladesh. A decrease in moist air supply from the Bay of Bengal due to a reduction of monsoon depression during the ENSO and IOD warm phases has been reported in several studies by Bhardwaj and Singh [27], Krishnamurthy and Krishnamurthy [30] and Dash et al. [31]. A decrease in thunderstorms triggering moist air circulation of the Bay of Bengal in ENSO and IOD–positive years may be the cause of a decrease in TS activities in Bangladesh.

Author Contributions: Conceptualization, M.W. and A.R.M.T.I.; methodology and analysis, M.W., A.R.M.T.I., J.-J.L., M.J.U., M.A.S. and S.M.S.; writing—original draft preparation, M.W. and A.R.M.T.I.; writing—review and editing, M.W. and S.S. and funding acquisition, M.W. and J.-J.L. All authors have read and agreed to the published version of the manuscript.

Funding: Project 42030605 supported by National Natural Science Foundation of China.

Conflicts of Interest: The authors declare no conflict of interest.

Appendix A

Table A1. Descriptive statistics of the annual average TS days over Bangladesh during 1975–2016.

Station	Mean	Median	Standard Deviation	Maximum	Minimum	1st Quartile	3rd Quartile	Variation per Decade
Dhaka	6.1	6.1	1.0	8.8	4.4	5.3	6.6	3.3
Madaripur	6.0	6.3	1.8	10.0	1.8	4.9	7.1	4.6
Faridpur	7.1	7.4	2.0	11.2	2.2	5.6	8.5	5.8
Mymensingh	8.2	8.3	1.6	11.8	4.1	7.3	8.9	5.5
Srimangal	9.8	9.9	1.6	13.1	6.0	8.7	11.1	6.5
Sylhet	11.8	11.9	1.6	16.2	9.1	10.5	12.8	7.5
Bogura	5.7	5.6	1.3	8.5	2.9	4.9	6.8	2.3
Rajshahi	6.3	6.4	1.2	9.0	4.4	5.2	7.2	3.5
Ishwardi	5.5	5.6	1.4	8.4	2.8	4.7	6.3	2.2
Dinajpur	4.6	4.7	1.7	7.6	1.0	3.3	6.0	3.2
Rangpur	6.3	6.3	1.6	9.0	2.7	5.3	7.3	4.2
Jassore	7.9	8.1	1.5	10.3	2.6	7.0	9.0	2.1
Khulna	5.2	5.2	1.4	8.2	1.8	4.3	6.1	2.5
Satkhira	5.1	5.5	1.6	8.5	1.8	4.1	6.0	2.3
Barishal	4.9	5.3	1.6	7.3	1.7	3.8	6.3	3.3
Bhola	4.3	4.2	1.0	6.4	2.5	3.6	5.1	1.1
Khepupara	4.4	4.6	1.8	8.5	1.3	3.1	5.5	1.2
Patuakhali	5.2	5.3	1.3	7.8	2.1	4.2	6.1	0.5
Chandpur	4.3	4.0	1.3	7.5	2.5	3.3	5.1	3.1
Chattogram	4.3	4.4	1.1	6.7	1.3	3.6	5.0	2.5
Comilla	3.9	4.0	1.3	6.8	1.4	3.0	4.6	3.5
Cox's Bazar	4.2	4.0	1.0	6.7	2.4	3.4	4.9	2.5
Feni	3.2	3.0	1.2	6.5	0.7	2.3	4.2	2.3
Hatiya	4.8	4.4	1.7	8.8	2.0	3.6	5.7	1.4
Kutubdia	3.1	3.2	1.2	5.5	0.7	2.1	3.8	1.9
M.court	3.2	3.5	1.4	6.3	0.3	2.1	4.2	3.3
Rangamati	4.5	4.8	1.5	7.1	0.6	3.6	5.5	1.2
Sandwip	3.3	3.1	1.4	6.9	0.6	2.4	4.3	1.0
Teknaf	2.4	2.5	1.2	5.2	0.3	1.4	3.2	0

Table A2. Probability of maximum number of TS days for different periods at different locations of Bangladesh.

Year	5	5	10	10	15	15	20	20
Station Name	Maximum	3rd Quartile	Maximum	3rd Quartile	Maximum	3rd Quartile	Maximum	3rd Quartile
Dhaka	>7	3	>19	10	>18	6	>42	4
Madaripur	>5	4	>5	3	>5	4	>5	4
Faridpur	>10	8	>12	9	>15	9	>17	8
Mymensingh	>22	16	>26	12	>28	12	>26	20
Srimangal	>9	8	>28	13	>20	12	>23	12
Sylhet	>4	3	>33	4	>46	15	>100	10
Bogra	>13	12	>14	12	>16	14	>18	13
Rajshahi	>8	2	>26	7	>13	6	>15	6
Ishurdi	>14	8	>14	10	>16	10	>18	12
Dinajpur	>7	5	>9	5	>8	5	>8	6
Rangpur	>9	5	>7	5	>7	5	>7	5
Jessore	>3	2	>3	2	>3	1	>4	2
Khulna	>12	8	>16	10	>17	17	>19	19
Satkhira	>2	2	>2	2	>2	2	>2	2
Barisal	>3	3	>4	2	>5	3	>4	2
Bhola	>13	7	>15	10	>18	14	>18	13
Khepupa	>13	5	>11	8	>16	7	>12	9
Patuakhali	>24	18	>28	25	>30	23	>28	24
Chandpur	>3	2	>6	5	>8	5	.>10	4
Chittagon	>11	7	>17	6	>15	6	>8	5
Comilla	>25	23	>27	25	>28	25	>26	24
Cox's Bazar	>8	6	>10	8	>16	8	>12	4
Feni	>24	17	>23	20	>24	22	>21	19
Hatiya	>7	6	>17	7	>12	8	>14	7
Kutubdia	>3	2	>3	2	>3	2	>3	2
M.court	>17	13	>20	12	>14	11	>18	12
Rangamat	>8	6	>18	8	>11	8	>16	7
Sandwip	>6	4	>14	5	>6	3	>12	5
Teknaf	>12	4	>13	2	>16	5	>12	5

References

1. Ahasan, D.M.N.; Quadir, D.; Khan, K.; Haque, M.S. Simulation of a thunderstorm event over Bangladesh using WRF–ARW model. *J. Mech. Eng.* **2015**, *44*, 124. [CrossRef]
2. Saha, T.R.; Quadir, D.A. Variability and trends of annual and seasonal thunderstorm frequency over Bangladesh. *Int. J. Climatol.* **2016**, *36*, 344–345. [CrossRef]
3. Wahiduzzaman, M.; Luo, J.J. A statistical analysis on the contribution of El Niño–Southern Oscillation to the rainfall and temperature over Bangladesh. *Meteorol. Appl. Phys.* **2020**. [CrossRef]
4. Wahiduzzaman, M.; Oliver, E.C.J.; Wotherspoon, S.J.; Luo, J.J. Seasonal forecasting of tropical cyclones in the North Indian Ocean region: The role of El Niño–Southern Oscillation. *Clim. Dyn.* **2020**, *54*, 1571–1589. [CrossRef]
5. Manohar, G.K.; Kandalgaonkar, S.S.; Tinmaker, M.I.R. Thundersytrom activity over India and the Indian Southwest Monsoon. *J. Geophys. Res.* **1999**, *104*, 41188–41694.
6. Kulkarni, M.K.; Revadekar, J.V.; Varikoden, H. About the variability in thunderstorms and rainfall activity over India and its association with El Niño and La Niña. *Nat. Hazards* **2015**, *63*, 2005. [CrossRef]
7. Allen, J.T.; Karoly, D.J. A climatology of Australian severe thunderstorm environments 1979–2011: Inter–annual variability and ENSO influence. *Int. J. Climatol.* **2014**, *34*, 81–97. [CrossRef]
8. Allen, J.T.; Tippett, M.; Sobel, A. Influence of El Niño/Southern Oscillation on tornado and hail frequency in the United States. *Nat. Geosci.* **2015**, *8*, 278–283. [CrossRef]
9. Pinto, O.; Ferro, M.A.S. A study of the long–term variability of thunderstorm days in southeast Brazil. *J. Geophys. Res.* **2013**, *118*, 5231–5246. [CrossRef]
10. Pinto, O. Thunderstorm climatology of Brazil: ENSO and tropical Atlantic connections. *Int. J. Climatol.* **2014**, *35*, 871–878. [CrossRef]
11. Yuan, T.; Di, Y. Variability of lightning flash and thunderstorm over eastern China and Indonesia on ENSO time scales. *Atmos. Res.* **2014**, *152*, 377–390.

12. Kunkel, K.E.; Karl, T.R.; Brooks, H.; Kossin, J.; Lawrimore, J.H.; Arndt, D.; Bosart, L.; Changnon, D.; Cutter, S.L.; Doesken, N.; et al. Monitoring and understanding trends in extreme storms: State of knowledge. *Bull. Am. Meteorol. Soc.* **2013**, *94*, 499–514. [CrossRef]
13. Mir, H.; Hussain, A.; Babar, Z.A. Analysis of thunderstorms activity over Pakistan during 19612–000. *J. Geophys. Res.* **2006**, *188*, 41694.
14. Kunz, M.; Sander, J.; Kottmeier, C. Recent trends of thunderstorm and hailstorm frequency and their relation to atmospheric characteristics in southwest Germany. *Int. J. Climatol.* **2009**, *29*, 22832–22897. [CrossRef]
15. Enno, S.E.; Briede, A.; Valiukas, D. Climatology of thunderstorms in the Baltic countries. *Theor. Appl. Climatol.* **2013**, *111*, 309–325. [CrossRef]
16. Zhang, Q.; Ni, X.; Zhang, F. Decreasing trend in severe weather occurrence over China during the past 50 years. *Sci. Rep.* **2017**, *7*, 42310. [CrossRef]
17. Singh, O.; Bhardwaj, P. Spatial and temporal variations in the frequency of thunderstorm days over India. *Weather* **2017**, *99*, 1–7. [CrossRef]
18. Araghi, A.; Adamowski, J.; Jagharg, M.R. Detection of trends in days with thunderstorms in Iran over the past five decades. *Atmos. Res.* **2016**, *15*, 242–243. [CrossRef]
19. Karmakar, S. Climatology of thunderstorm days over Bangladesh during the pre–monsoon season. *Bangladesh J. Sci. Tech.* **2001**, *33*, 1022–1031.
20. Islam, A.R.M.T.; Nafiuzzaman, M.; Rifat, J. Spatiotemporal variations of thunderstorm frequency and its prediction over Bangladesh. *Meteorol. Atmos. Phys.* **2020**. [CrossRef]
21. Shahid, S. Rainfall variability and the trends of wet and dryperiods in Bangladesh. *Int. J. Climatol.* **2010**, *3030*, 2299–2313. [CrossRef]
22. Karl, R.; Williams, C.N. An approach to adjusting climatological time series for discontinuous in homogeneities. *J. Clim. Appl. Meteorol.* **1987**, *2626*, 17441–17763.
23. Adamowski, K.; Bougadis, J. Detection of trends in annual extreme rainfall. *Hydrol. Process.* **2003**, *17*, 35473–35560. [CrossRef]
24. Wilks, D.S. *Statistical Methods in the Atmospheric Science*, 3rd ed.; Academic Press: Cambridge, MA, USA, 2011; p. 704.
25. Singh, C.; Mohapatra, M.; Bandyopadhyay, B.K.; Tyagi, A. Thunderstorm climatology over northeast and adjoining east India. *Mausam* **2011**, *6262*, 1631–1670.
26. Sahu, R.K.; Dadich, J.; Tyagi, B.; Visa, N.K.; Singh, J. Evaluating the impact of climate change in threshold values of thermodynamic indices during pre–monsoon thunderstorm season over Eastern India. *Nat. Hazards* **2020**, *102*, 1541–1569. [CrossRef]
27. Bhardwaj, P.; Singh, O. Spatial and temporal analysis of thunderstorm and rainfall activity over India. *Atmosfera* **2018**, *3131*, 2552–2584. [CrossRef]
28. Sahu, R.K.; Dadich, J.; Tyagi, B.; Vissa, N.K. Trends of thermodynamic indices thresholds over two tropical stations of north–east India during pre–monsoon thunderstorms. *J. Atmos. Solar Ter. Phys.* **2020**, *211*, 105472. [CrossRef]
29. Glazer, R.; Torres–Alavez, J.A.; Coppola, E.; Das, S.; Ashfaq, M.; Sines, T. Projected changes to Severe Thunderstorm environments as a result of 21st century warming from RegCM CORDEX–CORE simulations. *EGU Gen. Assem.* **2020**, *2020*, 970.
30. Krishnamurthy, L.; Krishnamurthy, V. Influence of PDO on South Asian monsoon and monsoon–ENSO relation. *Clim. Dyn.* **2013**, *42*, 91. [CrossRef]
31. Dash, S.K.; Jenamani, R.K.; Shekhar, M.S. On the decreasing frequency of monsoon depressions over the Indian region. *Curr. Sci. India* **2004**, *86*, 14041–14411.

Publisher's Note: MDPI stays neutral with regard to jurisdictional claims in published maps and institutional affiliations.

atmosphere

MDPI

Article

Spatial Characteristics of Precipitation in the Greater Sydney Metropolitan Area as Revealed by the Daily Precipitation Concentration Index

Kevin K. W. Cheung [1,*], Aliakbar A. Rasuly [2,*], Fei Ji [1] and Lisa T.-C. Chang [1]

1 Climate and Atmospheric Science, NSW Department of Planning, Industry and Environment, Lidcombe, NSW 2141, Australia; Fei.Ji@environment.nsw.gov.au (F.J.); LisaTzu-Chi.Chang@environment.nsw.gov.au (L.T.-C.C.)

2 Department of Earth and Environmental Sciences, Macquarie University, North Ryde, NSW 2109, Australia

* Correspondence: kkwc09@hotmail.com (K.K.W.C.); aarasuly@yahoo.com (A.A.R.)

Abstract: In this study; the spatial distribution of the Daily Precipitation Concentration Index (DPCI) has been analyzed inside the Greater Sydney Metropolitan Area (GSMA). Accordingly, the rainfall database from the Australian Bureau of Meteorology archive was utilized after comprehensive quality control. The compiled data contains a set of 41 rainfall stations indicating consistent daily precipitation series from 1950 to 2015. In the analysis of the DPCI across GSMA the techniques of Moran's Spatial Autocorrelation has been applied. In addition, a cross-covariance method was applied to assess the spatial interdependency between vector-based datasets after performing an Ordinary Kriging interpolation. The results identify four well-recognized intense rainfall development zones: the south coast and topographic areas of the Illawarra district characterized by Tasman Sea coastal regions with DPCI values ranging from 0.61 to 0.63, the western highlands of the Blue Mountains, with values between 0.60 and 0.62, the inland regions, with lowest rainfall concentrations between 0.55 and 0.59, and lastly the districts located inside the GSMA with DPCI ranging 0.60 to 0.61. Such spatial distribution has revealed the rainstorm and severe thunderstorm activity in the area. This study applies the present models to identify the nature and mechanisms underlying the distribution of torrential rains over space within the metropolis of Sydney, and to monitor any changes in the spatial pattern under the warming climate.

Keywords: precipitation concentration index; extreme rainfall; spatial inter-dependency

Citation: Cheung, K.K.W.; Rasuly, A.A.; Ji, F.; Chang, L.T.-C. Spatial Characteristics of Precipitation in the Greater Sydney Metropolitan Area as Revealed by the Daily Precipitation Concentration Index. *Atmosphere* **2021**, *12*, 627. https://doi.org/10.3390/atmos12050627

Academic Editor: Anita Drumond

Received: 7 April 2021
Accepted: 8 May 2021
Published: 13 May 2021

1. Introduction

The awareness of the importance of the spatial and temporal distribution of precipitation is important not only from a meteorological viewpoint, but also for its importance in different fields such as agriculture, hydrology, water resources and flood control. Estimation of the spatial and temporal distribution of precipitation is a complex undertaking, particularly in cases where detailed information concerning the impact of topography and land−use impacts on the prevailing atmospheric circulation is not quite available, such as the situation over southeastern Australia [1].

The concentration index (CI) is one of the indices that can be applied to characterize the temporal concentration of precipitation followed by spatial analysis [2]. A CI analysis makes it possible to characterize different spatial scales, which is of interest due to its effects on geo−hydrological processes and the analysis of erosion and soil loss [3]. Applying this type of analysis, interest is not only focused on climate but also on the effect of heavy rainfall on other areas of the environment and society [4,5]. The CI method was already applied in many different parts of the world [6–18]. While we focus on the most commonly applied index here, there are other indices that indicate different aspects of precipitation concentration, such as the relative cumulative precipitation, inequality concentration indices and the ordered version of the n index [2].

Many of the previous studies analyzed the CI based on monthly precipitation and examined the annual and seasonal CI values, which are mostly determined by the climatological and synoptic characteristics of a particular region. On a shorter timescale, daily analysis and prediction of the intensity of precipitation would help in in water resource planning and also identifying areas of high and low flash flooding potential. Likewise, it would facilitate the regulation of the flows from high−intensity areas towards low−intensity ones [19]. For example, a high precipitation concentration, represented by large percentages of the yearly total precipitation in a few very rainy days, has the potential to cause floods and also drought phenomena. This is exactly the scenario of precipitation extremes inside the Sydney region, such as rainstorms, ex-tropical cyclone remnants, east coast lows and severe thunderstorm events occurring over a few days that account for high percentages of the annual total. These events may potentially bring more frequent disasters for human society of the Greater Sydney metropolitan area (GSMA). In the past, specific attention has been paid to patterns of such uneven spatial variation of intense rainfalls using different statistics and mathematical methods (e.g., [20]). Nevertheless, a long time-span daily precipitation series has not been analyzed by applying a CI approach. Thus, this study examines the precipitation concentration in the area based on data with daily resolution. The DPCI is also good supplementary information to other extreme precipitation indices on similar timescale, such as the highest amount of daily precipitation (RX), the maximum consecutive 5-day precipitation (RX5D), number of days with precipitation $\geq$ 20 mm or above 50 mm (R20/D50MM) and days with precipitation > 95 percentile (D95P) recommended by the World Climate Research Programme's Expert Team on Climate Change Detection and Indices (ETCCDI, [21,22]).

The organization of the paper is as follows. Section 2 first introduces the climatological characteristics of the GSMA. Section 3 then depicts the methodology applied, including the daily CI and spatial correlation analysis. Results are discussed in Section 4. Finally an overall summary is given in Section 5 together with further discussion on implications and future work.

2. Climatology of the Study Area

The GSMA, which is located on the southeast coast of Australia in New South Wales and lies in the western part of the Tasman Sea (Figure 1), includes a highly populated area of approximately 3.8 million in population. The study area is bounded in the north by 33°30′ S latitude, extending to 150°30′ E longitude in the west, and to the southeast at 34°30′ latitude and 151°30′ longitude. The region is bowl-shaped with a low plain in the middle which is effectively walled in on three sides by hills. In general, the Sydney region enjoys a temperate climate and commonly the broad-scale wind pattern is westerly in the winter, and easterly in the summer. The climate of this region arises from a complex interaction of broad-scale, regional and local controls [23].

Rainfall over the GSMA may occur throughout the year but is highest between March and June. Also, precipitation is slightly higher during the first half of the year when easterly winds dominate (February–June), and lower in the second half; mainly from July to September. Rainfall can occur throughout the year with variation concerning altitude and distance from the coast, with wetter areas being closer to the coast or in higher altitudes. Due to the low predictability of rain as well as the well-known impacts from climate drivers to the region [24], the wettest and driest months change annually. Within the study area and surrounds, annual rainfall varies from around 700 mm to 1400 mm. More climatological information of the study area can be found in [25] and [26].

On the regional scale, rainfall in the GSMA is influenced by the synoptic weather systems in the region, such as fronts originated from the Southern Ocean, east coast lows, subtropical cyclones and ex-tropical cyclone remnants migrating to the higher latitudes. More locally, the GMSA is also known as one of the hotspots for severe thunderstorms in Australia [26]. As can be seen in the following analysis, the torrential rain from severe thunderstorms contribute significantly to the spatial characteristics of CI in the area.

Figure 1. The location map of the GSMA within New South Wales of Australia (upper). The lower panel shows the boundaries of the local government areas within the GSMA, with the names of some major local cities (dots).

3. Data and Method

3.1. Data

Daily rainfall data for forty−one (41) weather stations have been extracted from the Australian Bureau of Meteorology (BoM) online archives. Recording periods varied in duration for each station, but many data are available from 1950 to 2015. Rainfall data from the BoM has already been quality controlled with confirmation of the extremes with local reports and that observations from nearby stations do not disagree with each other. We have further verified such agreement among the stations after we downloaded the data. Most of the stations have complete data series in the study period—even the station with the least recorded data over the whole period has over 90% of coverage. The provided daily rainfall data, presenting relatively uniform coverage throughout the study area, have been carefully entered in a particular GIS database. Rainfall station characteristics are shown in Table A1 (Appendix A) and their spatial distribution is mapped in Figure 2. The only series with less than 10% missing data on an annual scale were used to calculate the DPCI indexes.

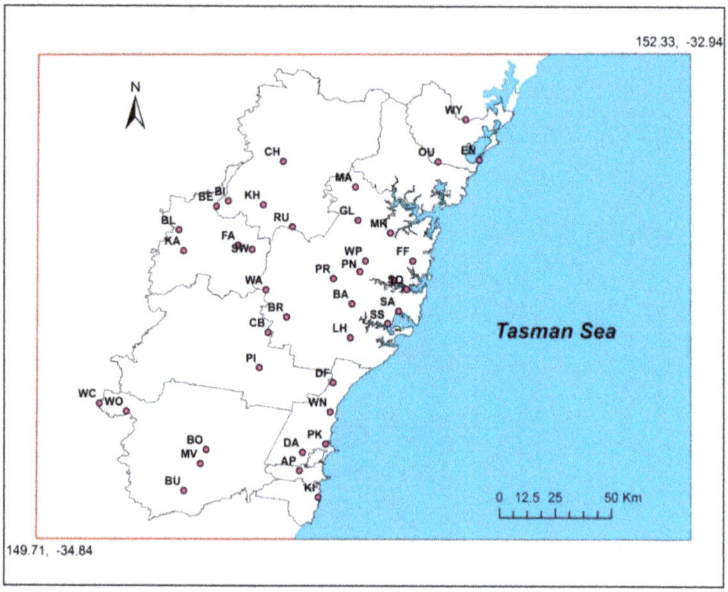

Figure 2. Spatial distribution of the rainfall stations with abbreviated codes as in Table A1 (Appendix A). Topography in the area can refer to the DEM in Figure 4.

To reach the main aims of the study, four interconnected techniques of Concentration Index, Moran's Spatial Autocorrelation, Ordinary Kriging interpolation, and a cross-covariance method were applied to assess the spatial dependence (covariance) between vector-based datasets. The last method was applied to reveal several of the inherent spatial inter-dependencies among dissimilar variables.

3.2. *The Daily Precipitation Concentration Index (DPCI)*

The DPCI, proposed by [27], is examined in this study. An example is illustrated for the Albion Park station data; this station recorded the highest rain during the 67 years from 1950 to 2015 (see Table A2 in Appendix A). In computing the DPCI, only observed daily precipitation values more than 1 mm were considered. The DPCI method was applied to the data based on the fact that the contribution of daily rainfall events to the total amount is generally well described by a negative exponential distribution [28].

The DPCI in this study consists of aggregating daily precipitation into increasing 10−mm categories and determining the relative impact of the different classes by analyzing the relative contribution (as a percentage) of the accumulated precipitation, Y, as a function of the accumulated percentage of occurrence frequency (X). Previous work such as [11] showed that such function can be based on Equation (1).

$$Y = aX.e^{bX} \tag{1}$$

where a and b are constants that can be determined through the least−squares method.

$$In\ a = \frac{\sum x_i^2 \sum \ln Y_i + \sum x_i \sum x_i ln\ x_i - \sum x_i^2 \sum lnx_i - \sum x_i \sum x_i \ln Y_i}{N \sum x_i^2 - N\left(\sum x_i\right)^2} \tag{2}$$

$$b = \frac{N \sum x_i lnY_i + \sum x_i \sum lnx_i - N \sum x_i lnx_i - \sum x_i \sum x_i inY_i}{N \sum x_i^2 - N\left(\sum x_i\right)^2} \tag{3}$$

where N is the number of classes. After determining the two constants a and b, the integral of the exponential curve (so−called Lorenz curve) between 0 and 100 shows the area S (Figure 3), which is given by:

$$S = \int_0^{100} \left[\frac{a}{b} e^{bx}\left(x - \frac{1}{b}\right)\right] dx \tag{4}$$

Based on S, the area S' compressed by the exponential curve, the equidistribution line and $X = 100$ is apparently the difference between 5000 (half of the total area) and the value of S:

$$S' = 5000 - S \tag{5}$$

Applying Equation (5) the DPCI value for each rainfall station is then a fraction of S' to the lower surface of the triangle bounded by the equidistribution line.

$$DPCI = \frac{S'}{5000} \tag{6}$$

Examples of the empirical curves or "concentration curves" of Y versus X for Albion Parks and Wombeyan stations are presented in Figure 3. The annual DPCI values for these two stations are 0.62 and 0.54, respectively (see Table 1). By definition, the value of the DPCI is always a number between 0 and 1, and geometrically it represents the percentage of the triangle area between the line $Y = X$ and the exponential curve. The DPCI is virtually equal to 0 when the contribution of each category of precipitation to the total is the same, and equal to 1 when precipitation falls into one category only and the exponential curve becomes the straight line $Y = 0$. Exponential curves of this type were calculated for all meteorological stations across the GSMA. As an example, different stages of calculating the above−mentioned parameters are given in Tables A1 and A2 (Appendix A).

Table 1. Values of the constants "a", "b", DPCI, 90% percentile of rain and maximum daily rainfall (mm) at each of the 41 stations (with full names given in Table A1).

Station	"a"	"b"	DPCI	90% Rain	Max Daily Rain
AP	0.033	0.032	0.619	52	536.4
BA	0.053	0.029	0.609	58	243
BE	0.056	0.028	0.583	57.5	248
BI	0.055	0.028	0.590	58	237.6
BL	0.053	0.029	0.582	58	245
BO	0.053	0.029	0.568	58.5	214.2
BR	0.062	0.027	0.572	59	203.2
BU	0.040	0.031	0.618	54	399.6
CA	0.058	0.028	0.564	58.5	231.1
CH	0.056	0.028	0.567	58	229

Table 1. *Cont.*

Station	"a"	"b"	DPCI	90% Rain	Max Daily Rain
DA	0.038	0.032	0.612	52	336.8
DF	0.037	0.032	0.612	53	415
FA	0.049	0.030	0.580	57	280.4
FF	0.048	0.030	0.583	56.5	248.4
GL	0.048	0.030	0.577	56	220
KA	0.047	0.030	0.621	57	285
KI	0.047	0.030	0.618	56	304.4
KH	0.049	0.030	0.579	56	283.7
LH	0.047	0.030	0.584	56	254.5
MA	0.059	0.028	0.565	58	325
MV	0.047	0.030	0.585	57	422
MK	0.055	0.028	0.574	58	243.2
OU	0.047	0.030	0.584	55	320
PN	0.047	0.030	0.590	57	293
PI	0.054	0.029	0.603	58	245.9
PK	0.039	0.032	0.622	54	322.5
PR	0.051	0.029	0.581	57	321
RI	0.052	0.029	0.570	58	210
RO	0.045	0.030	0.601	56	196.6
SS	0.045	0.030	0.602	57	239
SW	0.048	0.030	0.606	56	274.4
SA	0.051	0.029	0.583	57	216.2
SO	0.048	0.030	0.616	57	327.6
EN	0.054	0.029	0.598	59	246
WA	0.054	0.029	0.603	58	215.9
WP	0.041	0.031	0.597	54	409.8
WO	0.072	0.026	0.587	62	165
WC	0.071	0.026	0.544	61	230.6
WN	0.039	0.032	0.626	54	436.8
WY	0.050	0.029	0.595	57	256.2

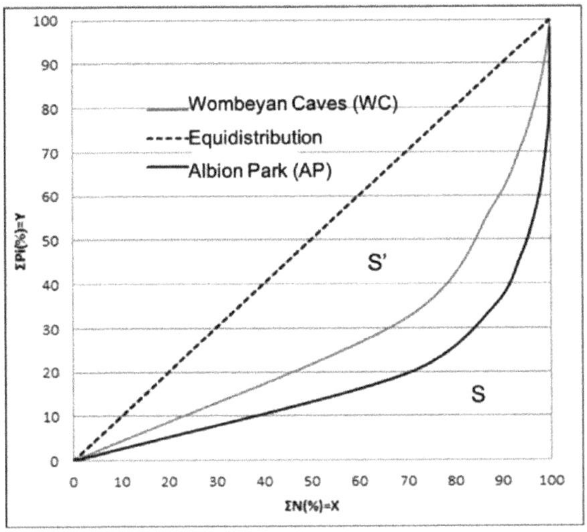

Figure 3. The empirical concentration curves for Albion Park and Wombeyan rain stations with the dash straight line the reference equidistribution line. The area S' is that bounded by the diagonal equidistribution line and the concentration curve, while area S is the remaining area underneath the equidistribution line.

3.3. Spatial Correlation

In the second stage of data analysis, a Moran's spatial autocorrelation technique was used to measure spatial autocorrelation based on rainfall station locations and DPCI values [29]. Given the set of 41 rainfall stations and associated DPCIs, it evaluates whether the pattern expressed is clustered, dispersed or random (Figure A1 in Appendix B). The tool calculates the Moran's I Index value and both a z-score and p-value to evaluate the significance of that Index. The Moran's I statistic for spatial autocorrelation is given by

$$I = \frac{N}{S_0} \frac{\sum_{i=1}^{n} \sum_{j=1}^{n} w_{i,j} z_i z_j}{\sum_{i=1}^{n} z_i^2} \tag{7}$$

where Z_i is the deviation of an attribute for feature (i.e., a particular rainfall station's DPCI) from its mean, $W_{i,j}$ is the spatial weight between stations i and j (designated as the significance, i.e., the p-value, of the correlation of rain between the two stations), N is the total number of stations and S_0 is the aggregate of the spatial weights by:

$$S_0 = \sum_{i=1}^{n} \sum_{j=1}^{n} w_{i,j} \tag{8}$$

For the current study, the Z_I score for the statistic is computed by applying the following equations.

$$Z_I = \frac{I - E[I]}{\sqrt{V[I]}} \qquad V[I] = E[I^2] - E[I]^2 \tag{9}$$

In which E is the expectation value and V the variance. Under the case of no spatial autocorrelation, $E[I] = -1/(N-1)$.

Subsequently, a spatial interpolation method, known as the Kriging technique, was applied to yield better results than other techniques ([30,31]). The Kriging technique assumes that the statistical surface to be interpolated has a certain degree of continuity ([32]). The technique applies moving averages and has the advantage of producing the standard error for the estimated values. Among all the Kriging methods, the ordinary mode was applied, as an advanced geostatistical procedure. This method was well fitted to all data layers to generate estimated DPCI surfaces from a re−projected set of point values [33]. The Kriging model is based on a statistical technique that includes autocorrelation; that is, the statistical relationships among the measured points. Potentially, geostatistical techniques not only have the capability to produce a prediction surface but also provide some measure of the certainty or accuracy of the predictions. Kriging tools weight the surrounding measured values to derive a prediction for each DPCI unmeasured location. There are variations of the techniques, such as the Ordinary Cokirging and those that consider topographical information, that can further improve the performance ([34,35]). The general formula for both interpolators is formed as a weighted sum of the data:

$$\hat{Z}(S_0) = \sum_{i=1}^{N} \lambda i \, Z\,(S_i) \tag{10}$$

where *Z (Si)* is regarded as the measured DPCI values at the *i*th location, λi shows an unknown weight for the measured value at the *i*th rainfall station location, *S0* specifies the prediction location and *N* indicates the number of stations. With the Kriging method, the weights are based not only on the distance between the measured points and the prediction location but also on the overall spatial arrangement of the measured points. To use the spatial arrangement in the weights, the spatial autocorrelation must be quantified. Thus, in ordinary Kriging, the weight, λi, depends on a fitted model to the measured points, the distance to the prediction location and the spatial relationships among the measured DPCI values around the prediction location. In the current study, an Ordinary Kriging formula is used to create maps of the prediction DPCI and "*b*" constant surfaces and associated accuracy models. Ordinary Kriging assumes the second−order trend removal model with no transformation type:

$$Z(S) = \mu + \varepsilon(s) \tag{11}$$

In the above equation, μ is an unknown constant whereas one of the main issues concerning ordinary Kriging is whether the assumption of a constant mean is reasonable. Sometimes there are good scientific reasons to reject this assumption. However, in this study, it was found that applying a second−order trend removal following an exponential Kernal Function (as a simple prediction method) gives remarkable flexibility in final interpolation method accuracy. Once again, the Kriging method was also applied to illustrate the variation and spatial distribution of the constant "*b*" values in the study area. This arbitrary way allows direct interstation comparison of the distribution of "*b*" value at each rainfall station across all districts.

To calculate the Pearson Overall Correlation Coefficient, a band collection statistic tool was furthermore computed among the DPCI and one of the other rainfall related parameters [36]. These parameters include the mean annual precipitation (AP), coefficient of variation (CV) of rainfall, the total number of rainfall days (TN), maximum rainfall observed (MxR) and the "*a*" and "*b*" constants taken from Equation (1). This tool was applied to provide statistics for the bivariate analysis of a set of raster bands by computing covariance and correlation for every event. The following equation was accordingly used to determine the covariance between layers i and j.

$$Cov_{ij} = \frac{\sum_{k=1}^{N} (Z_{ik} - u_i)\left(Z_{jk} - u_j\right)}{N - 1} \tag{12}$$

In the above equation, Z indicates for example DPCIs observed of a cell, i, j are layers of a stack, u is the mean of cells and N is the number of cells. The overall correlation between the rainfall datasets was then computed as:

$$Corr_{ij} = \frac{Cov_{ij}}{\sigma_i \sigma_j} \tag{13}$$

where the σ's are standard deviations. As usual, the correlation coefficient is between −1 and 1.

4. Results

4.1. Spatial Distribution of DPCI

By calculating the annual DPCI values it was found that they range greatly, between 0.54 and 0.63, and are spread across the study area represented by the 41 rainfall stations. This range is consistent with the global results in [2] that showed high values (>0.5) of the Gini Index (which has the same concept of the DPCI but without the assumed mathematical form of the Lorenz curve as in Equation (1), thus the Gini index is highly correlated with the DPCI) over eastern Australia. Table 1 indicates the DPCI values and the rainfall percentage contributed by 90% of the rainiest days for the 41 weather stations across GSMA from 1950 to 2015. Also values of the constants "*a*" and "*b*" (as the exponential curves are given by Equation (1)) and observed maximum daily rainfall are represented in the table. DPCI values present strongly different daily precipitation regimes, as Woonona station (0.63) is located in the southeast of the study area and precipitation there has a higher concentration and is more irregular than in Wombeyan station (0.54) which is located on the Tableland somewhere in the outlying southwest of the study area. The concentration can be considered a function of the relative separation of the equidistribution line, which is greater in Albion Park (with the highest maximum daily rainfall observed) than others (Figure 3).

Applying the Global Moran's I statistic it is possible to test an existing spatial autocorrelation based on rainfall station locations and DPCI values. The Spatial Autocorrelation tool returns five values: the Moran's I index, expected index, variance, z-score and *p*-value. Given the z-score of 3.55, there is a less than 1% likelihood that this clustered pattern could be the result of random chance, expressing the fact that there are spatially significant clusters of DPCI values among the existing dataset based on the spatial autocorrelation report.

The result of the Ordinary Kriging interpolator model is shown in Figure 4 after smoothing small errors depending on the measurement parameters overlaid with a Digital Elevation Model (DEM) of the study area. The maximum values of DPCI are crossing over the Kiama, Shellharbour and Wollongong districts located in the southeast of the study area. For example, Woonona station (34.34° S; 150.90° E) represents the highest value of 0.63, while the lowest values of DPCI could be seen in Wombeyan station (34.31° S; 140.97° E) with 0.54. The highest DPCIs were detected primarily in the Illawarra (along with the south coast) and Blue Mountains districts (Katoomba station with 0.62). Furthermore, the secondary maximum annual values of DPCI were found around the Sydney Metropolitan, mainly around the Central Business District (CBD). On the other hand, districts with the lowest values are located in the southwest Tablelands of the Wingecarribee and Hawkesbury districts. Meanwhile, the strongest gradient occurs between the west and east and between the northwest and southeast of GSMA, as coastlines meet the highest DPCI values.

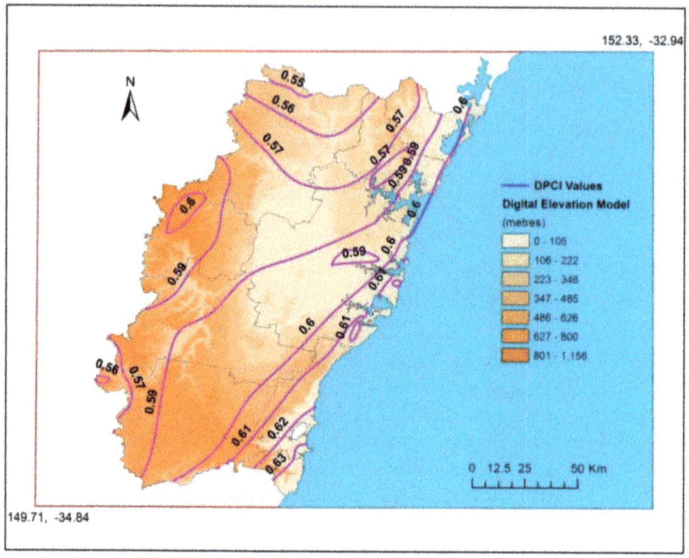

Figure 4. Spatial distribution of DPCI (contours) overlaid on a DEM inside the GSMA.

To find more about the nature of the spatial distribution of intense rains inside of the GSMA, the geographical distribution of the "*b*" constant, which from Equation (1) is the parameter to control the shape of the rainfall concentration curve and thus carries important information on the rainfall distribution, has been converted to classes of intensity level of rainfall occurrence (Figure 5). Very high intense amounts can be seen near the topography southeast of the study area, just over the Illawarra Escarpment. Besides, in some parts of the Sydney Metropolitan district, for example in the west of the City, and areas located in the northwestern corner of the Parramatta River, very intense "b" values can be observed. In comparison with the lowlands of the GSMA, over the Blue Mountains (Katoomba station), intense rain events are also relatively high. In contrast, non−intense classes of "b" values can be seen over the inland parts of the GSMA.

For comparison, the geographic position of the flash flood events (observed during 1989−2015 with a thundery−rain more than 50 mm) is overlaid on the distribution of the "*b*" constant map. In the GSMA, flash flood events are mostly induced by several weather systems, such as the local thunderstorms and east coast lows over the ocean. It can be seen that most of these flash flood events occurred in the areas with the high "*b*" values, which determine the shape of the concentration curve. The spatial pattern of "*b*" also highly resembles that of the severe thunderstorms, especially those with hail occurrence [26]. These facts indicate that the storm activity in the GSMA largely determine the CI pattern on the daily timescale.

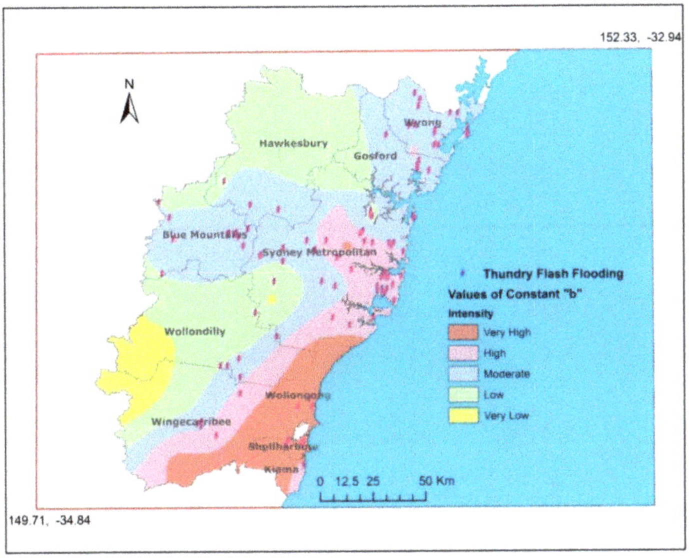

Figure 5. Spatial distribution of the constant "*b*" (shaded) with classification into four groups of intensity and occurrence locations of the thundery flash flooding days during 1989–2015 (lightning symbol) based on the BoM storm archive.

4.2. Spatial Correspondences

A cross−covariance model was presented to assess the spatial dependence (covariance) between two vector-based datasets. Here the first dataset is the DPCI, while the second is one of the important rainfall-related parameters such as mean annual rainfalls (AP), coefficient of variation (CV), the total number of rainfall days (TN), maximum rainfalls observed (MxR) and the "*a*" and "*b*" constants taken from Equation (1) that control the concentration curve. In the analysis, the attribute of one point (i.e., the DPCI) is correlated with the second attribute (i.e., one of the rain−related parameters) at another point, and this is repeated for all pairs of geographic points. The spatial distribution of the correlation (termed cross-covariance surface or cloud) can then be applied to examine the local characteristics of spatial correlation between the two attributes (datasets). The details of this cross-covariance model has been documented in Appendix C. This technique was applied to look for spatial shifts in existing correspondences between the DPCI and the other datasets throughout the GSMA.

A covariance surface with directional search capabilities was also involved in the modeling. For this reason, the values in the cross−covariance cloud were put into six bins based on the direction and distance separating a pair of locations [37]. These binned values were then averaged and smoothed to produce a cross−covariance surface (and associated correlations) for each pair of dataset throughout the study area (Figure 6). It can be seen that the DPCI possesses regions of high covariance with most of the rain parameters and also the "*a*" and "*b*" constants in the concentration curve, however, there is variability in the locations with the highest covariance. For example, the DPCI has high covariance with the climatological parameters AP and TN over the northwest. The highest covariance with the 'magnitude' of the concentration curve ("*a*") is also on the western side. This may be due to the topographic variation. However, the pattern of covariance with the two parameters directly related to the DPCI, namely the MxR and "*b*", has a southwest−northeast orientation. This is aligned with the distribution of the DPCI in Figure 4.

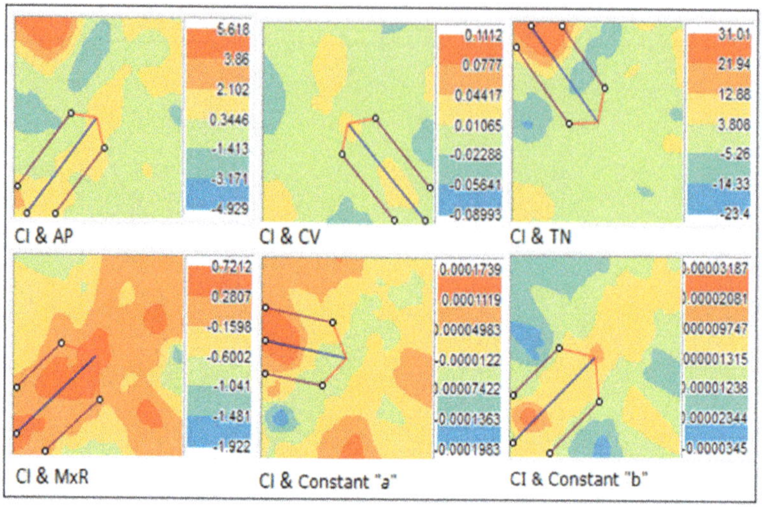

Figure 6. The result of cross−covariance surfaces for all pairs of variables: (CI,AP); (CI,CV); (CI,TN); (CI,MxR); (CI,a); (CI,b). CI is the same as the DPCI. Six bins have been set for the categories of covariance values. The arrows (with the blue and red lines) are examples of directional searches of high covariance values over the surfaces. See Appendix C for details.

In Table 2, the values of Pearson's correlation coefficient for the five pairs of variables are indicated. The Pearson overall correlations (r) for all rainfall related parameters, except the total number of rainy days (TN), are statistically significant at 0.95 and 0.99 levels, respectively. Correlation between TN and annual DPCI is nearly +0.24 ($p < 0.5$) and not significant; in other words high number of rain days is not a good indicator of high DPCI. The reason is that similar annual values could be achieved with different daily distributions.

Table 2. Values of the Pearson's correlation coefficient, significance level and category of cross−covariance spatial shifting for five pairs of variables: (CI,AP); (CI,CV); (CI,TN); (CI,MxR); (CI,a); (CI,b).

Parameter to Correlate with DPCI	Correlation Coefficient (r)	Significance Level (p)	Cross−Covariance Spatial Shifting
Mean annual precipitation (AP)	0.472	0.01	High
CV of annual rainfall	0.314	0.05	Medium
Total number of rainy days (TN)	0.239	Non−significant	Very low
Maximum daily rainfall (MxR)	0.508	0.01	High
"a"	−0.701	0.01	Very high
"b"	0.703	0.01	Very high

5. Conclusions and Discussion

5.1. Summary

In the current study, daily rainfall observations (1950–2015) from 41 rainfall stations inside the GSMA have been analyzed. According to the applied criteria and techniques used, the outcomes are summarized as follow:

- Within the GSMA, the essential features of climate in different districts are characterized by narrow rainfall zones close to the coast, under the combined influence of the Tasman Sea and the topography and land use patterns, leading to very different rainfall spatial distributions.
- The DPCI values in the Illawarra coastal elevated areas, parts of the Sydney Metropolitan area and the Blue Mountains are high, with concentration index values close to 0.60–0.63. This reflects the fact that very few rainy days could bring a high percentage of annual precipitation.
- The DPCI values obtained and distribution pattern of constant "*b*" are largely subject to influences from the topography and land use of the region. Generally, western and central regions inside GSMA are areas where rainfall is regular compared to eastern regions, while the southeastern districts and small parts of Metropolitan areas show the most aggressive DPCI values.
- Despite the significant variations in spatial cross−correlating models between the DPCI and 6 other rain−related parameters (AP, CV, TN, MxR, "*a*" and "*b*"), there are considerable positive relationships among data layers at 0.95 significance levels for most parts of the study area.
- The spatial patterns of the DPCI and "*b*" constant highlight the importance of catastrophic effects of such intense rainfall events, predominantly originating with severe thunderstorm and flash flood events.

5.2. Discussion

Inside of the Sydney Metropolitan area, daily precipitation is one of the factors in the processes of creating flash floods, and accordingly, differences in the spatial distribution of precipitation can lead to dissimilar precipitation regimes and various climatic conditions [38]. As was indicated in Table A1 (Appendix A), even if the annual total amounts are similar in many of the rainfall stations, precipitation processes may be different due to a different degree of concentrated rainfall in the time and space of the study area. Accordingly, the spatial distribution of precipitation can produce noticeably different impacts on natural and social processes across the GSMA—of particular interest for water management—flood control programs, and water availability for natural ecosystems. As the results show, the daily concentration of precipitation on an annual scale (expressed by the DPCI values in Table 1) is characterized by two different spatial gradients. One lies from the east to the west and the second is detectable from south to north, the latter characterized by the Tasman Sea coastal areas.

Overall, the spatial distribution of DPCIs follows a gradient between inland and the coastal areas, which may indicate approaching intense rainfall from different geographic directions. The results in this study have indicated that most parts inside GSMA are subject to severe rain, but with different likelihood of high DPCI (Figure 4). For example, the gigantic water resources of the Tasman Sea may influence the distribution of intense rainfall. On the other hand, a large proportion of rainfall comes from severe thunderstorms that occur over the northeast GSMA, the CBD and over the inner metropolitan area [26]. Also, the increased roughness associated with variation in topography and heat island phenomena may affect the spatial distribution of concentrated precipitation [39].

However, the pronounced dissimilar DPCI values and the subsequent cross−covariance surfaces (Figures 4 and 6) support the overall picture of multi−subjected developing areas and approaching weather systems from various directions in the region, which are under dissimilar synoptic patterns causing atmospheric instability [40,41]. It was found by previous studies that at least four types of weather patterns account for most of the rainfall in the region [42,43], and logically the amount, frequency, and intensity of precipitation events vary substantially in the region, as shown in the records during a long period from 1950–2015 [44,45]. Another weather pattern occurs in summer and involves the location of the Tropical Convergence Zone bringing torrential rainfalls [46]. Occasionally, weather systems from the southeast generate storms striking the region with torrential precipitation.

During the warm months (October to March) the prevailing easterly moist winds provide much of the moisture needed in the intensification of widespread and severe thunderstorm activity in the region [26]. Given the short duration of typical thunderstorm activity in terms of hours, likely it would contribute substantially to the DPCI.

Not all variations in the total precipitation and associated differences in the DPCI can be explained simply in terms of differences between dissimilar weather systems and the nature of the prevailing air masses [47]. The geographical distribution of DPCI and the *"b"* values illustrate that the coastal areas are subject to a high probability of intense rainfall (Figure 5). In the southwest extension of the coastal area, over the Illawarra Escarpment, topography has clear influences on the rainfall amounts. The high *"b"* distribution in the vicinity of elevated topography of the Illawarra Escarpment suggests an orographic enhancement of instability, particularly for sites facing the east (as indicated by Figure 4). Similarly, in the highland area west of Sydney, there appears at least two different patterns of intense rainfall events. The Blue Mountain ranges, located at the northwest of the study area, have some of the highest DPCI values, particularly in the summer months. Thus, the issue arises whether the limited number of rain stations, especially over the high mountains, can capture such topographic effect to extreme precipitation adequately. One way to improve is to extend the data sources representing rainfall distribution, which may include a radar−based estimate and gridded reanalysis dataset, the latter able to reduce the uncertainty during the spatial interpolation process. The other method is to incorporate theoretical topographic rainfall models (e.g., [48,49]) to improve the representation of extreme precipitation over high elevations.

Internationally, the values of CIs found across Europe are similar to those described in Iran by [6] and are lower than those offered by [7] in China. It has been proposed by [7] as a general explanation for differences between results from [27] in the Iberian Peninsula and China, that different climate systems and precipitation mechanisms were responsible for rainfall (such as a typhoon). Generally, it has been suggested that precipitation trends based on annual maximum daily events observed in most parts of the world have nearly the same signs. However, the trend of heavy precipitation is disproportionately larger than the trend of the total [50]. Some of the previous investigations and the more recent work of [5] demonstrated the prominence and precision of CI applications in different parts of the world. It was suggested that even without any change in total precipitation, there may be changes in the frequency of intense daily precipitation in a climate change context; a fact that would have led to meaningful variations in the precipitation concentration patterns [51–53].

Author Contributions: Conceptualization, K.K.W.C. and A.A.R.; methodology, K.K.W.C. and A.A.R.; software, A.A.R.; validation, K.K.W.C., A.A.R., F.J. and L.T.-C.C.; formal analysis, A.A.R.; investigation, K.K.W.C.; resources, F.J.; data curation, L.T.-C.C.; writing—original draft preparation, K.K.W.C. and A.A.R.; writing—review and editing, F.J. and L.T.-C.C.; visualization, A.A.R.; supervision, K.K.W.C. and A.A.R.; project administration, K.K.W.C. All authors have read and agreed to the published version of the manuscript.

Funding: This research received no external funding.

Informed Consent Statement: Not applicable.

Data Availability Statement: The data applied in this study is acquired from the Bureau of Meteorology, Australia and available from the authors.

Acknowledgments: A.A.R. would like to acknowledge the honorary fellowship from Macquarie University, Australia during part of this study.

Conflicts of Interest: The authors declare no conflict of interest.

Appendix A

The tables in this appendix document basic information of the rain stations and rainfall statistics in this study (Table A1), and parameters for computing the DPCI using Albion Park station as an example (Table A2).

Table A1. The geographic coordinates, study period, average annual rainfall (AP), coefficient of variation (CV) and total number of rainy days (TN) for the 41 rain stations across the GSMA.

Station Name (With Abbreviated Map Codes)	Latitude	Longitude	Altitude (m)	Study Period (year)	AP (mm)	CV (%)	TN
Albion Park (AP)	−4.57	150.78	8	1950–2015	1186.9	35.1	4811
Bankston Airport (BA)	−33.92	150.99	6.5	1969–2015	874.6	25.1	3910
Berambing (BE)	−33.54	150.44	792	1950–2015	1452.1	27.7	6774
Bilpin (BI)	−33.52	150.49	610	1950–2015	1363.2	26.2	7224
Blackheath (BL)	−33.63	150.29	1060	1950–2015	1240.6	28.3	6395
Bowral (BO)	−34.49	150.40	690	1962–2015	931.7	26.6	5041
Bringelly (BR)	−33.97	150.73	122	1950–2015	803.3	32.9	4496
Bundanoon (BU)	−34.65	150.31	688	1950–2015	1249.6	30.7	5894
Camden (CA)	−34.03	150.65	61	1950–2015	805.6	32.7	4964
Colo Heights (CH)	−33.36	150.71	320	1963–2015	1034.4	26.7	4991
Dapto (DA)	−34.50	150.79	10	1950–2015	1232.0	35.4	4822
Darkes Forest (DF)	−34.23	150.91	370	1950–2015	1558.4	31.1	6994
Faulconbridge (FA)	−33.69	150.53	460	1950–2015	1225.3	34.7	3351
Frenchs Forest (FF)	−33.75	151.23	158	1957–2015	1374.9	25.5	6012
Glenorie (GL)	−33.59	151.01	170	1950–2015	1002.7	27.7	5249
Katoomba (KA)	−33.71	150.31	1015	1950–2015	1449.9	27.8	7479
Kiama (KI)	−34.68	150.85	10	1950–2011	1332.9	31.1	5689
Kurrajong Heights (KH)	−33.53	150.63	460	1950–2015	1278.8	28.9	6347
Lucas Heights (LH)	−34.05	150.98	140	1958–2015	1021.9	26.6	5090
Maroota (MA)	−33.46	151.00	203	1950-2015	952.9	31.7	4062
Moss Vale (MV)	−34.54	150.38	675	1950–2015	961.6	30.1	6106
Mount Kuring-Gai (MK)	−33.64	151.14	215	1965–2015	1171.1	28.7	4560
Ourimbah (OU)	−33.36	151.33	195	1954–2015	1406.5	25.3	6183
Parramatta North (PN)	−33.79	151.02	55	1966–2015	970.6	26.9	4554
Picton (PI)	−34.17	150.61	165	1950–2015	886.5	32.8	4224
Port Kembla (PK)	−34.47	150.88	9	1964–2015	1119.9	30.4	4440
Prospect Reservoir (PR)	−33.82	150.91	61	1950–2015	936.4	28.6	5692
Richmond (RI)	−33.62	150.75	20	1950–2015	868.8	29.5	5287
Riverview Observatory (RO)	−33.83	151.16	40	1950–2015	1204.2	26.6	4946
Sans Souci (SS)	−33.99	151.13	9	1950–2015	1153.5	28.3	6728
Springwood (SW)	−33.71	150.58	320	1950–2015	1166.5	30.0	5727
Sydney Airport (SA)	−33.95	151.17	6	1950–2015	1123.0	27.5	6384
Sydney Observatory Hill (SO)	−33.86	151.21	39	1950–2015	1264.7	26.8	6607
The Entrance (EN)	−33.35	151.50	22	1950–2015	1176.8	25.6	5889
Wallacia (WA)	−33.86	150.64	50	1950–2015	890.9	33.0	5027
West Pennant Hills (WP)	−33.75	151.04	120	1950–2014	1115.2	30.8	4858
Wollondilly (WO)	−34.34	150.08	270	1974–2015	692.1	25.8	2781
Wombeyan Caves (WC)	−34.31	149.97	580	1952–2015	847.8	22.7	4466
Woonona (WN)	−34.34	150.90	45	1950–2015	1328.3	32.5	5679
Wyee (WY)	−33.20	151.44	40	1950–2015	1250.9	24.8	6293

Table A2. Frequency distribution (Ni, for rain >1 mm), total precipitation (Pi), relative cumulative frequencies (X) and percentage of total precipitation (Y) for the Albion Park station.

Classes	Midpoint	Ni	ΣNi	Pi	Σpi	ΣN (%) = X	Σpi (%) = Y
1–10	5	3037	3037	15,185	15,185	63.13	19.43
10.1–20	15	796	3833	11,940	27,125	79.67	34.70
20.1–30	25	367	4200	9175	36,300	87.30	46.44
30.1–40	35	187	4387	6545	42,845	91.19	54.81
40.1–50	45	109	4496	4905	47,750	93.45	61.09
50.1–60	55	71	4567	3905	51,655	94.93	66.08
60.1–70	65	52	4619	3380	55,035	96.01	70.41
70.1–80	75	38	4657	2850	57,885	96.80	74.05
80.1–90	85	22	4679	1870	59,755	97.26	76.45
90.1–100	95	30	4709	2850	62,605	97.88	80.09
100.1–110	105	20	4729	2100	64,705	98.30	82.78
110.1–120	115	12	4741	1380	66,085	98.55	84.55
120.1–130	125	8	4749	1000	67,085	98.71	85.82
130.1–140	135	12	4761	1620	68,705	98.96	87.90
140.1–150	145	13	4774	1885	70,590	99.23	90.31
150.1–200	175	25	4799	4375	74,965	99.75	95.91
200.1–250	225	8	4807	1800	76,765	99.92	98.21
250.1–300	275	2	4809	550	77,315	99.96	98.91
300.1–350	325	1	4810	325	77,640	99.98	99.33
500.1–550	525	1	4811	525	78,165	100.00	100.00
Sum		4811		78,165		1890.96	1507.27

Appendix B

Figure A1 in this appendix illustrates the physical interpretation of the Global Moran's I Statistic. Depending on the z-score value, the rainfall distribution changes from a dispersed pattern (negative extreme), random pattern (most of the z-score around the mean) to a clustered pattern (positive extreme). The values of the Moran's Index (−0.001755), z-score (3.553463) and the p-value (0.000380) based on the dataset in this study are given in the upper left corner of the figure.

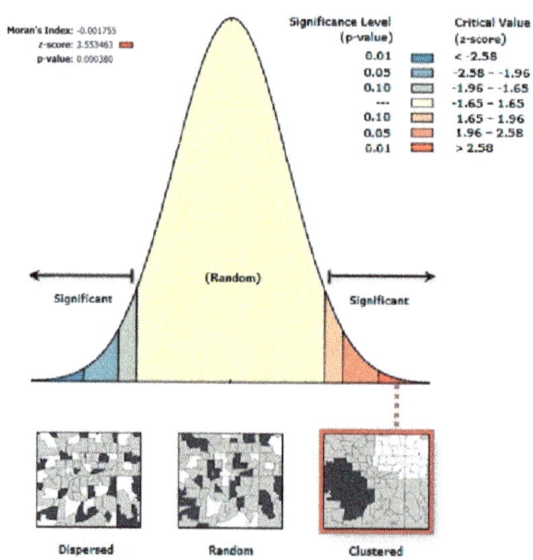

Figure A1. Illustration of the Global Moran's statistic and association with the dispersed, random and clustered precipitation patterns.

Appendix C

In this appendix the details in the procedure of performing spatial cross-covariance analysis between two attributes (datasets) are documented. An example of the DPCI (first attribute) and the constant "*b*" (second attribute) is illustrated in Figure A2. There are several steps in the analysis:

- The empirical cross-covariance for a pair of locations (NSW rainfall stations) between two datasets (DPCI and "*b*") is first plotted as a function of the distance between the two locations (Figure A2 upper panel). In this illustration, each red dot shows the empirical cross-covariance between the pair of stations, with the attribute of one station taken from the first dataset and the attribute of the second station taken from the second dataset. The Cross-covariance cloud can be used to examine the local characteristics of spatial correlation between two datasets, and it can be used to look for spatial shifts in the correlation between two datasets. A cross-covariance cloud looks something like the NSW example.
- The values in the cross-covariance cloud are put into bins based on the direction and distance separating a pair of locations. These binned values are then averaged and smoothed to produce a cross-covariance surface. The legend (Figure A2 **lower panel left**) shows the colors and values separating classes of covariance values.
- A covariance surface with search direction capabilities is also provided in the ArcGIS tool. The extent of the cross-covariance surface is controlled by the lag size and number of lags that are specified (Figure A2 lower panel right). The search direction and width are indicated by the blue and red lines over the cross-covariance surface. One example has been shown in the figure and these options can be modified.

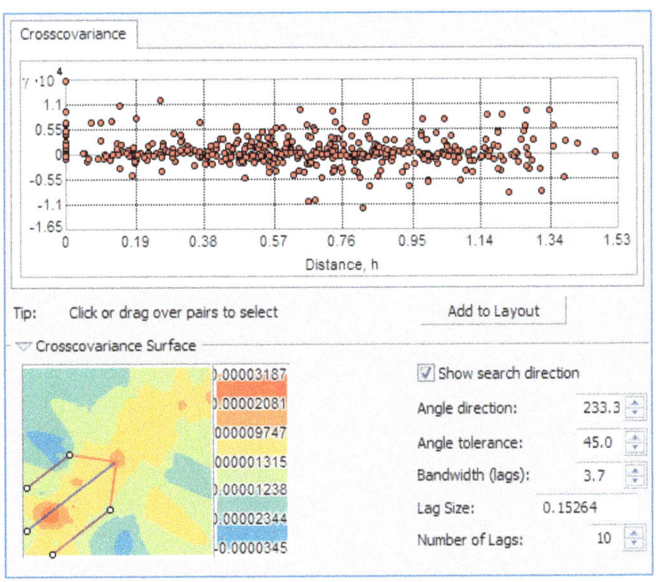

Figure A2. Cross-covariance surface or cloud between the DPCI and the constant "*b*" in the concentration curve. The upper panel has all the covariance values according to the distance between the two points for computing the covariance. The lower panel has the cross-covariance map and an example of the directional search arrow. The search arrow is set based on the parameters on the right hand side (such as the direction, lag size and number of lags).

References

1. Rasuly, A.A.; Cheung, K. The Spatial Distribution of Severe Thunderstorm Rainfall Events throughout the GMSTWA and Adjacent Tasman Sea. In *Proceedings of the the 7th Annual CAWCR Workshop, Observing, Estimating and Forecasting Rainfall: From Science to Applications, Melbourne, Australia, 21–23 October 2013*; Centre for Australian Weather and Climate Research: Melbourne, Australia, 2013.
2. Monjo, R.; Martin-Vide, J. Daily precipitation concentration around the world according to several indices. *Int. J. Climatol.* **2016**, *36*, 3828–3838. [CrossRef]
3. Máyer, P.; Jaén, M.V.M. Daily precipitation concentration and the rainy spells in the Canary Islands: Two risk factors. *Boletín de la Assoc. de Geogr. Espaanoles* **2014**, *65*, 463–468.
4. Martin-Vide, J. Geographical factors in the pluviometry of Mediterranean Spain: Drought and torrential rainfall. In *The U.S.-Spain Workshop on Natural Hazards, Iowa. Inst. Hydra Res.*; The University of Iowa: Iowa City, IA, USA, 1994; pp. 9–25.
5. Benhamrouche, A.; Boucherf, D.; Hamadache, R.; Bendahmane, L.; Martin-Vide, J.; Teixeira Nery, J. The spatial distribution of the daily precipitation concentration index in Algeria. *Nat. Hazards Earth Syst. Sci.* **2015**, *15*, 617–625. [CrossRef]
6. Alijani, B.; O'Brien, J.; Yarnal, B. Spatial analysis of precipitation intensity and concentration in Iran. *Theor. Appl. Climatol.* **2008**, *94*, 107–124. [CrossRef]
7. Zhang, Q.; Xu, C.Y.; Gemmer, M.; Chen, Y.Q.; Liu, C.L. Changing properties of precipitation concentration in the Pearl River basin, China. *Stoch. Env. Res. Risk Assess.* **2009**, *23*, 377–385. [CrossRef]
8. Suhaila, J.; Jemain, A.A. Spatial analysis of daily rainfall intensity and concentration index in Peninsular Malaysia. *Theor. Appl. Climatol.* **2012**, *108*, 235–245. [CrossRef]
9. Cortesi, N.; Gonzalez-Hidalgo, J.C.; Brunetti, M.; Martin-Vide, J. Daily precipitation concentration across Europe 1971–2010. *Nat. Hazards Earth Syst. Sci.* **2012**, *12*, 2799–2810. [CrossRef]
10. Coscarelli, R.; Caloiero, T. Analysis of daily and monthly rainfall concentration in Southern Italy (Calabria region). *J. Hydrol.* **2012**, *416–417*, 145–156. [CrossRef]
11. Caloiero, T. Analysis of daily rainfall concentration in New Zealand. *Nat. Hazards* **2014**, *72*, 389–404. [CrossRef]
12. Voskresenskaya, E. Precipitation inequality over Ukraine. *J. Sci. Res. Rep.* **2014**, *3*, 384–396.
13. Sarricolea, P.; Martín-Vide, J. Spatial analysis of rainfall daily trends and concentration in Chile. *Investig. Geogr.* **2014**, *47*, 53–66. [CrossRef]
14. Zubieta, R.; Saavedra, M.; Silva, Y.; Gira'Idez, L. Spatial analysis and temporal trends of daily precipitation concentration in the Mantaro River basin: Central Andes of Peru. *Stoch. Env. Res. Risk Assess.* **2017**, *31*, 1305–1318. [CrossRef]
15. AL-Shamarti, H.K.A. The variation of annual precipitation and precipitation concentration index of IRAQ. *IOSR J. Appl. Phys.* **2016**, *8*, 36–44. [CrossRef]
16. Ezenwaji, E.E.; Nzoiwu, C.P.; Chima, G.N. Analysis of precipitation concentration index (PCI) for Awka urban area, Nigeria. *Hydrol. Current Res.* **2017**, *8*, 4. [CrossRef]
17. Nandargi, S.S.; Aman, K. Precipitation concentration changes over India during 1951 to 2015. *Sci. Res. Essays* **2018**, *13*, 14–26. [CrossRef]
18. Zhang, K.; Yao, Y.; Qian, X.; Wang, J. Various characteristics of precipitation concentration index and its cause analysis in China between 1960 and 2016. *Int. J. Climatol.* **2019**, *39*, 4648–4658. [CrossRef]
19. Patel, N.R.; Shete, D.T. Analyzing Precipitation Using Concentration Indices for North Gujarat Agro Climatic Zone, India. In *Proceedings of the International Conference on Water Resources, Coastal and Ocean Engineering (ICWRCOE), Mangalore, Karnataka, Indian*; Elsevier Procedia: Amsterdam, The Netherlands, 2015; Volume 4, pp. 917–924.
20. Rasuly, A.A. The spatial variation and distribution of thunderstorm rainfall in the Greater Sydney Region. In Proceedings of the International Conference on Storms, Brisbane, Australia, 4–9 July 2004.
21. Zhang, X.; Alexander, L.; Hegerl, G.C.; Jones, P.; Klein Tank, A.; Peterson, T.C.; Trewin, B.; Zwiers, F.W. Indices for monitoring changes in extremes based on daily temperature and precipitation data. *WIREs Clim. Chang.* **2011**, *2*, 851–870. [CrossRef]
22. Kalyan, A.V.S.; Ghose, D.K.; Thalagapu, R.; Guntu, R.K.; Agarwal, A.; Kurths, K.; Rathinasamy, M. Multiscale spatiotemporal analysis of extreme events in the Gomati River basin, India. *Atmosphere* **2021**, *12*, 480. [CrossRef]
23. Griffiths, D.J.; Colquhoun, J.R.; Batt, K.L.; Casinader, T.R. Severe thunderstorms in New South Wales: Climatology and means of assessing the impact of climate change. *Clim. Chang.* **1993**, *25*, 369–388. [CrossRef]
24. Risbey, J.S.; Pook, M.J.; McIntosh, P.C.; Wheeler, M.C.; Hendon, H.H. On the remote drivers of rainfall variability in Australia. *Mon. Wea. Rev.* **2009**, *137*, 3233–3253. [CrossRef]
25. Rasuly, A.; Cheung, K. Applying a climatologically oriented GIS in comparison of TRMM estimated severe thunderstorm rainfalls with ground truth in Sydney metropolitan area. *J. Appl. Hydrol.* **2014**, *1*, 1–13.
26. Rasuly, A.A.; Cheung, K.; McBurney, B. Hail events across the Greater Metropolitan Severe Thunderstorm Warning Area. *Nat. Haz. Earth Syst. Sci.* **2015**, *15*, 973–984. [CrossRef]
27. Martin-Vide, J. The spatial distribution of a daily precipitation concentration index in Peninsular Spain. *Int. J. Climatol.* **2004**, *24*, 959–971. [CrossRef]
28. Brooks, C.E.P.; Carruthers, N. Handbook of Statistical Methods in Meteorology. Her Majesty's Stationery Office, M.O.538.: London, UK, 1953; p. 412.
29. Mitchell, A. *The ESRI Guide to GIS Analysis*; ESRI Press: Redlands, CA, USA, 2005; Volume 2.

30. Karnieli, A. Application of Kriging technique to areal precipitation mapping in Arizona. *GeoJournal* **1990**, *22*, 391–398. [CrossRef]
31. Ly, S.; Charles, C.; Degre, A. Different methods for spatial interpolation of rainfall data for operational hydrology and hydrological modeling at watershed scale: A review. *Biotechnol. Agron. Soc. Environ.* **2013**, *17*, 392–406.
32. Yang, X.; Xie, X.; Liu, D.L.; Ji, F.; Wang, L. Spatial interpolation of daily rainfall data for local climate impact assessment over Greater Sydney region. *Adv. Meteorol.* **2015**. [CrossRef]
33. Cressie, N. The origins of Kriging. *Math. Geol.* **1990**, *21*, 239–252. [CrossRef]
34. Adhikary, S.K.; Muttil, N.; Yilmaz, A.G. Cokriging for enhanced spatial interpolation of rainfall in two Australian catchments. *Hydrol. Process.* **2017**, *31*, 2143–2161. [CrossRef]
35. Usowicz, B.; Lipiec, J.; Lukowski, M.; Sloninski, J. Improvement of spatial interpolation of precipitation distribution using cokriging incorporating rain-gauge and satellite (SMOS) soil moisture data. *Remote Sens.* **2021**, *13*, 1039. [CrossRef]
36. Zhang, P.; Huang, Y.; Shekhar, S.; Kumar, V. Exploiting spatial autocorrelation to efficiently process correlation-based similarity queries. In *Lecture Notes in Computer Science*; Springer: Berlin/Heidelberg, Germany, 2003; Volume 2750.
37. Haining, R. *Spatial Data Analysis: Theory and Practice*; Cambridge University Press: Cambridge, UK, 2003.
38. Johnson, F.; White, C.J.; van Dijk, A.; Ekstrom, M.; Evans, J.P.; Jakob, D.; Kiem, A.S.; Leonard, M.; Rouillard, A.; Westra, S. Natural hazards in Australia: Floods. *Clim. Chang.* **2016**, *139*, 21–35. [CrossRef]
39. Walsh, K.; White, C.J.; McInnes, K.; Holmes, J.; Schuster, S.; Richter, H.; Evans, J.P.; Luca, A.D.; Warren, R.A. Natural hazards in Australia: Storms, wind and hail. *Clim. Chang.* **2016**, *139*, 55–67. [CrossRef]
40. Hopkins, L.C.; Holland, G. Australian Heavy-rain days and associated east coast cyclones: 1958–1992. *J. Clim.* **1997**, *10*, 621–635. [CrossRef]
41. Pepler, A.S.; Dowdy, A.J.; Rensch, P.; Rudeva, I.; Catto, J.L.; Hope, P. The contributions of fronts, lows and thunderstorms to southern Australian rainfall. *Clim. Dyn.* **2020**, *55*, 1489–1505. [CrossRef]
42. Speer, M.; Wiles, P.; Pepler, A. Low pressure systems off the New South Wales coast and associated hazardous weather: Establishment of a database. *Aust. Meteorol. Oceanogr. J.* **2009**, *58*, 29–39. [CrossRef]
43. Risbey, J.S.; McIntosh, P.C.; Pook, M. Synoptic components of rainfall variability and trends in southeast Australia. *Intl. J. Climatol.* **2013**, *33*, 2459–2472. [CrossRef]
44. Bureau of Meteorology. *Stormy Weather, A Century of Storms, Flood and Drought in New South Wales*; Australian Government: Melbourne, Australia, 2011; p. 40.
45. Dare, R.A.; Davidson, N.E.; McBride, J.L. Tropical cyclone contribution to rainfall over Australia. *Mon. Wea. Rev.* **2012**, *140*, 3606–3619. [CrossRef]
46. Lavender, S.L.; Abbs, D.J. Trends in Australian rainfall: Contribution of tropical cyclones and closed lows. *Clim. Dyn.* **2013**, *40*, 317–326. [CrossRef]
47. Pepler, A.; Timbal, B.; Rakich, C.; Coutts-Smith, A. Indian ocean dipole overrides ENSO's influence on cool season rainfall across the Eastern seaboard of Australian. *J. Clim.* **2014**, *27*, 3816–3826. [CrossRef]
48. Chang, L.T.-C.; Cheung, K.K.W.; McAneney, J. Case Study of TRMM satellite rainfall estimation for landfalling tropical cyclones: Issues and challenges. *Trop. Cycl. Res. Rev.* **2013**, *2*, 109–123.
49. Abbate, A.; Papini, M.; Longoni, L. Extreme rainfall over complex terrain: An application of the linear model of orographic precipitation to a case study in the Italian pre-alps. *Geosciences* **2021**, *11*, 18. [CrossRef]
50. Easterling, D.R.; Evans, J.L.; Groisman, P.Y.; Karl, T.R.; Kunkel, K.E.; Ambenje, P. Observed variability and trends in extreme climate events: A brief review. *Bull. Amer. Meteor. Soc.* **2000**, *81*, 417–426. [CrossRef]
51. Allan, R.J.; Haylock, M.R. Circulation Features Associated with the winter rainfall decrease in Southwestern Australia. *J. Clim.* **1993**, *6*, 1356–1367. [CrossRef]
52. Alexander, L.V.; Wang, X.L.L.; Wan, H.; Trewin, B. Significant decline in storminess over southeast Australia since the late 19th century. *Aust. Meteorol. Oceanogr. J.* **2011**, *61*, 23–30. [CrossRef]
53. Dey, R.; Lewis, S.; Abram, N.; Arblaster, J. A review of past and projected changes in Australia's rainfall. *Wiley Interdiscip. Rev.* **2019**, *10*. [CrossRef]

atmosphere
MDPI

Article

Multiscale Spatiotemporal Analysis of Extreme Events in the Gomati River Basin, India

AVS Kalyan [1,2], Dillip Kumar Ghose [2], Rahul Thalagapu [1], Ravi Kumar Guntu [3], Ankit Agarwal [3,4], Jürgen Kurths [5,6,7] and Maheswaran Rathinasamy [1,*]

[1] Department of Civil Engineering, MVGR College of Engineering, Vizianagaram 535005, Andhra Pradesh, India; kalyanprime@mvgrce.edu.in (A.K.); rahulthalagapu222@gmail.com (R.T.)
[2] Department of Civil Engineering, National Institute of Technology Silchar, Silchar 788010, Assam, India; dillipghose_2002@yahoo.co.in
[3] Department of Hydrology, Indian Institute of Technology Roorkee, Roorkee Uttarakhand 247667, India; guntu_r@hy.iitr.ac.in (R.K.G.); agarwal@pik-potsdam.de (A.A.)
[4] GFZ German Research Centre for Geosciences, Section 4.4: Hydrology, Telegrafenberg, 14473 Potsdam, Germany
[5] Research Domain Transdisciplinary Concepts and Methods, Potsdam Institute for Climate Impact Research, Telegrafenberg, 14412 Potsdam, Germany; juergen.kurths@pik-potsdam.de
[6] Institute of Physics, Humboldt Universität zu Berlin, 10117 Berlin, Germany
[7] Institute of Information, Technology, Mathematics and Mechanics, Lobachevsky University of Nizhny Novgorod, 603950 Nizhnij Novgorod, Russia
* Correspondence: maheswaran27@yahoo.co.in

check for updates

Citation: Kalyan, A.; Ghose, D.K.; Thalagapu, R.; Guntu, R.K.; Agarwal, A.; Kurths, J.; Rathinasamy, M. Multiscale Spatiotemporal Analysis of Extreme Events in the Gomati River Basin, India. *Atmosphere* **2021**, *12*, 480. https://doi.org/10.3390/atmos12040480

Academic Editor: Peter Hoffmann

Received: 23 February 2021
Accepted: 7 April 2021
Published: 9 April 2021

Abstract: Accelerating climate change is causing considerable changes in extreme events, leading to immense socioeconomic loss of life and property. In this study, we investigate the characteristics of extreme climate events at a regional scale to -understand these events' propagation in the near future. We have considered sixteen extreme climate indices defined by the World Meteorological Organization's Expert Team on Climate Change Detection and Indices from a long-term dataset (1951–2018) of 53 locations in Gomati River Basin, North India. We computed the present and future spatial variation of theses indices using the Sen's slope estimator and Hurst exponent analysis. The periodicities and non-stationary features were estimated using the continuous wavelet transform. Bivariate copulas were fitted to estimate the joint probabilities and return periods for certain combinations of indices. The study results show different variation in the patterns of the extreme climate indices: D95P, R95TOT, RX5D, and RX showed negative trends for all stations over the basin. The number of dry days (DD) showed positive trends over the basin at 36 stations out of those 17 stations are statistically significant. A sustainable decreasing trend is observed for D95P at all stations, indicating a reduction in precipitation in the future. DD exhibits a sustainable decreasing trend at almost all the stations over the basin barring a few exceptions highlight that the basin is turning drier. The wavelet power spectrum for D95P showed significant power distributed across the 2–16-year bands, and the two-year period was dominant in the global power spectrum around 1970–1990. One interesting finding is that a dominant two-year period in D95P has changed to the four years after 1984 and remains in the past two decades. The joint return period's resulting values are more significant than values resulting from univariate analysis (R95TOT with 44% and RTWD of 1450 mm). The difference in values highlights that ignoring the mutual dependence can lead to an underestimation of extremes.

Keywords: extreme climate indicators; bi-variate copula; Hurst exponent; wavelet transform

1. Introduction

The adverse effects of climate change associated with global warming have been disrupting various natural processes with a visible impact on ecological, economic, and social aspects. Particularly, the extreme climate events have become sensitive to the climate change and have become more hazardous in the recent decades. Several studies have shown that there is a significant increase in the frequency and intensity of the extreme climate events [1–4]. Numerous studies [5–12] have concluded that there is a considerable increase in the hot days, warm nights at several places across the globe. Similarly, there has been considerable evidence [13–20] wherein spatio-temporal changes in extreme precipitation are observed in addition to the increasing temperature which have led to extreme droughts, flood conditions and heat waves.

In this context, observation, detection, and detailed analysis of extreme climate events have become imperative. The detection of extreme climate changes requires the use of specific indices [16]. Different extreme indices have evolved and are now focused on relative thresholds related to the tails of distributions of meteorological variables [17]. Out of the several indices, the Expert Team on Climate Change Detection and Indices (ETCCDI) have selected around 27 indices as prominent ones [5].

Studies on extreme climate events/change using these indices proposed by ETCCDI are highly relevant and increasing. Garcia-Cueto et al., [18] observed 4–5 °C and 2–3 °C rise of summer temperature increases in Northwestern Mexico under RCP 4.5 and RCP8.5, respectively. Wang et al., [19] examined the spatiotemporal trends of temperature and precipitation indices proposed by ETCCDI in the Loess Plateau region of China and observed significant warming and drying trends at all temperature and precipitation extremes, respectively. The application of ETCCDI indices in India's upper Tapi river basin has exhibited increasing trends in hottest and coldest days and decreasing trends in coldest nights in most portions of the basin [20]. In contrast, mixed trends were observed in 1-day, five-day precipitation totals. Sharma et al., [20] analyzed the spatiotemporal variation in extreme precipitation and temperature at daily scale across India using the ETCCDI indices. They observed that the number of 'warm days' per year increased significantly, while the number of 'cold days', 'warm nights' and 'cold nights' per year decreased significantly at several locations.

Even though previous studies have dealt with extreme events, there have been only very few studies based on joint probability characteristics of extreme events. It is imperative to investigate the joint occurrence of these extreme events in the context of global warming.

The Gomati River Basin has a monsoonal climate with a high variation in precipitation. Abeysingha et al. [21] showed a significant decrease in precipitation and an increase in temperature during the last century, making it one of India's hotspots region. Even though they investigated the spatiotemporal trends in the total precipitation and temperature, there is no investigation on extreme events. Therefore, in this study, we have analyzed the joint distribution of bivariate climate extreme events which is of great significance for water resources management. Further, in this study, selected extreme indices' spatiotemporal characteristics were investigated using Sen's slope, the Hurst exponent, and the wavelet transform. The results will provide a scientific basis for future prediction of extreme events and help in disaster mitigation and prevention.

2. Study Area and Data

2.1. Gomati Basin

The Gomati river basin is bounded between longitude 80°0′10″ E-83°11′4″ E and latitude 25°31′16″ N-28°53′17″ N in the Doab region of the Ganga and Ghaghara river basin covering a total area of 30,437 km^2 [21]. It is one of the main tributaries of River Ganga originating from Fulhaar Jheel lake near Mainkot of about 30 km east of the Pilibhit town of the Uttar Pradesh (see Figure 1). The river basin is mainly comprised of the alluvial stratum. The river is the primary source of supply for the cities of Lucknow and Jaunpur in Uttar Pradesh. The entire basin's climate varies from semi-arid to sub-humid tropical with an average annual precipitation of 850–1100 mm. The basin receives about 75% of the total annual precipitation between June and September due to the southwest monsoon.

The basin has a gentle southeasterly slope of maximum and minimum elevations varying between 58 m and 238 m [22].

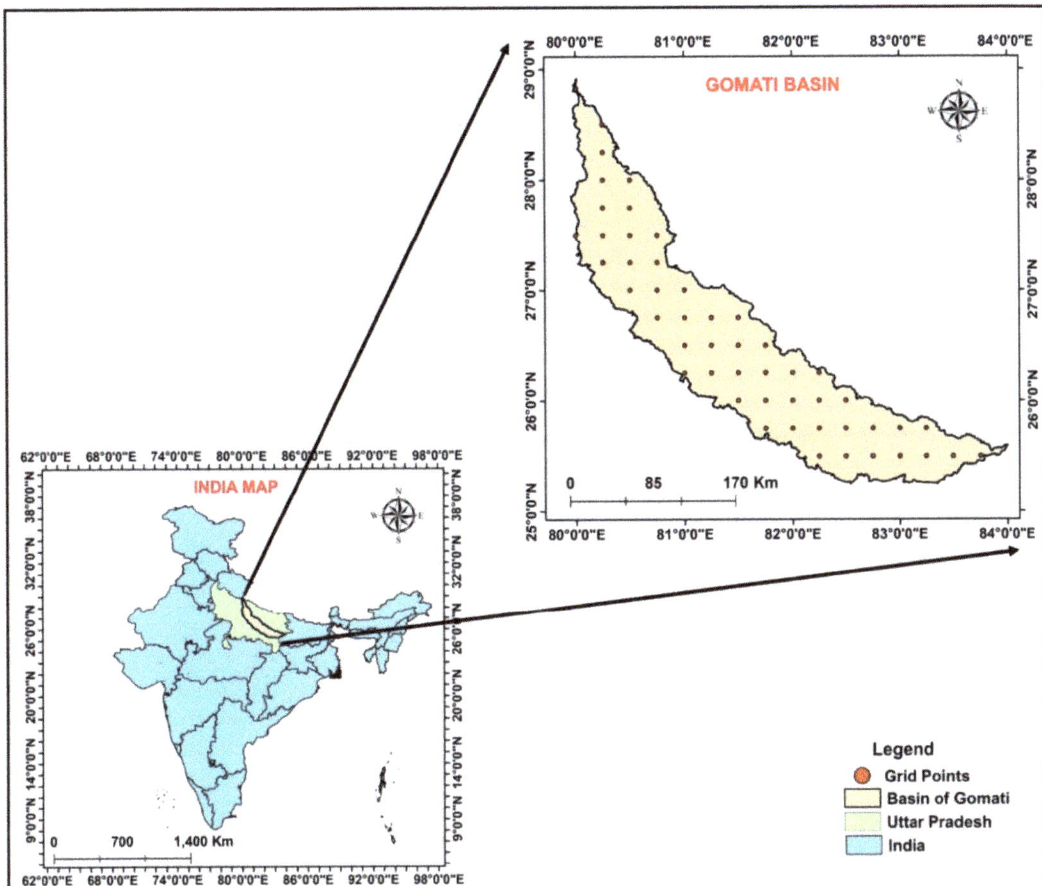

Figure 1. Indian map showing the Uttar Pradesh state and the Gomati river basin boundary in the green and orange color, respectively. Dots in the Gomati river basin show selected precipitation grid points.

2.2. Data

Daily data for the period 1951–2018 is extracted from the gridded datasets of temperature and precipitation developed by Pai et al., [23] and Srivastava et al., [24] for a spatial domain of 80° E to 84° E and 25° N to 29° N covering the Gomati basin. The spatial resolution of the precipitation and temperature data is $0.25° \times 0.25°$ and $1° \times 1°$, respectively. Both datasets have been extensively used in earlier studies [25–27]. It shows that the data is highly accurate and capable of capturing the spatial distribution of precipitation and temperature over the country. For more information on the gridded products, see [23,24] for precipitation and temperature, respectively. The most commonly used ETCCDI indices are estimated for each grid location within the study region from the gridded dataset. Table 1 provides a brief description of the selected indices used in the present study (see [17,28]).

Table 1. Brief description of the Expert Team on Climate Change Detection and Indices (ETCCDI) used in the present study.

S.NO.	INDICATOR	DESCRIPTIVE NAME	DEFINITION	UNIT
1	R50MM	Very heavy precipitation days	Number of days with precipitation above 50 mm.	Days
2	D95P	Very wet days	Days with precipitation > 95p.	Days
3	DD	Dry days	Days with precipitation less than 1 mm.	Days
4	R20MM	Heavy precipitation days	Days with daily precipitation amount ≥ 20 mm.	Days
5	R95TOT	Percentage precipitation of very wet days	Precipitation at days exceeding the 95th percentile divided by total precipitation expressed in percentage.	Percent
6	R99TOT	Precipitation fraction extremely wet days	Precipitation at days exceeding the 99th percentile divided by total precipitation expressed in percentage.	Percent
7	RTWD	Total precipitation wet days	Precipitation amount on days with RR ≥ 1 mm.	mm
8	RX	Maximum precipitation	The highest amount of daily precipitation.	mm
9	RX5D	Maximum 5 days R	Maximum consecutive five-days precipitation.	mm
10	CD	Percentage of cold days	Percentage of days with TX lower than the 10th percentile.	Days
11	VCD	Very cold days	Days with TN < 1st percentile.	Days
12	WN	Warm nights	Percentages of days with TN higher than the 90th percentile.	Days
13	CSD	Maximum consecutive summer days	Maximum number of consecutive summer days (TX > 25 Celsius).	Days
14	VWD	Very warm days	Days with TX > 99th percentile per year.	Days
15	WD	Warm days	Total no. of days with TX higher than the 90th percentile.	Days
16	SUD	Summer days	Number of days with TX > 25 Celsius.	Days

Precise definitions are given at http://etccdi.pacificclimate.org/list_27_indices.shtml accessed on 11 December 2020.

3. Methods

3.1. Sen's Slope Estimator Test

Sen's slope estimator test, developed by [29], adopts a basic non-parametric method for estimating the magnitude of a trend, if any, in a time series. It is widely used for determining the magnitude of a trend in hydro-meteorological time-series datasets [30–32]. The Sen's slope value indicates the degree of a slope and its trend in the time series, where a positive value shows an increasing trend, while a negative one indicates a decreasing trend.

In this method, the slopes (X_{ij}) of all data pairs (i, j) are first calculated by:

$$X_{ij} = \frac{(Y_j - Y_i)}{t_j - t_i} \ (i = 1, 2, .., N) \tag{1}$$

where $X'_{ij}s$ are the slopes of the lines connecting each pair of points (t_i, Y_i) and (t_j, Y_j), Y_j and Y_i are data values at times j and i ($j > i$), respectively. The median of these N values of $X'_{ij}s$ is Sen's slope, which is calculated as follows:

$$\beta = \begin{cases} X_{\frac{N+1}{2}}, & N \text{ is odd} \\ \frac{1}{2}\left(X_{\frac{N}{2}} + X_{\frac{N+2}{2}}\right), & N \text{ is even} \end{cases} \tag{2}$$

A positive value of β indicates an upward (increasing) trend and a negative one refers to a downward (decreasing) trend in the time series.

3.2. Rescaled Range Analysis

The rescaled range (R/S) analysis is a method to determine the variability of a time series introduced by Harold Edwin Hurst in the mid-20th century while investigating the Nile River's discharge time series. Hydrological time series exhibit some structure, unlike a common random series. For instance, consecutive values of hydrological time series are dependent on each other [33]. Future climate trends can be predicted using the R/S analysis method. Many improvements were done on the R/S analysis since then and applied to many fields, such as climate change, hydrology, population and economic analysis. The R/S for a time series $\{\xi(t) = 1, 2\ldots\}$ for any positive integer $\tau \geq 1$ consists of the following steps

(i). To define a mean sequence:

$$\langle\xi\rangle_\tau = \frac{1}{\tau}\sum_{t=1}^{\tau}\xi(t)\ (t = 1,\ 2\ldots.\tau) \tag{3}$$

(ii). To create a cumulative deviation series:

$$X(t,\tau) = \sum_{\mu=1}^{t}(\xi(\mu) - \langle\xi\rangle_\tau)\ (1 \leq t \geq \tau) \tag{4}$$

(iii). To create a range series:

$$R(\tau) = \max X(t,\tau) - \min X(t,\tau) \tag{5}$$

(iv). To create a standard deviation series:

$$S(\tau) = \sqrt{\frac{1}{t}\sum_{t=1}^{\tau}(\xi(t) - \langle\xi\rangle_\tau)^2} \tag{6}$$

For $R(\tau)/S(\tau) \triangleq R/S$,

If a power-law relationship is given in Equation (7), it indicates that the time series exhibits the Hurst phenomena and H is called the Hurst exponent.

$$\frac{R}{S}\alpha\tau^H \tag{7}$$

The nature of the trend can be determined based on the value of the Hurst exponent, with $0 < H < 0.5$ indicating that the future change is opposite to that of the past, $0.5 < H < 1$ that the change is sustainable and H = 0.5 indicating that the change is random [34]

3.3. Wavelet Transform

The Wavelet transform decomposes the original signal into proxies representing the original time series' inherent features in the time-frequency domain [35]. In general, a Wavelet transform can be classified into a continuous wavelet transform (CWT) and a discrete wavelet transform. CWT has been used for analyzing the time-frequency characteristics of climate and hydrological parameters [36–39]. In this study, we have used the Morlet wavelet as the mother wavelet for analysis, which is defined as:

$$\psi(t) = \pi^{-\frac{1}{4}}e^{iw_0t}e^{-t\frac{2}{2}} \tag{8}$$

where w_0 is the dimensionless frequency, t is a non-dimensional time parameter; i is the unit of an imaginary number [40,41].

The CWT of a discrete signal $\xi(t)$ with a Morlet wavelet $\psi(t)$ is given as:

$$W_f(a,b) = \frac{1}{a}\int_R \xi(t)\psi*\left(\frac{t-b}{a}\right)dt \tag{9}$$

where $W_f(a,b)$ is the transform; a and b are the scale parameter and translation parameter, respectively; $\psi*$ is the complex conjugate and t is the time scale [42].

The CWT has edge artefacts that result on account of the finite-length of the time series used and, understandably, errors are expected at the beginning and end of the wavelet power spectrum. This may be avoided by padding up the end of the time series with zeroes prior to its transformation and later removing them. However, padding with zeroes introduces discontinuities at the endpoints and, as one goes to larger scales, decreases the amplitude near the edges as more zeroes enter the analysis. The cone of influence is the region of the wavelet spectrum in which edge effects become important and is defined here as the *e*-folding time for the autocorrelation of the wavelet power at each scale, and the peaks within these regions are reduced in magnitude presumably due to zero padding. To test the significance of any statistic of a given wavelet power spectrum, a background reference spectrum is used as a basis for comparison. Based on a Monte Carlo simulation study by [41], the latter authors have recommended the use of a white or a red noise background spectrum based on an AR (1) (lag – 1) autoregressive process.

3.4. Bivariate Copula Functions

A copula is a joint distribution function derived from two or more marginal distributions of random variables. The Sklar theorem is the foundation of the mathematical framework of copulas [43]. Copulas are very useful in unravelling random variables' behavior, thereby deriving the joint distribution of two or more random variables [44]. Bi-variate copulas are used to derive the return periods of climate indices to describe climate indices' characteristics. Over the years, several families of copulas have been derived. Out of these Archimedean copulas (Frank, Clayton, and Gumbel) are widely used for hydrological applications (climate extreme analysis) due to their simple implementation, construction and flexibility in applying for both positively or negatively correlated variables [16,45]. The mathematical description of Frank, Clayton, and Gumbel copulas' distribution functions and their parameters are provided in Table 2. For details on the mathematical description, see [46,47].

Table 2. Mathematical description of copula functions and their parameters.

Copula Function	C(u,v)	Parameter Range
Frank	$-\frac{1}{\Theta}ln\left[1+\frac{(e^{-\theta u}-1)(e^{-\theta v}-1)}{(e^{-\theta}-1)}\right]$	$\theta \in (-\infty,\infty)\backslash\{0\}$
Clayton	$\max\left[\left(u^{-\theta}+v^{-\theta}-1\right)^{-\frac{1}{\theta}},0\right]$	$\theta \in [-1,\infty]\backslash\{0\}$
Gumbel	$exp\left\{-\left[(-\ln u)^{\theta}+(-\ln v)^{\theta}\right]^{\frac{1}{\theta}}\right\}$	$\theta \in [1,\infty]$

The estimation of return periods using copulas goes as follows:

1. Fit a marginal distribution for the climate indices (for example, RTWD and R95TOT) and develop their cumulative distribution functions (CDF). We used 15 marginal distributions (1) Weibull distributions, (2) gamma, (3) extreme value, (4) exponential, (5) Birnbaum–Saunders, (6) generalized extreme value, (7) inverse Gaussian, (8) log-normal, (9) Nakagami, (10) logistic, (11) log-logistic, (12) t location-scale, (13) Rayleigh, (14) normal, and (15) Rician,
2. The Bayesian Information Criterion (BIC) is applied to identify each climate indices' best marginal distribution. The marginal distributions parameters are calculated using a maximum likelihood algorithm, and the X^2 goodness-of-fit is used for statistical significance.
3. The correlation was calculated between different climate indices to evaluate the mutual dependence between the two climate indices using Kendall's Tau Correlation coefficient.
4. Obtain the joint CDF of two variables by estimating the copula parameters (see Table 2)
5. Finally, a summary report is generated containing copulas from best fitted to the lowest one. The best copula function was selected based on the minimum principle of Akaike Information criterion (AIC) and RMSE following [48].

4. Results and Discussion

Here, we first present the spatiotemporal trend of precipitation at different timescales followed by spatiotemporal variability of extreme climate indices for the period from 1951–2018. Rescaled range analysis is employed to predict the future trends of the extreme climate indices in the basin. Regularity of climate indices and its response to climate change are unraveled using wavelet analysis. Furthermore, investigates the joint occurrence of extreme events in the context of global warming.

4.1. Spatio-Temporal Variability of Precipitation

The spatio-temporal trend analysis of precipitation for the entire period is carried out at monthly, seasonal and annual time scale (Figure 2). The additional supporting analysis is shown in the Supplementary Materials for brevity purposes (Figures S1–S9). It is evident that there has been a significant reduction in the precipitation amount over the basin, especially during the months of June-September (Figure 2). Mixed positive and negative trends are observed in June with negative trends concentrated toward the basin's southern region. Slightly positive and significant negative trends are seen in July towards the basin's northern and southern regions with maximum slopes at 0.4821 and −0.9073. August showed negative trends at all the stations over the basin with southern region exhibiting significant negative trends with a maximum slope of −1.2952. Significant negative trends with a maximum slope of −0.7229 are observed at almost all the stations over the basin except for a few stations in the northern region for September. Seasonal precipitation also showed negative trends with the summer season being statistically significant with a maximum slope of −2.1514. A similar pattern of significant negative trends at all stations is observed for annual precipitation in the basin's central and southern regions with a maximum slope of −3.2934. It can be inferred from the monthly, seasonal and annual precipitation trends that there has been a significant reduction in precipitation during the period (1951–2018), especially in the basin's central and southern region. The decreasing in precipitation agree with the previous studies concerning precipitation trends across India at various spatial (regional, sub-basin, basin and meteorological sub-divisions, homogeneous regions) and temporal scales (monthly, seasonal, and annual) [48,49].

(a)

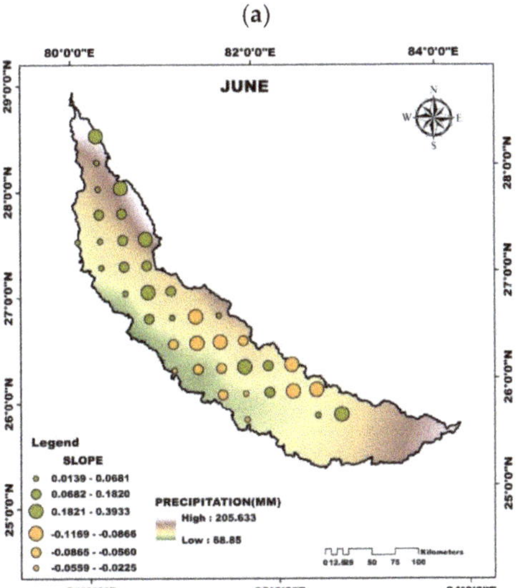

(b)

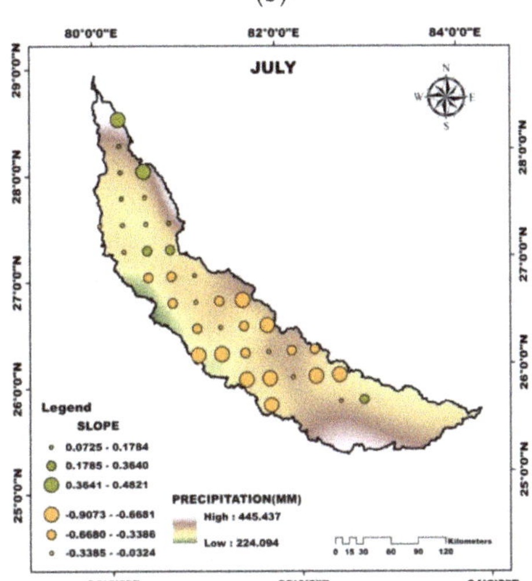

Figure 2. *Cont.*

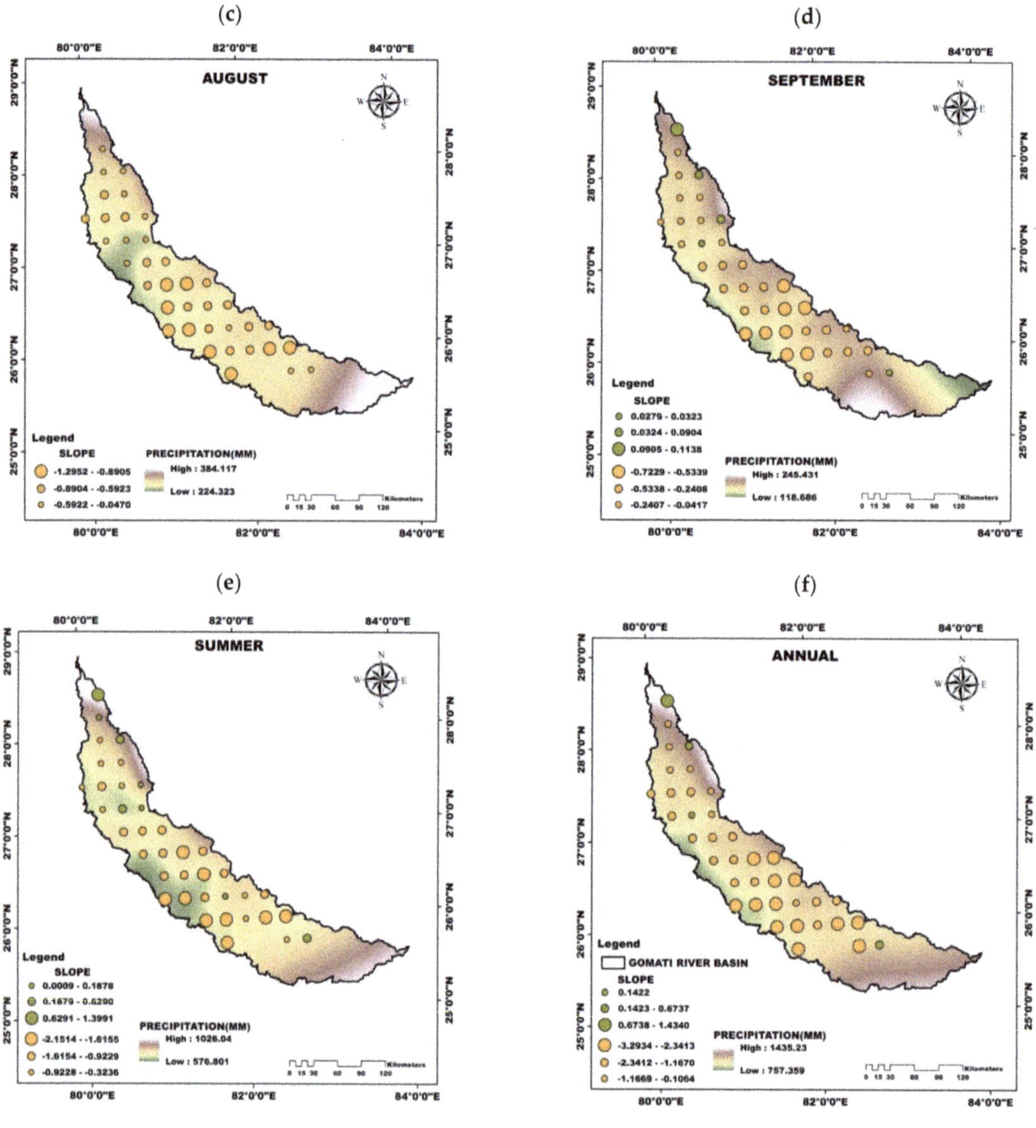

Figure 2. Spatio Temporal trends in the total precipitation at different time scales, (**a**–**d**) monthly, (**e**) seasonal, and (**f**) annual. The figure provides the Sen slope of the trend in each grid, which has significant trends.

4.2. Spatio-Temporal Variability of Extreme Climate Indices

The spatio-temporal trends in 16 selected extreme climate indices for the entire period is presented in Figure 3. Figure 4 provides a summary of the number of stations showing significant trends of extreme climate indices. The results show a substantial reduction in the precipitation amount over the entire basin, barring few exceptions. This significant negative trend is visible in RTWD, highlighting a considerable decrease during the period with a maximum slope of −3.3002 at the basin's central region. Although the trends of all the stations are decreasing, 19 stations were statistically significant. R50MM for all stations showed decreasing trends, 12 of which were statistically significant, with higher trend slopes of −0.6613 in the basin's southern region. D95P showed negative trends for all stations over the basin, out of which 20 were significant and the maximum slope was found to be equal to −1.5150 in the southern region. Further, 17 stations displayed statistically significant negative trends for the index R20MM with maximum slope (−1.6938) being observed at the southern region. A similar pattern of negative trends can be observed with the other precipitation indices such as R95TOT, RX5D, and RX for almost all the stations. The number of dry days (DD) showed positive trends during the period over the basin at 36 stations out of which 17 stations are statistically significant with a maximum slope of 2 at the southern region of the basin. It is evident from the negative trends of various precipitation indices that there is a significant reduction of precipitation over the basin. The index DD exhibiting positive trends over the basin during the same period indicates that the basin has moved towards drier weather.

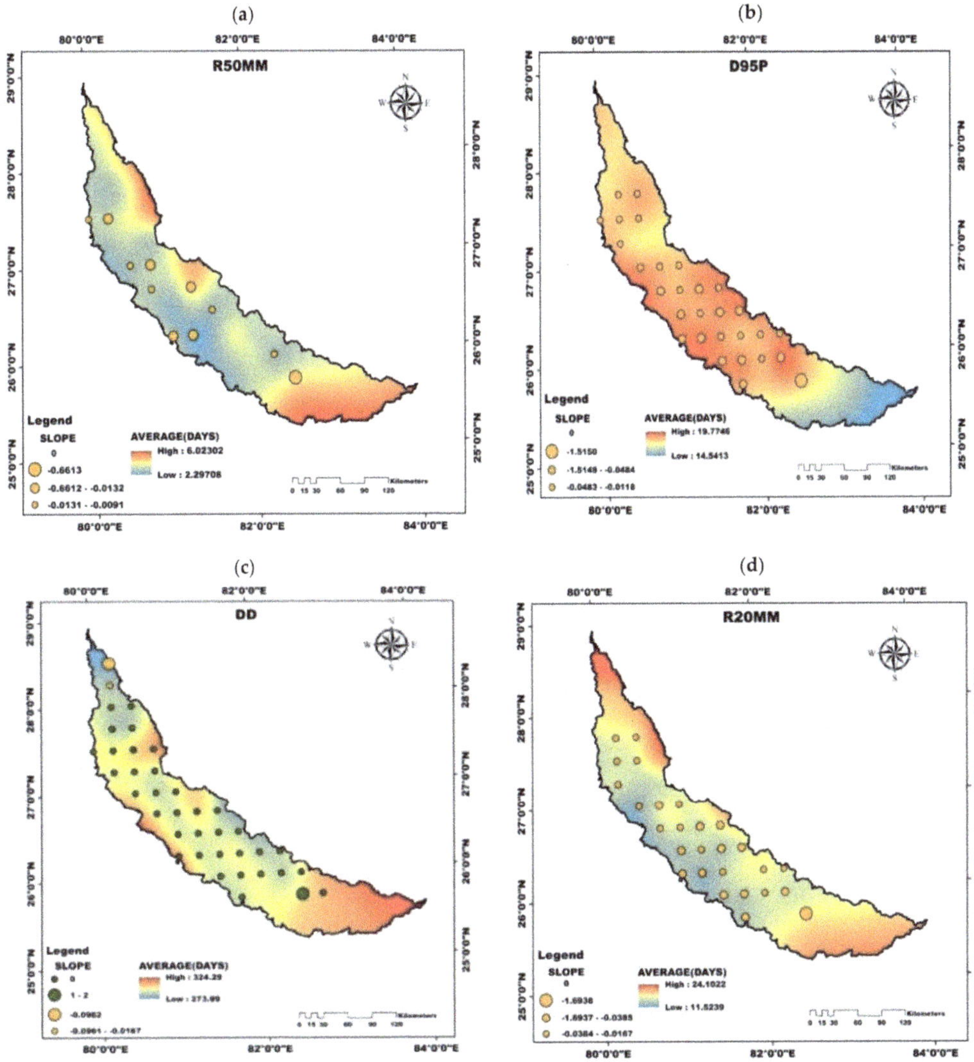

Figure 3. *Cont.*

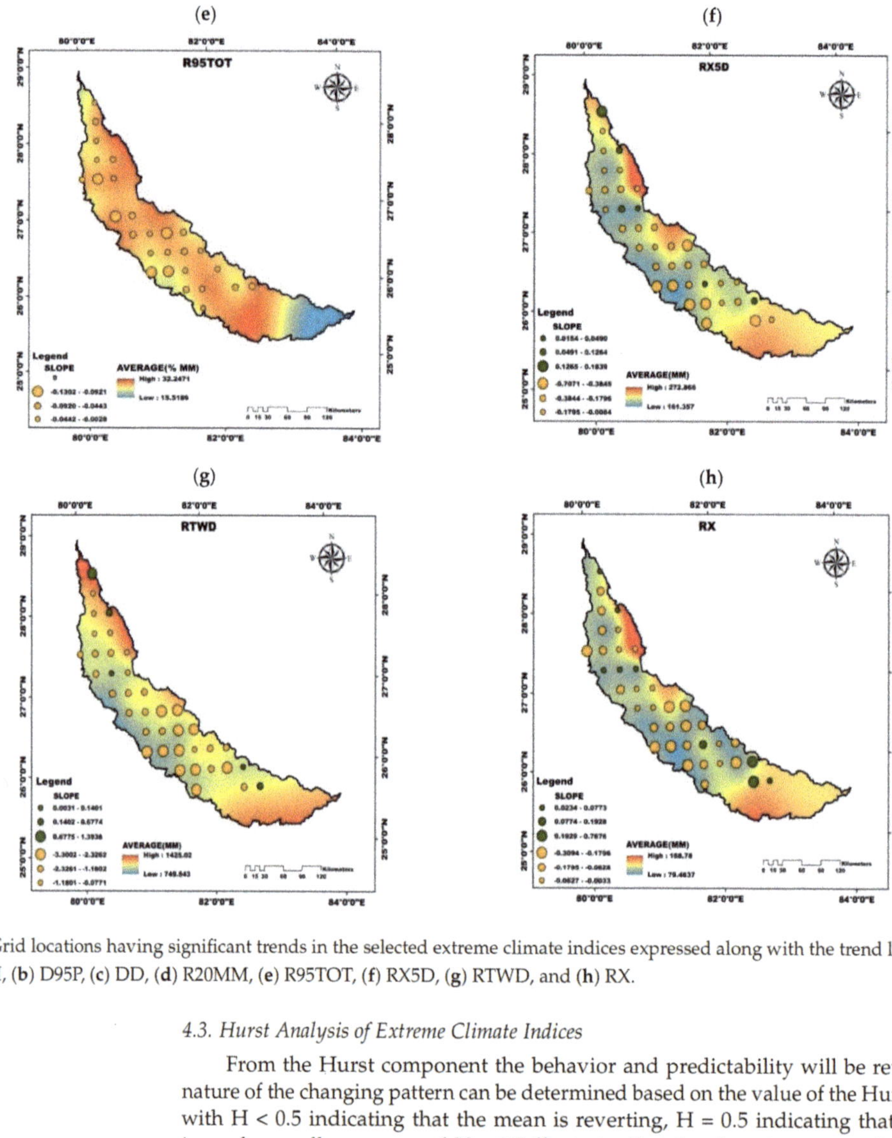

Figure 3. Grid locations having significant trends in the selected extreme climate indices expressed along with the trend line's slope for (**a**) R50MM, (**b**) D95P, (**c**) DD, (**d**) R20MM, (**e**) R95TOT, (**f**) RX5D, (**g**) RTWD, and (**h**) RX.

4.3. Hurst Analysis of Extreme Climate Indices

From the Hurst component the behavior and predictability will be revealed. The nature of the changing pattern can be determined based on the value of the Hurst exponent, with $H < 0.5$ indicating that the mean is reverting, $H = 0.5$ indicating that the change is random walk process and $H > 0.5$ illustrates that the changing pattern is sustainable. Therefore, by knowing the value of H, the intrinsic nature (mean reverting or trending) of the time series can be described and it plays a significant role with implications for forecast skill, low frequency variations [34]. The spatial distribution map of H values for each of the extreme climate indices is presented in Figure 5. The future trends can be determined based on the value of the Hurst exponent obtained for the index. R50MM shows mixed increasing and decreasing future trends all over the basin with the Hurst exponent value around 0.43 and 0.67.

A sustainable decreasing trend is observed for D95P at all stations over the basin, indicating a reduction in precipitation in the future. The number of dry days (DD) exhibits a sustainable decreasing trend at almost all the stations over the basin barring a few exceptions indicating that the basin is turning drier. For R95TOT, a sustainable decreasing trend is displayed at almost all the stations except for five stations that interestingly exhibited the opposite trend of the present pattern. All the stations showed a sustainable decreasing trend for the index RTWD in the basin, indicating that the number of wet days in the basin will reduce in the future. The indices RX, RX5D, and R20MM displayed sustainable decreasing trends over most of the basin stations except for 2–3 stations whose Hurst exponents too are closer to 0.50 indicating that the future trend over the entire basin can be considered as sustainable. Generally, processes with H > 0.5 indicate a persistent long memory effect and the trend is sustainable [16]. Since in the present study, most of the stations show H > 0.5, it shows that this region would be having a persisting phenomenon in terms of decreasing trend in precipitation and increasing trend in temperature in future. After observing the spatial distribution map of extreme climate indices (Figure 5), we can summarize that almost all the extreme precipitation indices exhibit a sustainable decreasing trend except a few stations. Our analysis reveals that precipitation will continue to decrease in the future meaning that the basin is inching towards a drier climate.

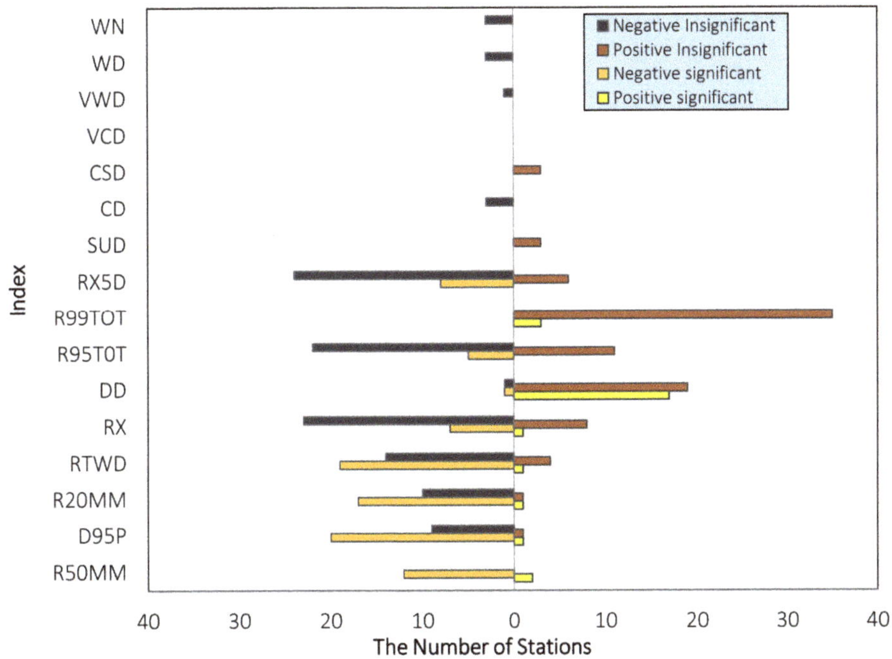

Figure 4. Number of stations where extreme climate indices have shown specific trends in the Gomati river basin during 1901–2018.

4.4. Periodic Oscillation Analysis

The variability of the climate indices is multifaceted, and often conveyed as multi-frequency and quasi-periodic in the time and frequency domain. Therefore, it is helpful to use wavelet methods for analyzing the variations in the frequency of climate indices at multiple scales averaged over 38 grid points in the Gomati river basin during 1951–2018. In doing so, the regularity of any climate index and its response to climate change can be better understood. Figure 6 shows the wavelet power spectrum for the climate indices, in which contours enclosing light yellow regions have larger power. The cone of influence is represented by a V shape with a black line to distinguish between non-significant periodic characteristics (edge effect artefacts) and significant periodic characteristics (within the cone of influence) at 95% confidence level [42]. Additionally, the global power spectrum and its statistical significance are depicted beside each climate index. The results show that the two-year period was dominant in the global wavelet spectrum of R50MM, with significant wavelet power found during 1975–1980 (Figure 6a). The wavelet power spectrum for D95P showed a significant power distributed across the 2–16-year bands, and the two-year period was dominant in the global power spectrum around 1970–1990 (Figure 6b). One exciting finding unraveled in Figure 6b is that the dominant two-year period of D95P has changed to the four-year after 1984 and remains there in the past two decades. The wavelet power spectrum of R20MM is following a similar pattern of D95P (Figure 6c). Figure 6c indicates that frequency of R20MM is 1 in two years until 1984, and later the frequency has changed to 1 in four years. Figure 6d shows the wavelet power in the 2–4-year band around 1955–1965, 1975–1983, 1990–2002, and 2007–2012 and in the 4–6-year band during 1955–1995, and in the 8–16-year band around 196–2000 of R99TOT. The 8-year period was found in the global wavelet of WD and a 2–4-year band in 1965–1975, 1995–2000 (Figure 6e). The global wavelet power spectrum of CD (Figure 6f) showed that the 4–6-year period was significant at the 95% confidence level during 1970–1980. WN was dominated by four- and eight-year periods, with a 4–8-year band, found in 1994–2014 (Figure 6g).

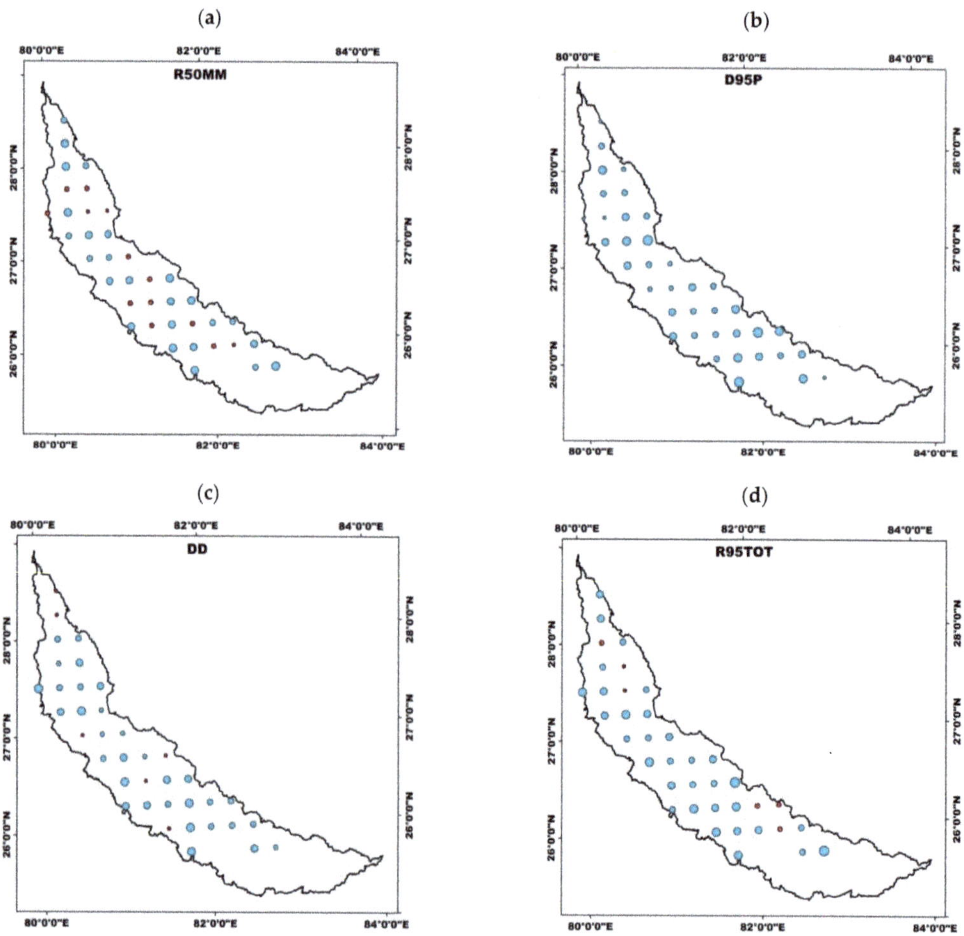

Figure 5. *Cont.*

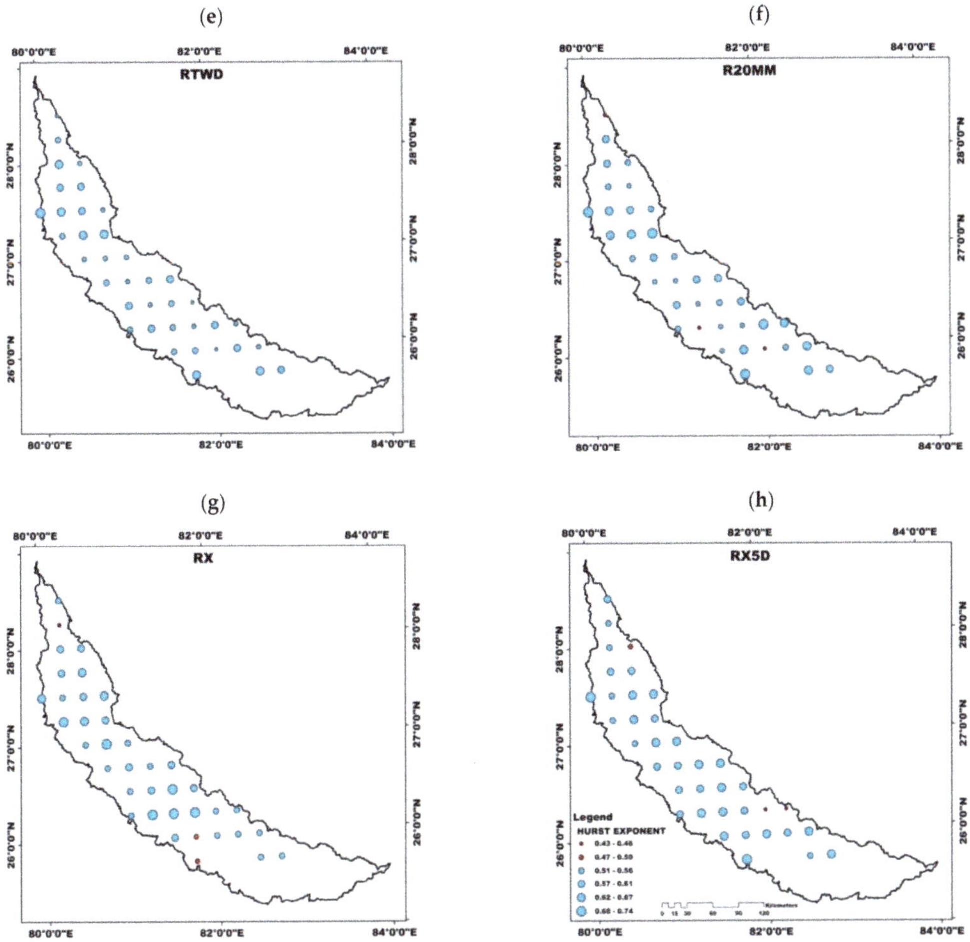

Figure 5. Spatial distribution map of extreme climate indices (**a**) R50MM, (**b**) D95P, (**c**) DD, (**d**) R95TOT, (**e**) RTWD, (**f**) R20MM, (**g**) RX, and (**h**) RX5D in Gomati Basin based on the Hurst Exponent.

Overall, the wavelet spectra of most of the extreme climate indices showed variability within the 4–8 years and for some, the inter-decadal variability is prominent. Earlier study by Rathinasamy et al. [50] have shown there is a linkage between extreme precipitation in the Indian subcontinent and the global climate teleconnections such as ENSO and PDO. Following that study, the reason for the presence of dominant variability observed in the extreme climate indices could be linked to global teleconnection patterns. However, in-depth analysis is required to understand the dynamics of this linkage which can be considered in the further analysis.

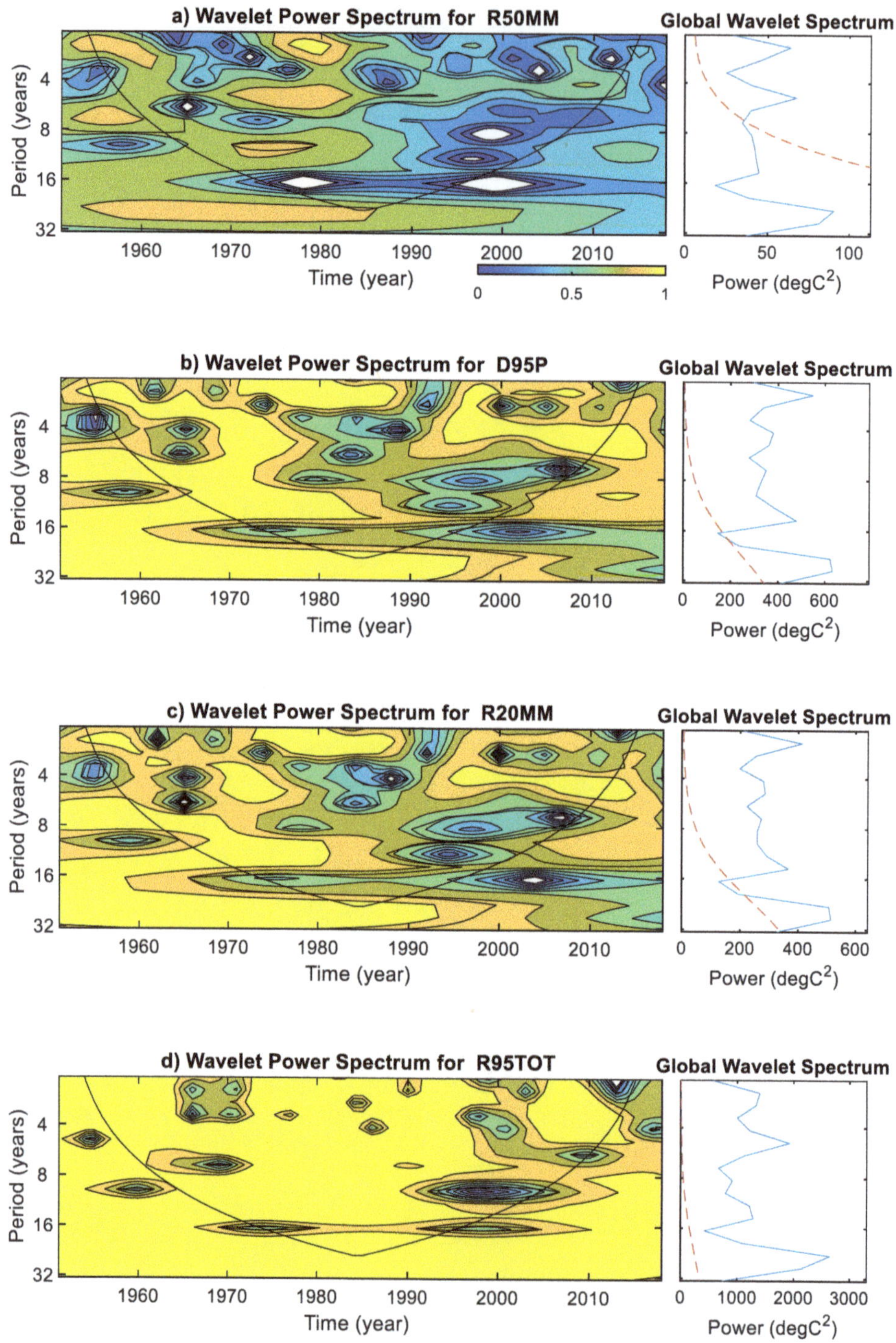

Figure 6. *Cont.*

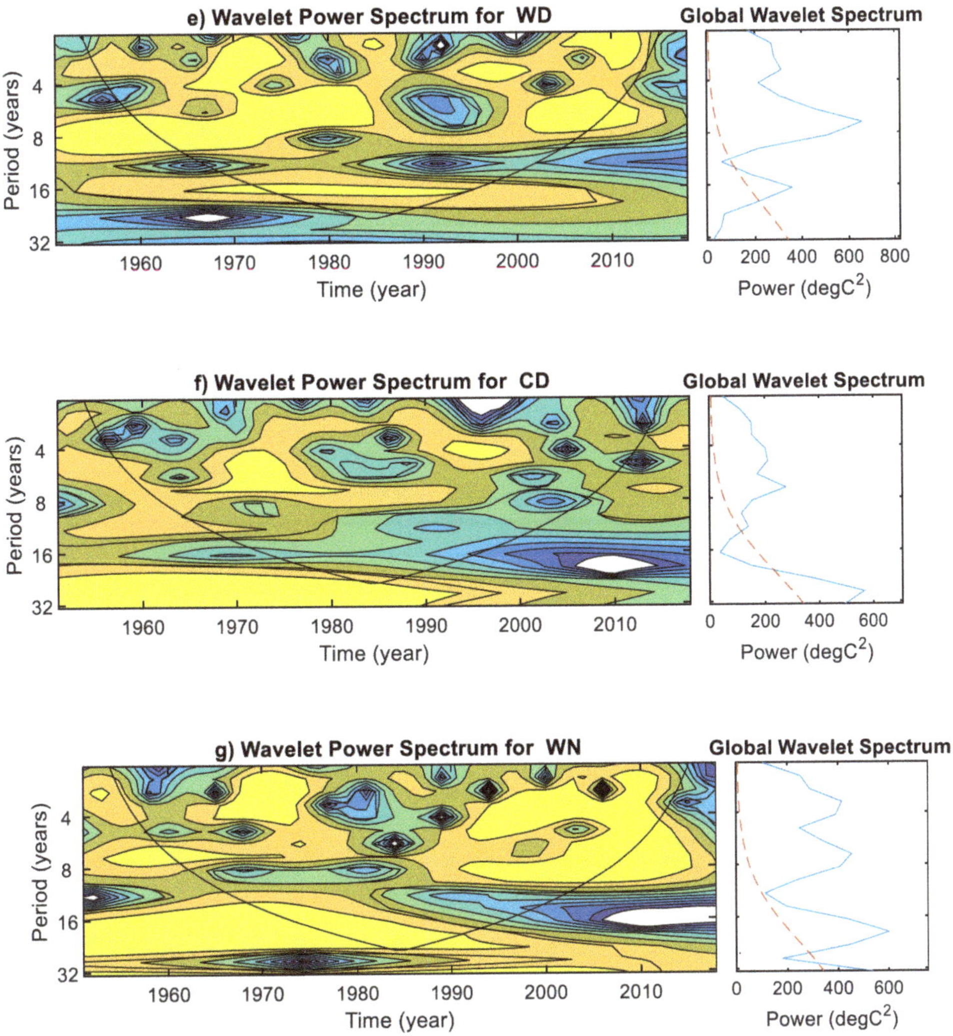

Figure 6. Continuous wavelet transformation for climate indices (**a**) R50MM, (**b**) D95P, (**c**) R20MM, (**d**) R95TOT, (**e**) WD, (**f**) CD, and (**g**) WN averaged over 38 grid points in the Gomati river basin during 1951–2018. The V shape represents the cone of influence and regions enclosed with light yellow are periods with significant power. The red color dashed line distinguishes between significant and non-significant periods. If the global wavelet spectrum curve is higher than the dashed line at a particular scale, then the period is statistically significant at the 95% confidence level.

4.5. Bivariate Joint Probability and Return Period Analysis

The mutual dependence between the hydro-climatic variables helps to understand extreme climate characteristics over time that individual indices are limited to cover [39]. Therefore, to detect and analyze the mutual dependence among the extreme climate in the Gomati river basin bi-variate copula functions are used. Bi-variate copula is efficient to compute the bi-variate joint probabilities and return periods of climate indices. First, the probability distributions of climate indices in the Gomati river basin were calculated using 15 marginal distribution functions (listed in Section 3.4), and then Bayesian Information criterion was used to find the optimal fitting function. The thumb rule is that a smaller BIC value point to better-fitting, with the marginal distribution function to the lowest value considered the best fitted distribution function. Table 3 shows the results of the calculation of BIC values for each distribution for every climate index. Our results show that R50MM, RTWD and RX are inverse Gaussian distributed, D95P is gamma, R20MM is Birnbaum-Saunders, RX5D is log-logistic, DD and WN is Rician, R95TOT is Log-normal, R99TOT is Rayleigh, CSD is Weibull, WD is normal, CD and SUD is Nakagami distributed. As presented in Figure 7, the optimal distribution function revealed above was more effective.

Table 3. BIC values and selection of distribution function of the Expert Team on Climate Change Detection and Indices (ETCCDI).

Climate Index	Type of Distribution Function									
	Weibull	Gamma	Birbaum-Saunders	Inverse Gaussian	Log Normal	Nakagami	Log-Logistic	Rayleigh	Normal	Rician
R50MM	241.5	230.6	226.9	***226.8***	227.0	237.0	228.9	242.8	247.7	243.8
D95P	399.3	***393.8***	394.0	394.0	394.2	394.5	397.0	446.7	396.6	396.4
R20MM	379.4	373.0	***372.8***	372.8	373.0	374.2	375.7	416.9	377.1	376.8
RTWD	940.1	928.6	927.2	***927.2***	927.3	931.0	929.1	985.7	934.8	934.5
RX	628.4	614.7	612.1	***612.0***	612.2	617.9	615.4	679.0	622.1	621.9
RX5D	725.1	709.2	706.7	706.6	706.6	713.0	***706.1***	769.1	717.9	717.6
DD	512.3	505.8	505.9	505.9	505.9	505.7	506.7	822.2	505.6	***505.6***
R95TOT	470.9	462.2	462.0	462.0	***462.0***	463.7	462.8	514.0	466.6	466.4
R99TOT	410.5	410.1	413.5	414.8	413.9	410.5	416.7	***406.3***	421.0	410.5
CSD	***576.6***	581.5	582.3	582.3	582.4	580.8	583.8	811.5	580.1	580.1
WD	371.8	385.9	433.2	437.5	409.7	376.1	385.9	385.7	***368.0***	368.9
CD	342.9	343.0	347.5	347.7	347.0	***341.4***	344.3	375.0	341.8	341.6
WN	378.2	383.5	397.5	398.7	393.6	379.2	385.8	387.7	378.6	***377.9***
SUD	512.6	505.0	505.1	505.1	505.1	***505.0***	507.7	823.0	505.1	505.0

Italic bold font indicates the selected distribution function.

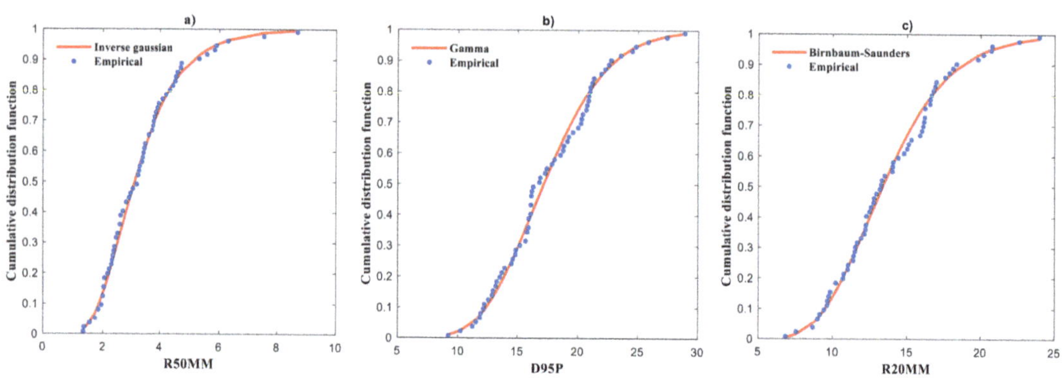

Figure 7. *Cont.*

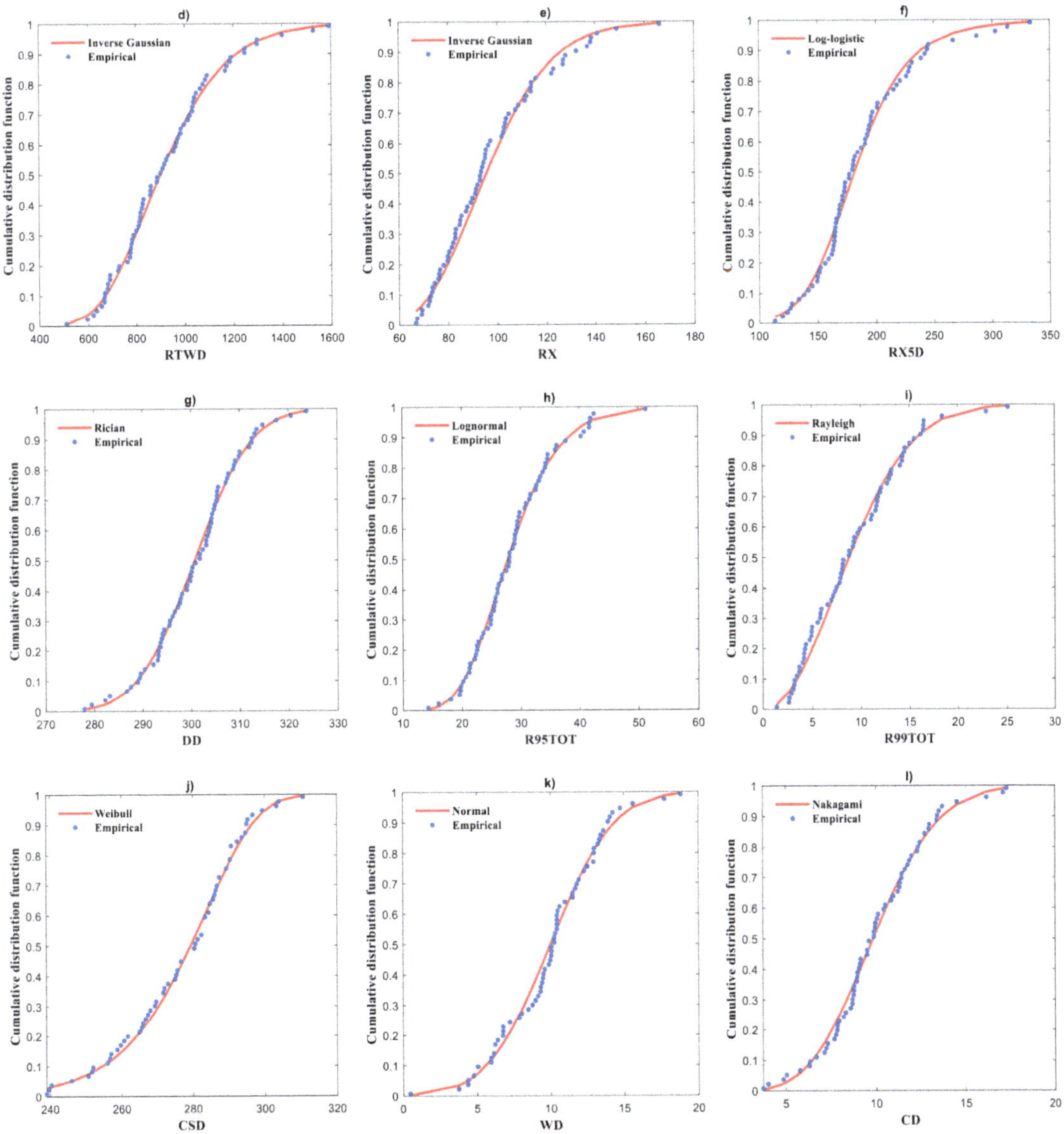

Figure 7. *Cont.*

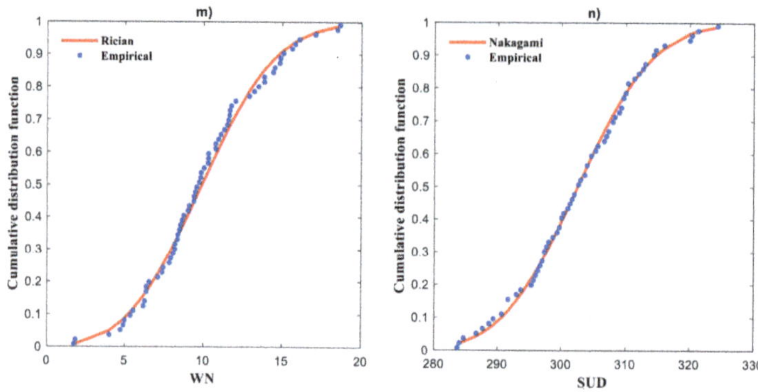

Figure 7. Observed data (blue dots) and fitted marginal distribution (red line) of the climate indices (**a**) R50MM, (**b**) D95P, (**c**) R20MM, (**d**) RTWD, (**e**) RX, (**f**) RX5D, (**g**) DD, (**h**) R95TOT, (**i**) R99TOT, (**j**) CSD, (**k**) WD, (**l**) CD, (**m**) WN, and (**n**) SUD averaged over 36 grid points in the Gomati river basin during 1951–2018.

The selected 16 climate indices could be combined into 120 combinations; however, a few combinations lack physical meaning. We attempt two combinations for the Gomati river basin to understand the combination of precipitation amount and intensity (modelling the flood drivers). Firstly, the R95TOT and RTWD combination is selected to calculate the joint probability and return period of extremes contribution to total precipitation and precipitation amount from wet days. Next, the R95TOT and D95P combination is selected to calculate the joint probability and return period of extreme precipitation contribution and number of days. The best copula function was selected based on the minimum principle of AIC and RMSE following [51]. The results are given in Table 4 and the joint probability and return period for two combinations averaged over 53 grid points during 1951–2018 are shown in Figures 8 and 9.

Table 4. Ranking of copula families based on RMSE and AIC values.

Climate Index Combination	Copula Family	RMSE	AIC	Rank
RTWD&R95TOT	Clayton	0.4175	−403.707	3
	Frank	0.2214	−490.013	2
	Gumbell	0.1956	−506.819	1
D95P&R95TOT	Clayton	0.3863	−414.272	3
	Frank	0.2253	−487.603	2
	Gumbell	0.1995	−504.131	1

We first calculated the mutual dependence of these climate indices (R95TOT and RTWD) in the 68-year record (68 pairs) using three approaches, i.e., Kendal's Tau rank ($r = 0.4934$), Spearman's rank correlation ($r = 0.6613$), and Pearson correlation coefficient ($r = 0.7428$), of which three methods illustrate a statistically significant dependence between R95TOT and RTWD. Hence, the bi-variate copula analysis is applied to describe the interdependence among them.

We first select the optimal marginal distribution functions, to the R95TOT and RTWD based on the BIC goodness-of-fit measure. Figure 7d,h show the fitted distributions (red line) compared to the observed (blue dots) for R95TOT and RTWD. We chose an inverse Gaussian and lognormal distributions to fit both variables (refer to Table 3). Then we evaluated three Archimedean copula families (Clayton, Frank, and Gumbel) using the MvCAT toolbox in MATLAB [51]. The parameters of the copula and their posterior distribution are inferred using a Bayesian analysis and Monte-Carlo simulations. For both combinations, Gumbel is selected as best copula family to describe the dependence structure (see Table 4). Figure 8 shows the isolines of the joint probability (Figure 8c) and return period (Figure 8f) based on the Gumbel copula. We then use the Gumbel copula model to derive non-exceedance probabilities and return periods and analyze the compound event. For the most likely design scenario, the values of R95TOT and RTWD are 48% and 1500 mm, respectively. The joint return period's resulting values are greater than values resulting from univariate analysis (R95TOT with 44% and RTWD of 1450 mm). The difference in values highlights that ignoring the mutual dependence can lead to a substantial underestimation of extremes. Similarly, for another combination (R95TOT and D95P), the same procedure is followed to derive the joint return periods and displayed in Figure 9. Intuitively, consideration of mutual dependence improves the accuracy of the return period of climate extremes.

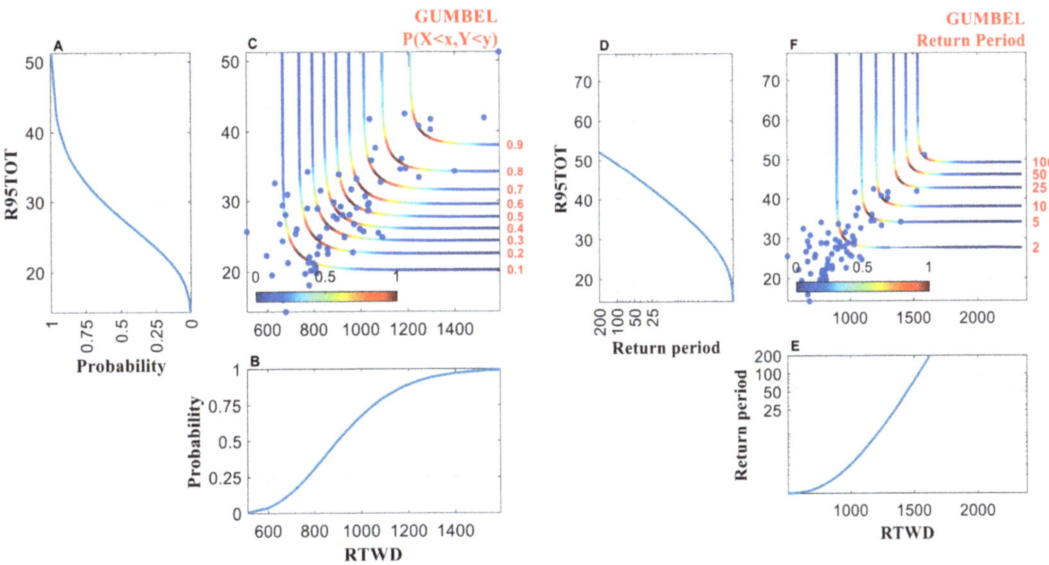

Figure 8. Cumulative distributive functions of (**A**) R95TOT and (**B**) RTWD, and their (**D**) and (**E**) are associated univariate periods (*y*-axis indicated in log scale). Joint probability iso lines derived from the Gumbell copula are displayed in Figure (**C**), and the associated return period isolines are illustrated in Figure (**F**). The colorbar [0 1] represents the joint density levels of joint probability and return periods, where 0 represents lower and one represents higher density level. Blue dots represent the observed pairs of R95TOT and RTWD. The figure is plotted using the MvCAT toolbox [51] in MATLAB (Version 9.4 R 2020b).

Overall, from the entire analysis, spatio-temporal variability and changing pattern of extreme climate indices indicate that the basin will experience a reduction in precipitation and an increase in the number of dry days pushing the basin towards drier weather. This finding is in line with prior studies dealt globally [17,52,53] as well as regionally [54–58]. In particular, Abeysingha et al. [56], after analyzing rainfall and temperature trends in the Gomati River basin, concluded that there had been a significant reduction in rainfall, consequently leading to decline in the streamflow coupled with increasing temperature results in dryness in the basin. Sachidanand et al. [58] concluded that the number of 'warm days' per year increased significantly, whereas the number of 'cold days', 'warm nights', and 'cold nights' per year decreased significantly at several locations in India. On the other side, a decreasing trend in precipitation is observed at some Uttar Pradesh locations, including the Gomati basin, highlighting the possibility of dryness in Northern India.

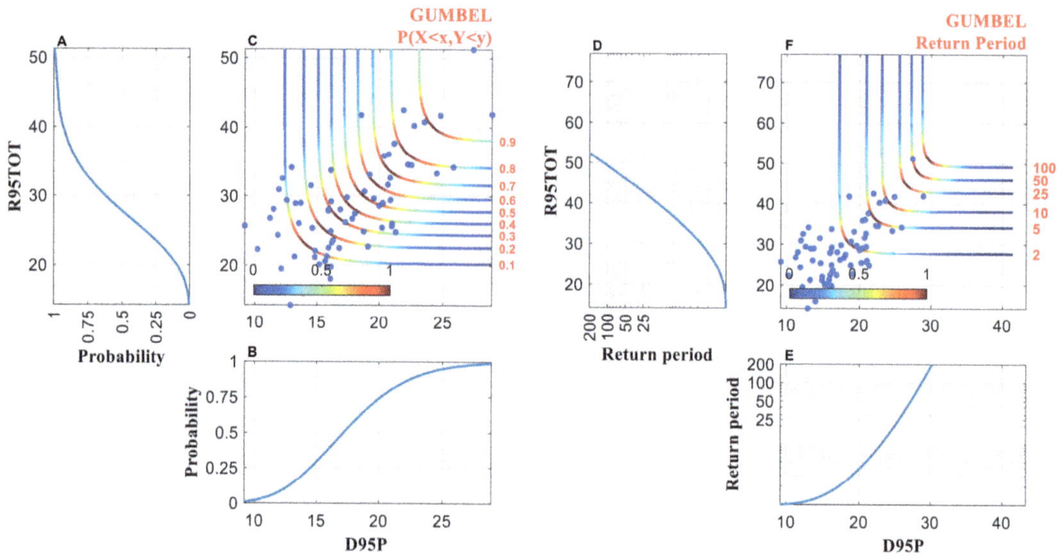

Figure 9. Same as Figure 8 but for a combination of climate indices R95TOT and D95P.

Numerous climate models have observed that there has been an increase in surface temperatures over the 20th century [52]. This increase in temperatures will lead to increased evaporation and surface drying, which increases the intensity and duration of drought events. On the other hand, for every 1 °C warming, the moisture-holding capacity of air is increased by about 7%, thereby leading to an increase in water vapor in the atmosphere, which in turn leads to intensifying the water cycle [59]. Therefore, a warmer climate will increase the risk of drought due to surface dryness and floods due to intensified water cycle but at different times and/or places. A similar pattern is observed in the present study. There has been a significant reduction in precipitation along with an increase in the number of dry days. The future trends based on the Hurst exponent indicate that this trend is likely to continue in the near future. Further copula modelling is applied to analyze compound extremes. Intuitively, the multivariate framework can better represent the risk due to the consideration of mutual dependence.

5. Conclusions

We investigated the long-term spatiotemporal variation of precipitation and the extreme climate indices in the Gomati River Basin in India. The extreme climate indices were characterized through a trend analysis, dominant periodicities and joint probability. The main conclusions from the study are as follows.

(a) Different extreme precipitation indices (D95P, R95TOT) show significant decreasing trends whereas the extreme temperature indices show an increasing trend indicating that the basin is experiencing a warm and dry climate compared to the past. Further, the positive persistence present in these variables show that the trend would continue in the future resulting in warmer and drier climate. The results presented in the study is in congruence with the other studies wherein the climate has become warmer

(b) The presence of the significant periodicities (4–8 years and decadal timescale) in the precipitation extreme indicate a possibility of connection with some of the well-known global atmospheric patterns such as ENSO and PDO. However, interestingly, the extreme temperature indices did not exhibit significant periodicities.

(c) Further, the analysis from the CWT of the extreme precipitation indices indicate that there is a significant non-stationarity indicating that the changes in the precipitation patterns after the 1980s.
(d) Further copula modelling is applied to analyze compound extremes. The multivariate framework better represented the risk due to the consideration of mutual dependence. Interestingly, the copula modelling shows that the joint probabilities for higher annual precipitation and precipitation receiving from very wet days are smaller than the marginal exceedance probabilities. Intuitively, consideration of mutual dependence improves the compound risk of climate extremes.

Supplementary Materials: The following are available online at https://www.mdpi.com/article/10.3390/atmos12040480/s1, Figure S1. Spatio Temporal trends in the total precipitation for different months (January(a)–December(l)). Figure S2. Spatio Temporal trends in the total precipitation for different seasons (Autumn(a)–Winter(d)). Figure S3. Spatio Temporal trends in the total precipitation for annual time scale. Figure S4. Spatio temporal variation of all the 16 extreme climate indices over the basin. Figure S5. Spatial distribution map of all the 16 extreme climate indices based on the Hurst exponent. Figure S6. Spatio Temporal trends in the total temperature(MIN) for different months(January–December). The figure provides the Sen slope of the trend in each grid, which has significant trends. Figure S7. Spatio Temporal trends in the total temperature(MIN) for different seasons(Autumn–Winter) & Annual time scale. The figure provides the Sen slope of the trend in each grid, which has significant trends. Figure S8. Spatio temporal variation of all the 16 extreme climate indices over the basin. Figure S9. Spatial distribution map of all the 16 extreme climate indices based on the Hurst exponent.

Author Contributions: Conceptualization, M.R.; Data curation, A.K. and R.T.; Formal analysis, A.K., R.T., R.K.G. and M.R.; Funding acquisition, M.R.; Investigation, R.K.G. and M.R.; Methodology, A.A. and M.R.; Project administration, A.A.; Resources, J.K.; Supervision, D.K.G. and J.K.; Validation, R.T., R.K.G. and A.A.; Visualization, A.K. and R.T.; Writing—original draft, A.K. and R.K.G.; Writing—review & editing, D.K.G., A.A., J.K. and M.R. All authors have read and agreed to the published version of the manuscript.

Funding: This research was supported by the Early Career Research Award (Dr.RM) under SERB, India under grant no. ECRA/2016/01721. R.K.G. was supported by the Inspire fellowship award under DST, India under grant No. IF 190581. A.A. acknowledges the funding support provided by the COPREPARE project funded by UGC and DAAD under the IGP 2020-2024. J.K. was supported by the Russian Ministry of Science and education agreement no. 075-15-2020-808.

Institutional Review Board Statement: Not applicable.

Informed Consent Statement: Not applicable.

Data Availability Statement: Not applicable.

Conflicts of Interest: The authors declare no conflict of interest.

References

1. Stocker, T.F.; Qin, D.; Plattner, G.-K.; Alexander, L.V.; Allen, S.K.; Bindoff, N.L.; Bréon, F.-M.; Church, J.A.; Cubasch, U.; Emori, S.; et al. Technical summary. In *Climate Change 2013: The Physical Science Basis. Contribution of Working Group I to the Fifth Assessment Report of the Intergovernmental Panel on Climate Change*; Cambridge University Press: Cambridge, UK, 2013; pp. 33–115.
2. Mutiibwa, D.; Vavrus, S.J.; McAfee, S.A.; Albright, T.P. Recent Spatiotemporal Patterns in Temperature Extremes across Conterminous United States. *J. Geophys. Res. Atmos.* **2015**, *120*, 7378–7392. [CrossRef]
3. Shrestha, U.B.; Shrestha, B.B. Climate Change Amplifies Plant Invasion Hotspots in Nepal. *Divers. Distrib.* **2019**, *25*, 1599–1612. [CrossRef]
4. Van Wilgen, B.W.; Fill, J.M.; Baard, J.; Cheney, C.; Forsyth, A.T.; Kraaij, T. Historical Costs and Projected Future Scenarios for the Management of Invasive Alien Plants in Protected Areas in the Cape Floristic Region. *Biol. Conserv.* **2016**, *200*, 168–177. [CrossRef]
5. Zhou, Y.; Ren, G. Change in Extreme Temperature Event Frequency over Mainland China, 1961–2008. *Clim. Res.* **2011**, *50*, 125–139. [CrossRef]
6. Shukla, R.; Chakraborty, A.; Sachdeva, K.; Joshi, P.K. Agriculture in the Western Himalayas—An Asset Turning into a Liability. *Dev. Pract.* **2018**, *28*, 318–324. [CrossRef]

7. El-Zein, A.; Tonmoy, F.N. Assessment of Vulnerability to Climate Change Using a Multi-Criteria Outranking Approach with Application to Heat Stress in Sydney. *Ecol. Indic.* **2015**, *48*, 207–217. [CrossRef]
8. Bahinipati, C.S.; Venkatachalam, L. Role of Climate Risks and Socio-Economic Factors in Influencing the Impact of Climatic Extremes: A Normalisation Study in the Context of Odisha, India. *Reg. Environ. Chang.* **2016**, *16*, 177–188. [CrossRef]
9. Shukla, R.; Agarwal, A.; Sachdeva, K.; Kurths, J.; Joshi, P.K. Climate Change Perception: An Analysis of Climate Change and Risk Perceptions among Farmer Types of Indian Western Himalayas. *Clim. Chang.* **2019**, *152*, 103–119. [CrossRef]
10. Zhang, J.; Terrones, M.; Park, C.R.; Mukherjee, R.; Monthioux, M.; Koratkar, N.; Kim, Y.S.; Hurt, R.; Frackowiak, E.; Enoki, T.; et al. Carbon Science in 2016: Status, Challenges and Perspectives. *Carbon* **2016**, *98*, 708–732. [CrossRef]
11. Stocker, T.F.; Clarke, G.K.C.; Le Treut, H.; Lindzen, R.S.; Meleshko, V.P.; Mugara, R.K.; Palmer, T.N.; Pierrehumbert, R.T.; Sellers, P.J.; Trenberth, K.E.; et al. Physical Climate Processes and Feedbacks. In *IPCC, 2001: Climate Change 2001: The Scientific Basis. Contribution of Working Group I to the Third Assessment Report of the Intergovernmental Panel on Climate Change*; Houghton, J.T., Ding, Y., Griggs, D.J., Noguer, M., van der Linden, P.J., Dai, X., Maske, K., Eds.; Cambridge University Press: Cambridge, UK, 2001; pp. 417–470, ISBN 978-0-521-01495-3.
12. Easterling, D.R. Climate Extremes: Observations, Modeling, and Impacts. *Science* **2000**, *289*, 2068–2074. [CrossRef]
13. Farajzadeh, H.; Matzarakis, A. Quantification of Climate for Tourism in the Northwest of Iran: Quantification of climate for tourism in the northwest of iran. *Meteorol. Appl.* **2009**, *16*, 545–555. [CrossRef]
14. Keggenhoff, I.; Elizbarashvili, M.; Amiri-Farahani, A.; King, L. Trends in Daily Temperature and Precipitation Extremes over Georgia, 1971–2010. *Weather Clim. Extrem.* **2014**, *4*, 75–85. [CrossRef]
15. Ruml, M.; Gregorić, E.; Vujadinović, M.; Radovanović, S.; Matović, G.; Vuković, A.; Počuča, V.; Stojičić, D. Observed Changes of Temperature Extremes in Serbia over the Period 1961–2010. *Atmos. Res.* **2017**, *183*, 26–41. [CrossRef]
16. Guo, E.; Zhang, J.; Wang, Y.; Quan, L.; Zhang, R.; Zhang, F.; Zhou, M. Spatiotemporal Variations of Extreme Climate Events in Northeast China during 1960–2014. *Ecol. Indic.* **2019**, *96*, 669–683. [CrossRef]
17. Zhang, X.; Alexander, L.; Hegerl, G.C.; Jones, P.; Tank, A.K.; Peterson, T.C.; Trewin, B.; Zwiers, F.W. Indices for Monitoring Changes in Extremes Based on Daily Temperature and Precipitation Data: Indices for Monitoring Changes in Extremes. *WIREs Clim. Chang.* **2011**, *2*, 851–870. [CrossRef]
18. García-Cueto, O.R.; Cavazos, M.T.; de Grau, P.; Santillán-Soto, N. Analysis and Modeling of Extreme Temperatures in Several Cities in Northwestern Mexico under Climate Change Conditions. *Theor. Appl. Climatol.* **2014**, *116*, 211–225. [CrossRef]
19. Wang, H.; Zhang, G.; Li, N.; Zhang, B.; Yang, H. Soil Erodibility as Impacted by Vegetation Restoration Strategies on the Loess Plateau of China: Effect of Vegetation Restoration on Soil Erodibility. *Earth Surf. Process. Landf.* **2019**, *44*, 796–807. [CrossRef]
20. Sharma, P.J.; Loliyana, V.D.; Resmi, S.R.; Timbadiya, P.V.; Patel, P.L. Spatiotemporal Trends in Extreme Rainfall and Temperature Indices over Upper Tapi Basin, India. *Theor. Appl. Climatol.* **2018**, *134*, 1329–1354. [CrossRef]
21. Abeysingha, N.; Singh, M.; Sehgal, V.; Khanna, M.; Pathak, H. Analysis of Rainfall and Temperature Trends in Gomti River Basin. *J. Agric. Phys.* **2014**, *14*, 56–66.
22. Dutta, V.; Kumar, R.; Sharma, U. Assessment of Human-Induced Impacts on Hydrological Regime of Gomti River Basin, India. *Manag. Environ. Qual. Int. J.* **2015**, *26*, 631–649. [CrossRef]
23. Pai, D.; Sridhar, L.; Rajeevan, M.; Sreejith, O.; Satbhai, N.; Mukhopadhyay, B. Development of a New High Spatial Resolution (0.25 × 0.25) Long Period (1901–2010) Daily Gridded Rainfall Data Set over India and Its Comparison with Existing Data Sets over the Region. *Mausam* **2014**, *65*, 1–18.
24. Srivastava, A.K.; Rajeevan, M.; Kshirsagar, S.R. Development of a High Resolution Daily Gridded Temperature Data Set (1969–2005) for the Indian Region. *Atmos. Sci. Lett.* **2009**, *10*, 249–254. [CrossRef]
25. Guntu, R.K.; Maheswaran, R.; Agarwal, A.; Singh, V.P. Accounting for Temporal Variability for Improved Precipitation Regionalization Based on Self-Organizing Map Coupled with Information Theory. *J. Hydrol.* **2020**, *590*, 125236. [CrossRef]
26. Guntu, R.K.; Rathinasamy, M.; Agarwal, A.; Sivakumar, B. Spatiotemporal Variability of Indian Rainfall Using Multiscale Entropy. *J. Hydrol.* **2020**, *587*, 124916. [CrossRef]
27. Pingale, S.M.; Khare, D.; Jat, M.K.; Adamowski, J. Spatial and Temporal Trends of Mean and Extreme Rainfall and Temperature for the 33 Urban Centers of the Arid and Semi-Arid State of Rajasthan, India. *Atmos. Res.* **2014**, *138*, 73–90. [CrossRef]
28. Alexander, L.V.; Fowler, H.J.; Bador, M.; Behrangi, A.; Donat, M.G.; Dunn, R.; Funk, C.; Goldie, J.; Lewis, E.; Rogé, M.; et al. On the Use of Indices to Study Extreme Precipitation on Sub-Daily and Daily Timescales. *Environ. Res. Lett.* **2019**, *14*, 125008. [CrossRef]
29. Sen, P.K. Robustness of Some Nonparametric Procedures in Linear Models. *Ann. Math. Stat.* **1968**, *39*, 1913–1922. [CrossRef]
30. Lettenmaier, D.P.; Wood, E.F.; Wallis, J.R. Hydro-Climatological Trends in the Continental United States, 1948–1988. *J. Clim.* **1994**, *7*, 586–607. [CrossRef]
31. Yue, S.; Hashino, M. Long term trends of annual and monthly precipitation in japan[1]. *JAWRA J. Am. Water Resour. Assoc.* **2003**, *39*, 587–596. [CrossRef]
32. Partal, T.; Kahya, E. Trend Analysis in Turkish Precipitation Data. *Hydrol. Process.* **2006**, *20*, 2011–2026. [CrossRef]
33. Koutsoyiannis, D. Hydrologic Persistence and the Hurst Phenomenon. In *Water Encyclopedia*; Lehr, J.H., Keeley, J., Eds.; John Wiley & Sons, Inc.: Hoboken, NJ, USA, 2005; p. sw434. ISBN 978-0-471-47844-7.
34. Graves, T.; Gramacy, R.; Watkins, N.; Franzke, C. A Brief History of Long Memory: Hurst, Mandelbrot and the Road to ARFIMA, 1951–1980. *Entropy* **2017**, *19*, 437. [CrossRef]

35. Maheswaran, R.; Khosa, R. Comparative Study of Different Wavelets for Hydrologic Forecasting. *Comput. Geosci.* **2012**, *46*, 284–295. [CrossRef]
36. Agarwal, A.; Maheswaran, R.; Kurths, J.; Khosa, R. Wavelet Spectrum and Self-Organizing Maps-Based Approach for Hydrologic Regionalization—A Case Study in the Western United States. *Water Resour. Manag.* **2016**, *30*, 4399–4413. [CrossRef]
37. Tamaddun, K.A.; Kalra, A.; Ahmad, S. Wavelet Analyses of Western US Streamflow with ENSO and PDO. *J. Water Clim. Chang.* **2017**, *8*, 26–39. [CrossRef]
38. Nicolay, S.; Mabille, G.; Fettweis, X.; Erpicum, M. 30 and 43 Months Period Cycles Found in Air Temperature Time Series Using the Morlet Wavelet Method. *Clim. Dyn.* **2009**, *33*, 1117–1129. [CrossRef]
39. Narasimha, R.; Bhattacharyya, S. A Wavelet Cross-Spectral Analysis of Solar–ENSO–Rainfall Connections in the Indian Monsoons. *Appl. Comput. Harmon. Anal.* **2010**, *28*, 285–295. [CrossRef]
40. Farge, M. Wavelet Transforms and Their Applications to Turbulence. *Annu. Rev. Fluid Mech.* **1992**, *24*, 395–458. [CrossRef]
41. Torrence, C.; Compo, G.P. A Practical Guide to Wavelet Analysis. *Bull. Am. Meteorol. Soc.* **1998**, *79*, 61–78. [CrossRef]
42. Yeditha, P.K.; Venkatesh, K.; Maheswaran, R.; Ankit, A. Forecasting of extreme flood events using different satellite precipitation products and wavelet-based machine learning methods. Chaos Interdiscip. *J. Nonlinear Sci.* **2020**, *30*, 063115.
43. Sklar, M. Fonctions de Repartition an Dimensions et Leurs Marges. *Publ. Inst. Statist. Univ. Paris* **1959**, *8*, 229–231.
44. Ganguli, P.; Merz, B. Extreme Coastal Water Levels Exacerbate Fluvial Flood Hazards in Northwestern Europe. *Sci. Rep.* **2019**, *9*, 1–14. [CrossRef]
45. Anandalekshmi, A.; Panicker, S.T.; Adarsh, S.; Muhammed Siddik, A.; Aloysius, S.; Mehjabin, M. Modeling the Concurrent Impact of Extreme Rainfall and Reservoir Storage on Kerala Floods 2018: A Copula Approach. *Model. Earth Syst. Environ.* **2019**, *5*, 1283–1296. [CrossRef]
46. Nelsen, R.B. *An Introduction to Copulas*; Springer Science & Business Media: Berlin/Heidelberg, Germany, 2007.
47. Genest, C.; Rivest, L.-P. Statistical Inference Procedures for Bivariate Archimedean Copulas. *J. Am. Stat. Assoc.* **1993**, *88*, 1034–1043. [CrossRef]
48. Ghosh, S.; Vittal, H.; Sharma, T.; Karmakar, S.; Kasiviswanathan, K.S.; Dhanesh, Y.; Sudheer, K.P.; Gunthe, S.S. Indian Summer Monsoon Rainfall: Implications of Contrasting Trends in the Spatial Variability of Means and Extremes. *PLoS ONE* **2016**, *11*, e0158670. [CrossRef]
49. Singh, D.; Ghosh, S.; Roxy, M.K.; McDermid, S. Indian Summer Monsoon: Extreme Events, Historical Changes, and Role of Anthropogenic Forcings. *WIREs Clim. Chang.* **2019**, *10*. [CrossRef]
50. Rathinasamy, M.; Agarwal, A.; Sivakumar, B.; Marwan, N.; Kurths, J. Wavelet Analysis of Precipitation Extremes over India and Teleconnections to Climate Indices. *Stoch. Environ. Res. Risk Assess.* **2019**, *33*, 2053–2069. [CrossRef]
51. Sadegh, M.; Ragno, E.; AghaKouchak, A. Multivariate C Opula A Nalysis T Oolbox (MvCAT): Describing Dependence and Underlying Uncertainty Using a B Ayesian Framework. *Water Resour. Res.* **2017**, *53*, 5166–5183. [CrossRef]
52. Solomon, S. IPCC (2007): Climate Change the Physical Science Basis. In Proceedings of the Agu Fall Meeting Abstracts, San Francisco, CA, USA, 10–14 December 2007; Volume 2007, p. U43D-01.
53. Alexander, L.V.; Zhang, X.; Peterson, T.C.; Caesar, J.; Gleason, B.; Klein Tank, A.M.G.; Haylock, M.; Collins, D.; Trewin, B.; Rahimzadeh, F.; et al. Global Observed Changes in Daily Climate Extremes of Temperature and Precipitation. *J. Geophys. Res.* **2006**, *111*, D05109. [CrossRef]
54. Arora, M.; Goel, N.K.; Singh, P. Evaluation of Temperature Trends over India/Evaluation de Tendances de Température En Inde. *Hydrol. Sci. J.* **2005**, *50*, 12. [CrossRef]
55. Kothawale, D.; Munot, A.; Krishna Kumar, K. Surface Air Temperature Variability over India during 1901–2007, and Its Association with ENSO. *Clim. Res.* **2010**, *42*, 89–104. [CrossRef]
56. Guntu, R.K.; Agarwal, A. Investigation of Precipitation Variability and Extremes Using Information Theory. *Environ. Sci. Proc.* **2021**, *4*, 14. [CrossRef]
57. Kurths, J.; Agarwal, A.; Marwan, N.; Rathinasamy, M.; Caesar, L.; Krishnan, R.; Merz, B. Unraveling the spatial diversity of Indian precipitation teleconnections via nonlinear multi-scale approach. *Nonlinear Process. Geophys.* **2019**, *26*, 251–266. [CrossRef]
58. Kumar, S.; Chanda, K.; Pasupuleti, S. Spatiotemporal Analysis of Extreme Indices Derived from Daily Precipitation and Temperature for Climate Change Detection over India. *Theor. Appl. Climatol.* **2020**, *140*, 343–357. [CrossRef]
59. Trenberth, K. Changes in Precipitation with Climate Change. *Clim. Res.* **2011**, *47*, 123–138. [CrossRef]

atmosphere

Article

Recognizing the Aggregation Characteristics of Extreme Precipitation Events Using Spatio-Temporal Scanning and the Local Spatial Autocorrelation Model

Changjun Wan [1], Changxiu Cheng [1,2], Sijing Ye [1,*], Shi Shen [1] and Ting Zhang [1]

[1] State Key Laboratory of Earth Surface Processes and Resource Ecology, Beijing Normal University, Beijing 100875, China; wancj@mail.bnu.edu.cn (C.W.); chengcx@bnu.edu.cn (C.C.); shens@bnu.edu.cn (S.S.); zhangting_bnu@mail.bnu.edu.cn (T.Z.)

[2] National Tibetan Plateau Data Center, Beijing 100101, China

* Correspondence: yesj@bnu.edu.cn; Tel.: +86-134-8881-1751

Abstract: Precipitation is an essential climate variable in the hydrologic cycle. Its abnormal change would have a serious impact on the social economy, ecological development and life safety. In recent decades, many studies about extreme precipitation have been performed on spatio-temporal variation patterns under global changes; little research has been conducted on the regionality and persistence, which tend to be more destructive. This study defines extreme precipitation events by percentile method, then applies the spatio-temporal scanning model (STSM) and the local spatial autocorrelation model (LSAM) to explore the spatio-temporal aggregation characteristics of extreme precipitation, taking China in July as a case. The study result showed that the STSM with the LSAM can effectively detect the spatio-temporal accumulation areas. The extreme precipitation events of China in July 2016 have a significant spatio-temporal aggregation characteristic. From the spatial perspective, China's summer extreme precipitation spatio-temporal clusters are mainly distributed in eastern China and northern China, such as Dongting Lake plain, the Circum-Bohai Sea region, Gansu, and Xinjiang. From the temporal perspective, the spatio-temporal clusters of extreme precipitation are mainly distributed in July, and its occurrence was delayed with an increase in latitude, except for in Xinjiang, where extreme precipitation events often take place earlier and persist longer.

Keywords: spatio-temporal patterns; spatio-temporal scanning; local spatial autocorrelation; extreme precipitation; climate change

Citation: Wan, C.; Cheng, C.; Ye, S.; Shen, S.; Zhang, T. Recognizing the Aggregation Characteristics of Extreme Precipitation Events Using Spatio-Temporal Scanning and the Local Spatial Autocorrelation Model. *Atmosphere* **2021**, *12*, 218. https://doi.org/10.3390/atmos12020218

Academic Editor: Ankit Agarwal

Received: 23 December 2020

Accepted: 2 February 2021

Published: 5 February 2021

1. Introduction

Precipitation is a climatic variable with high spatio-temporal variability, playing an important role in the eco-hydrological cycle [1]. An abnormal increase or decrease of precipitation will lead to an imbalance in surface runoff and soil moisture content, causing severe catastrophic effects on socioeconomic development, ecological environmental system and life safety [2,3]. Due to global climate change, the frequency and intensity of extreme precipitation events has increased in most regions [4–7]. Many studies have performed extreme precipitation trends analysis (extreme precipitation, precipitation intensity, precipitation distribution patterns, etc.) under global warming [8–10], while little research has been conducted on the regionality and the persistence of extreme precipitation events [11]. Extreme precipitation events are often more destructive if the intensity and frequency are relatively high within a certain spatial scope and temporal range [12,13].

Identifying regional persistence and aggregation characteristics of extreme precipitation events has mainly gone through three stages. Early studies were mostly based on extreme precipitation indicators, such as the widely used ETCCDMI (Expert Team on Climate Change Detection Monitoring and Indices) [14–16]. Those studies can reflect moderate disasters but cannot effectively characterize the disaster's severity [17,18]. Whereas, extreme precipitation events with long durations, which are often more destructive, have been overlooked.

In the second stage, related researches began to introduce time series to precipitation, the extreme precipitation events were artificially divided into N days (1 day, 3 days or 5 days). Biondi et al. defined days with precipitation values above a certain threshold as a complete and continuous extreme precipitation process, achieving good results [19]. Min et al. used the percentile threshold method to define extreme precipitation events and strong precipitation events by the area index method, then used a single or several uninterrupted extreme precipitation events to discover the persistent extreme precipitation events (1-day, 2-day, 3-day, etc.) [18]. A regional formulation of Intensity–Duration–Frequency curves of point rainfall maxima in a scale-invariant generalized extreme value (GEV) framework was proposed by Blanchet, under the assumptions that extreme daily rainfall is GEV-distributed, and extremes of aggregated daily rainfall follow simple-scaling relationships [20]. Gentilucci et al. [21], using GEV, successfully forecast extreme precipitation events.

In the third stage, space-time interaction becomes an essential feature for identifying extreme precipitation events. Extreme precipitation events are not only related to the duration of extreme precipitation but are also affected by the scope of coverage. Events possessing both persistent and regional characteristics often cause the most serious damage. Jing proposed an intensity–area–duration (IAD) analysis method [22], which was improved from the severity–area–duration (SAD) method created by Andreadis et al. [23], to define an extreme precipitation event when considering both time period and spatial continuity. The temporal range is established when the effective precipitation exceeds the corresponding extreme precipitation threshold on a certain time scale (1 day, 3 days, 5 days, or 7 days). The continuous spatial extent is established when adjacent grid points exceed the threshold during the same temporal range. Although the IAD method can accurately identify the spatial extent and temporal range of regional extreme precipitation events, the values of time scale are determined subjectively, separating the continuity of the rainfall process [14,24]. Meanwhile, these extreme precipitation events are mainly concerned with short-term multi-day extreme precipitation events, but relatively stable continuous extreme precipitation events are likely to cause greater destructive power [25]. Chen et al. redefined regional persistent extreme precipitation events (PEPE) with severe disasters by time intervals and spatial adjacency based on multi-day and single-station PEPEs [25].

Coupling temporal processes and spatial patterns, we can better identify extreme precipitation events [26–28]. However, the above syntheses just superimpose different time segments onto spatial pattern, separating space and time. The spatio-temporal scanning model (STSM) was developed by introducing time dimension into the spatial dynamic scanning window [29–31], and has been broadly applied in infectious diseases, criminology, economics, and geography [32–34]. It considered both spatial extent and temporal range through a scanning window, thus can be used to determine the boundary of extreme precipitation accumulation areas with significant spatial aggregation characteristics.

The definition of extreme precipitation thresholds also plays an important role in understanding the spatial and temporal aggregation characteristics of extreme precipitation events. Early definitions of extreme precipitation thresholds mainly adopted the absolute critical value method, giving one extreme precipitation threshold, which cannot reflect the actual distribution of precipitation extremes [35,36]. The catastrophic extreme precipitation events are related not only to the physical properties, but also to the ecological carrying capacity, which are highly regional. Regions with small spatial scales and similar climate characteristics can use absolute thresholds; regions with large spatial scales should use the percentile method. To better reflect the spatio-temporal characteristics of extreme precipitation events in China, the percentile method is more suitable to be adopted to define extreme precipitation [37,38].

In this paper, we combine the spatio-temporal scanning model (STSM) and the local spatial autocorrelation model (LSAM) to explore the spatio-temporal aggregation characteristics of extreme precipitation. First, we used daily precipitation data from the China Meteorological Forcing Data during 1979 to 2018 as input data; then integrated the 31 × 40 time sliding window and the 95% percentile threshold to extract the extreme precipitation threshold in July 2016. Second, the STSM was applied to detect the spatio-temporal extreme precipitation events in China. The spatio-temporal aggregation characteristics were evaluated by log likelihood ratio (LLR) and the relative risk (RR). Last, the local spatial autocorrelation model (LSAM) was integrated to discover the internal distribution of extreme precipitation in the spatio-temporal accumulation area.

2. Materials and Methods

2.1. Data

Ground-based rain gauge records are stable and have the longest historical precipitation observation data, widely used in hydrology and climate research [39–41]. However, ground-based rain gauge records are sparse at point scale, and there are not enough to develop a reliable high-resolution global dataset and capture the spatio-temporal variation characteristics of precipitation [42]. The China Meteorological Forcing Data [43] has provided long-standing, globally covered precipitation data through the fusion of remote sensing products, reanalysis dataset and in-situ observation data at weather stations, improving the extraction accuracy of extreme precipitation events [43–47]. The forcing dataset is expected to be better with more stations as the input observation dataset. A large number of stations were used to generate the China Meteorological Forcing Data, which allowed it to show superior quality. Two ground-based observation data sources are used in the China Meteorological Forcing Data: China Meteorological Administration's China Meteorological Data Service Center, approximately 700 stations; and the National Oceanic and Atmospheric Administration (NOAA)'s National Centers for Environmental Information (NCEI), approximately 300–400 weather stations in China [45]. This paper uses these daily datasets at 0.1° × 0.1° (longitude, latitude), covering China from 1979 to 2018. The effective precipitation value of each grid point is defined as greater than 1 mm.

2.2. Extreme Precipitation Threshold Extraction Method

The percentile method calculates a percentile as the extreme value for each grid. The detailed calculation is as followed: for each grid, we obtain a 31-day monthly time series of effective precipitation using a forward sliding window of 15 days and a backward sliding window of 15 days; then, we apply the 31-day monthly time range of each grid point from 1979 to 2018 to obtain a 31 × 40 yearly precipitation sequence. Finally, the 95th percentile of the yearly precipitation sequence is extracted as the extreme precipitation threshold for the grid. For example, to calculate the extreme precipitation threshold of a certain grid point on 16 July, the effective precipitation sequence from 1–31 July from 1979 to 2018, and the 95th percentile of the effective sequence is the extreme precipitation threshold for the grid point.

2.3. Spatio-Temporal Scanning Model

STSM selects an event in the scanning area as the center of the bottom surface of the dynamic cylinder scanning window, continuously enlarging the radius of the bottom surface (the upper limit of the radius is generally set less than or equal to 50% of the total number of points in the research area) and the height of the cylinder (the upper limit of the time is generally set greater than or equal to 50% of the maximum time sequence) until reaching the upper limit. This scanning process will repeat for each event in study area (Figure 1). Then, the LLR and RR are calculated based on the actual number of events and expected number of events inside and outside the scanning window. Finally, the sample data of the scanning area are simulated multiple times using the Monte Carlo randomization method to obtain a confidence value of the aggregated regions. The window of the spatio-temporal clustering area needs to satisfy both the LLR greater than 0 (LLR > 0) and the ratio of extreme precipitation events in the window greater than outside (RR > 1).

Among all the clustering areas, the clustering area with the maximum LLR is the first-level accumulation area, indicating that the clustering area has the highest occurrence probability of extreme precipitation.

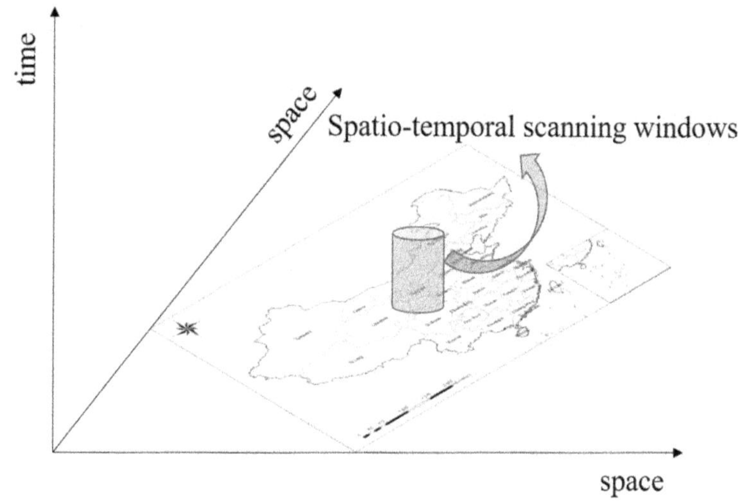

Figure 1. Schematic diagram of the spatio-temporal scanning model (STSM).

There are only two states for the daily precipitation extreme precipitation event data (either extreme or not extreme precipitation events), which are suitable for the Bernoulli distribution model, also known as a typical 0–1 distribution. Both the shape and size of the scanning windows should be considered: based on the structure of the scanning regions and the spatial characteristics of China, we adopted a circular scanning window. The principle of the STSM is as followed:

The probability of extreme precipitation events (P) for the entire study area G can be calculated by Equation (1), where m_z is the actual number of extreme precipitation events in the scanning window Z, $\mu(Z)$ is the total number of events, $\mu(G)$ is the total number of extreme precipitation events in all regions G, m_s is the actual number of extreme precipitation events, $p \in [0,1]$ is the probability of extreme precipitation events in the scanning window, and $q \in [0,1]$ is the probability of extreme precipitation events outside the scanning window.

$$P = p^{m_z}(1-p)^{\mu(Z)-m_z} q^{m_g-m_z}(1-q)^{((\mu(G)-\mu(Z))-(m_g-m_z)} \tag{1}$$

Assuming L(Z) is the likelihood function value of the spatio-temporal scanning window Z, then Equation (1) can be expressed as Equation (2):

$$L(Z) = \left(\frac{m_z}{\mu(Z)}\right)^{m_z}\left(1-\frac{m_z}{\mu(Z)}\right)^{\mu(Z)-m_z}\left(\frac{m_g-m_z}{\mu(G)-\mu(Z)}\right)^{m_g-m_z}\left(1-\frac{m_g-m_z}{\mu(G)-\mu(Z)}\right)^{((\mu(G)-\mu(Z))-(m_g-m_z)} \tag{2}$$

Based on the null assumption, the likelihood function L_0 is given as Equation (3):

$$L_0 = \left(\frac{m_g}{\mu(G)}\right)^{m_g}\left(\frac{\mu(G)-m_G}{\mu(G)}\right)^{\mu(G)-m_g} \tag{3}$$

K is an indication function. If the probability of an extreme precipitation event in the spatio-temporal scanning window is greater than the outside of the window, K = 1; otherwise, it is 0. The maximum LLR in Z has the expression in Equation (4):

$$\max \mathrm{LLR} = \max\log\left[\frac{L(Z)}{L_0}\right] K\left(\frac{m_z}{\mu_z} > \frac{m_g - m_z}{\mu(G) - \mu(Z)}\right) \tag{4}$$

LLR is mainly used to characterize the probability of the spatio-temporal accumulation area; RR is mainly used to characterize the spatio-temporal persistence through measuring the number of extreme precipitation grids inside and outside the spatio-temporal clusters. RR is defined as Equation (5), where n_i and E_i represent the number of extreme precipitation events and the number of expected extreme precipitation events, observed in the spatio-temporal scanning window i, respectively; N represents the total number of extreme precipitation grids in the study area; E represents the total expected number of extreme precipitation grids in the study area. N is equal to E to meet the data requirements of the spatio-temporal scanning.

$$\mathrm{RR_i} = \frac{n_i / E_i}{(N - n_i)/(E - E_i)} \tag{5}$$

To ensure the results are statistically significant ($p <= 0.001$), this experiment randomly generates M data sets according to the Monte Carlo test.

2.4. Local Spatial Autocorrelation Model

Spatial autocorrelation is one of the most commonly used models for spatial aggregation [48,49]. It can measure the correlation of the same object in different spatial locations. Spatial autocorrelation is divided into global spatial autocorrelation and local spatial autocorrelation. The global spatial autocorrelation assumes that the space is homogeneous and one trend exists in the entire region. Its value is between −1.0 and 1.0 through normalization. Moran's I > 0 indicates a positive spatial correlation in the spatial unit; the larger the value, the more aggregative the units. Moran's I < 0 indicates a negative spatial correlation in the spatial unit; the smaller the value, the sparser the units. Moran's I = 0 indicates that the spatial unit does not have spatial autocorrelation and is randomly distributed. However, it can only detect the global spatial aggregation, and cannot locate the specific accumulation area [50–52]. Therefore, it is necessary to introduce local spatial autocorrelation to analyse the local aggregation characteristics aggregation, such as LISA (local indicators of spatial association) and Moran's I scatter plot.

LISA applies the Moran index to each regional unit, describing the similarities between the spatial units and its neighborhood. Moran scatter plots use a two-dimensional coordinate system to visually describe observed variables and spatial lag vectors. The x-axis represents the normalized observations, and the y-axis represents the spatial lag vector (the weighted average of the observations around the observation). The coordinate system is divided into four quadrants (HH, HL, LL, and LH) according to the order of the combination of high and low, which represent the spatial relationship between a certain research area and the adjacent area. Among them, the first quadrant indicates a high-value aggregation (HH), that the Moran index is positive, the z score is positive, and the LISA value is positive (aggregate). We chose Queen's case as the spatial weight matrix with eight neighborhoods. At the same time, the LISA value is calculated according to Equation (6), where S is the cumulative precipitation difference, t_i and t_j are the precipitation at i and j, respectively, t is the mean precipitation, and $Z(I_i)$ obeys the standard normal distribution. $Z(I_i)$ is calculated by normalizing the local autocorrelation index I_i to obtain the significance p of each grid point. This paper considers p-values that do not exceed 0.001 to be statistically significant.

$$\begin{gathered} S = \begin{cases} \frac{1}{n}\sum_{i=1}^{n}(t_i - t), & t_i \geqq t \\ 0, & t_i < t \end{cases}; \\ I_i = \frac{(t_i - \bar{t})}{S\sum_{j=1}^{n} w_{i,j}} \sum_{j=1}^{n} w_{i,j}(t_j - \bar{t}); \\ Z(I_i) = \frac{I - E[I_i]}{\sqrt{V[I_i]}}; \end{gathered} \tag{6}$$

2.5. Experimental Process

The main experimental steps in this paper are shown in Figure 2: first, the extreme precipitation threshold based on the daily precipitation grid data from 1979 to 2018 by percentile threshold method was extracted using the 31 × 40 time sliding window. Then, the extreme precipitation threshold and daily precipitation value were compared to select the extreme precipitation grid. Second, the spatio-temporal clusters of extreme precipitation in July 2016 were extracted using the Bernoulli distribution's STSM. Third, the LSAM was used to detect the hot spots with accumulated precipitation differences to achieve fine positioning. Finally, the extracted spatio-temporal clusters of extreme precipitation were compared with the historical extreme precipitation events to evaluate the extraction accuracy of the STSM.

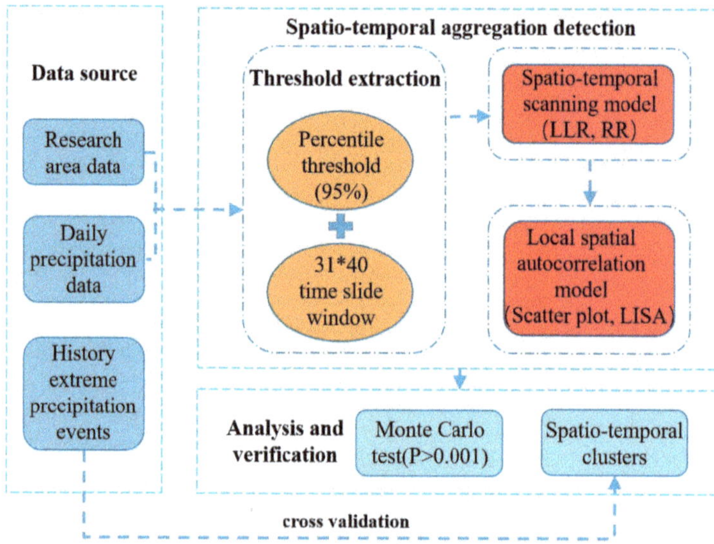

Figure 2. Spatio-temporal aggregation detection flow chart for extreme precipitation. LLR: log likelihood ratio; RR: relative risk; LISA: local indicators of spatial association.

3. Results

3.1. Spatio-Temporal Aggregation Characteristics of Extreme Precipitation Events

When using STSM, the parameters have a great influence on the results. In this paper, we set the spatial scanning window threshold parameter to 5%, 10%, 15%, and 20%, then applied them to extract the spatio-temporal clusters of extreme precipitation events in China July 2016.

As Figure 3 shows, when the spatial scanning window parameter is set to 5%, the scanning result is too fragmented. Almost the whole of China has accumulation areas, and many accumulation areas are small and dense, which cannot effectively represent the spatio-temporal aggregation of extreme precipitation events. When the parameter is set to 10%, the accumulation area starts to have a good degree of discrimination, but the number of accumulation areas is still too high. When the spatial scanning window is set to 15% and 20%, the positions and sizes of the extreme precipitation accumulation areas tend to be similar. Although the accumulation areas of 20% coincides with the accumulation areas of 15%, and both show obvious spatio-temporal aggregation characteristics; the number of accumulation areas are reduced when the parameter is set to 20%, which are relatively rough, overlooking small areas with extreme precipitation, and may include spurious accumulation areas where extreme precipitation events did not exist. The spatial scan window parameter is better set to 15% (an appropriate level of detail and better consistency) with the advantage of a stable calculation result. The time scan window parameter is set to an empirical value of 50% [53]. The results of the 15% scanning window are shown in Table 1 and Figure 4.

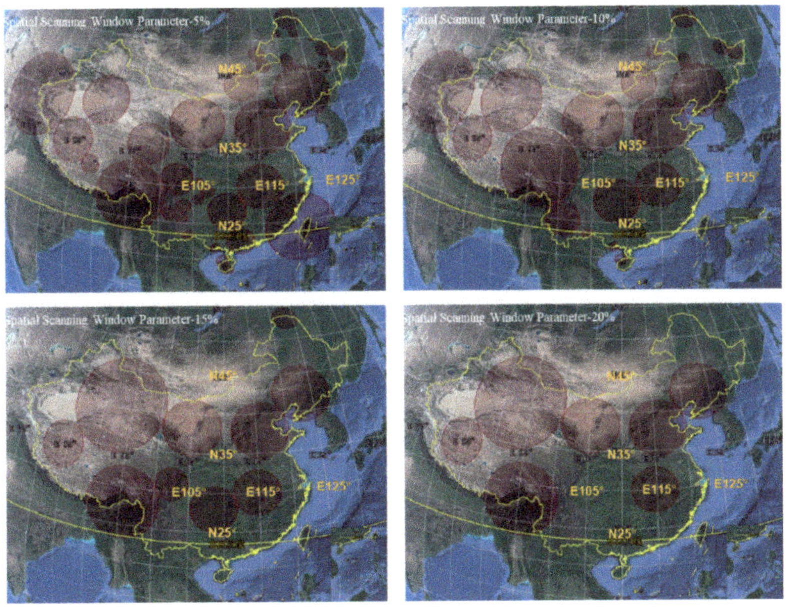

Figure 3. The results for scanning window thresholds set to 5%, 10%, 15%, and 20%.

Table 1. Spatio–temporal scan results of China from 1 to 31 July 2016 (15%).

Cluster Date	Cluster Region	Number of Events	Expected Events	Log Likelihood Ratio	Relative Risk	p-Value
2016/07/19–2016/07/20	1	6000	707.01	9763.16	8.69	0.001
2016/07/28–2016/07/28	2	3598	287.49	8634.29	12.70	0.001
2016/07/08–2016/07/08	3	6703	1079.89	8130.93	6.36	0.001
2016/07/01–2016/07/04	4	5819	931.85	7091.63	6.38	0.001
2016/07/20–2016/07/25	5	8874	2201.44	6724.74	4.15	0.001
2016/07/21–2016/07/25	6	6505	1903.84	3937.85	3.49	0.001
2016/07/10–2016/07/10	7	2474	390.88	3041.10	6.39	0.001
2016/07/26–2016/07/26	8	677	57.90	1493.53	11.73	0.001
2016/07/05–2016/07/05	9	879	122.00	1221.12	7.23	0.001
2016/07/13–2016/07/13	10	1047	232.35	895.07	4.52	0.001

The STSM detected ten statistically significant (through confidence tests) spatio-temporal clusters, which better reflected the distribution of extreme precipitation events in time and space (Figure 4). Cluster 1 is centered on 37.95 N, 115.65 E with a radius of 393.84 km. From the provincial perspective, it is mainly centered on Hebei, Shanxi, Shandong and Henan. The accumulation lasted two days from 19 July to 20 July. Its corresponding LLR is 9763.16, which is 1.13 times the aggregation degree of cluster 2. It's RR also reaches the third highest value (8.69). Cluster 2 centered on 33.95 N, 83.95 E with a radius of 350.26 km, covering the northern Xizang and southern Xinjiang, with the highest RR (12.52). Cluster 8 reaches the second highest value (11.69). Other statistically significant clusters ($p < 0.001$) have relatively small RR values that gradually reduce with a more stable gradient. Cluster 6, 7, 8, 9, and 10's LLR are relatively smaller than cluster 1, 2, 3, 4, 5.

Cluster 5′s LLR is 1.7 times than cluster 6. The extreme precipitation duration is short, and the detected clusters over three days only include cluster 4, cluster 5, and cluster 6.

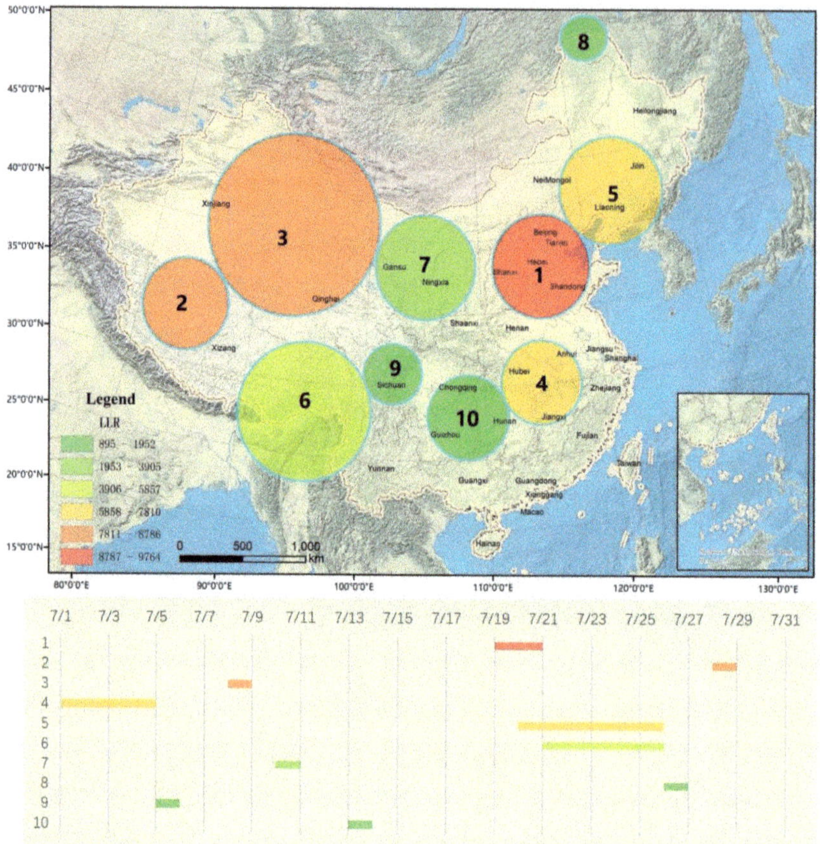

Figure 4. Spatio-temporal aggregation pattern of extreme precipitation events in July 2016, China.

3.2. *Internal Spatio-Temporal Aggregation Characteristics with the Local Spatial Autocorrelation Model*

Combined with the LSAM to further explore the internal aggregation characteristics of the spatio-temporal accumulation area of extreme precipitation, we selected the largest LLR (cluster 1) and largest RR (cluster 8) as examples. The difference between the daily maximum precipitation value and the extreme precipitation threshold was accumulated, then hot spots in the extreme precipitation areas were extracted by GeoDa. Cluster 1 starts from 19 July 2016 to 20 July 2016, lasting for two days. The Moran's I scatter plot (left) and p-values (right) in space are shown in Figure 5.

China's cumulative precipitation difference from 19 to 20 July 2016 has a significant correlation. Moran's I correlation index reaches 0.984636. The high-high value regions (extreme precipitation events surrounded by extreme precipitation events) are mainly distributed in Hebei, Shanxi, Henan, Hubei.

By overlapping the high-high value regions ($p > 0.001$) and the spatio-temporal clusters in Figure 4, we found that cluster 1 had obvious hotspots; cluster 4, cluster 5, and cluster 10 also had some hotspots (Figure 6). They decreased outwards from one of the internal regions, surrounded by low-value regions. The possible reason for this distribution is that the cumulative precipitation difference is selected from 19 July 2016, to 20 July 2016, where extreme precipitation events occurred in these regions; other regions of China are more stable. These results suggest that the combined use of LSAM is helpful for the exploration of the internal aggregation characteristics of these clusters.

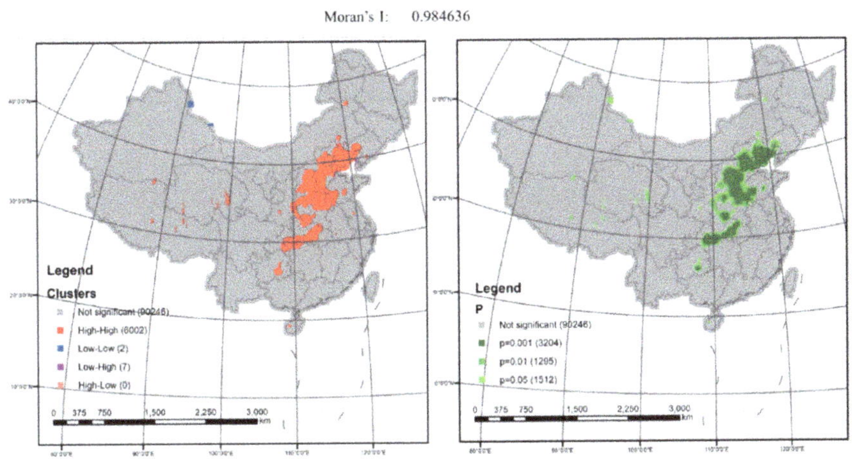

Figure 5. Moran's I scatter plot (**left**) and *p*-values (**right**) in China from 19 to 20 July 2016.

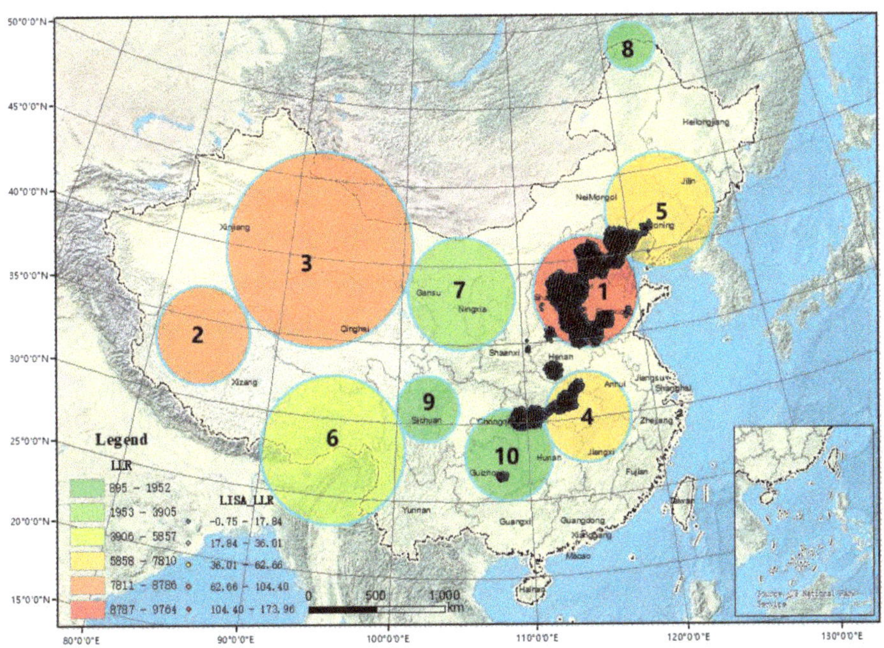

Figure 6. Overlap between the high–high value areas and the clusters of the STSM from 19 to 20 July 2016.

China's cumulative precipitation difference on 28 July 2016 also has a significant correlation (Figure 7). Moran's I correlation index reaches 0.91667, slightly smaller than the highest LLRs Moran's I, perhaps because of sparser precipitation. The high-high values are mainly distributed in cluster 2, cluster 5 and cluster 1, which occurred near 28 July 2016 (Figure 8). In addition, the LISA value on 28 July 2016, which reached 423.97, is larger than the LISA value on 19 to 20 July 2016, which reached 173.96. This indicates that RR is more suitable for characterizing the actual spatio-temporal persistence of the accumulation areas, which is also more catastrophic; LLR is more suitable for characterizing the most likely spatio-temporal accumulation areas. Although the region with a larger RR has a higher probability of occurrence (LLR) of extreme precipitation events, there is no clear positive correlation; that is, the region most likely to have extreme precipitation events does not necessarily have the strongest RR, and the assessment of catastrophic ability must consider the local natural environment.

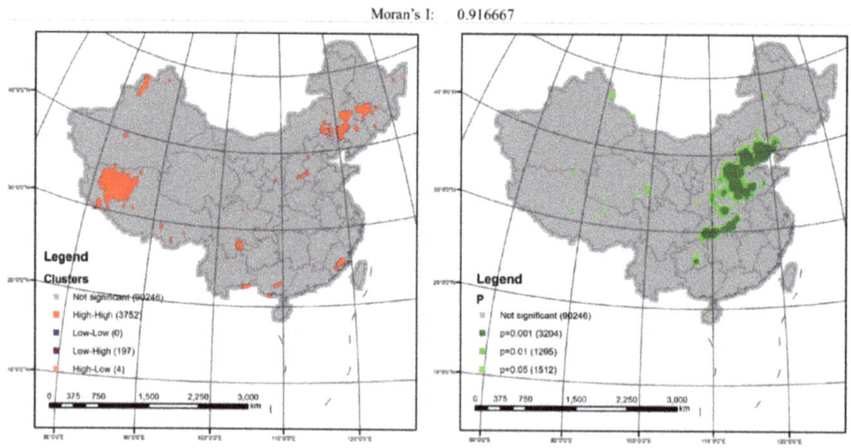

Figure 7. Moran's I scatter plot (**left**) and *p*-values (**right**) in China on 28 July 2016.

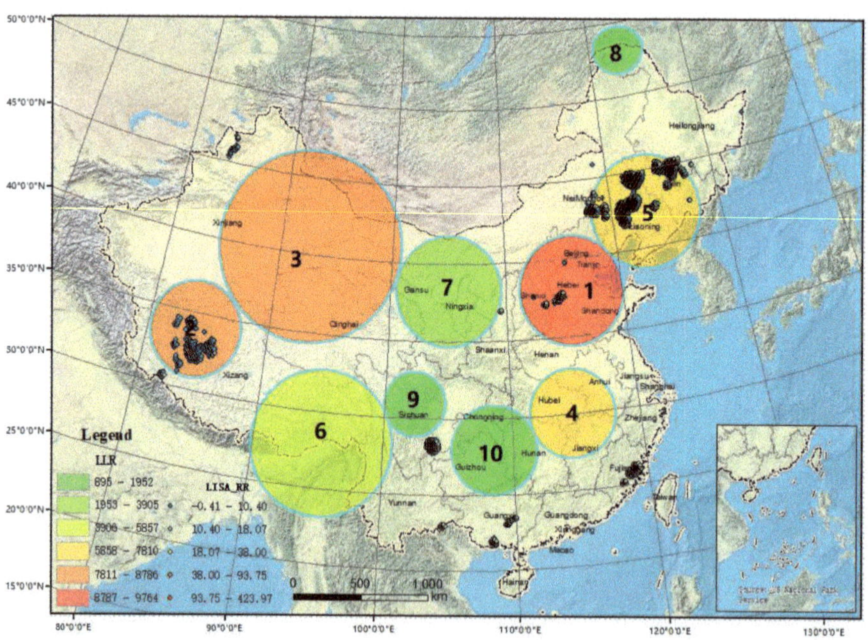

Figure 8. Overlap between the high–high value areas and the clusters of the STSM on 28 July 2016.

Taken together, these results suggest that extreme precipitation events are more likely to occur in eastern and northern China with a significant aggregation, e.g., North China Plain centered on Hebei, north-western China centered on Gansu, and the Tianshan Mountains.

4. Discussion

Our research, coupled the spatial and temporal properties of extreme precipitation events using STSM, successfully discovered the spatio-temporal clusters of extreme precipitation events. To optimize the spatial positions of the extreme precipitation events, LSAM was used to further detect the internal distribution of extreme precipitation clusters. According to the meteorological reports provided by the National Information Centre, extreme precipitation in 2016 was mainly concentrated in East China and North China, and the North China Plain is the rainstorm center far beyond the same period of national precipitation from 18 to 20 July 2016 [54,55]. Zhou et al. used 2016 rainwater and typhoon information collected by the Water Resources and Hydrology Bureau of China to analyze the extreme precipitation events, finding that the extreme precipitation process began early, was long lasting, and widely covered China in 2016 [56]. Twenty-eight provinces and cities were influenced by these extreme precipitation events: the first echelon affected by extreme precipitation events were along the Yangtze river: Jiangxi, Hunan, Zhejiang, Guangdong, and Fujian; the second echelon (Guangxi, Shanghai, Anhui, Chongqing, Hubei, Jiangsu, and Guizhou) was basically located on the outskirts of the first echelon [56]. They are highly coincident with economically developed regions, causing a major impact on China's economic development. Extreme precipitation events have also occurred in the inland areas, e.g., Xinjiang and Gansu corresponded to cluster 2 and cluster 3. These studies are highly consistent with our findings, which showed that STSM combined with LSAM is useful in recognizing the aggregation characteristics of extreme precipitation events. Particularly, this method contributes to a decrease in subjective and an increase in objective information when determining the location and range of extreme precipitation areas through coupling the spatial and temporal scale, and enables the quantitative evaluation of these areas with LLR and RR.

There are also some limitations in this study. First, although the percentile threshold method can reduce the influence of spatio-temporal heterogeneity and climate multi-deformation, it cannot eliminate the influence of the threshold divided subjectivity by human. Second, the spatial resolution ($0.1^\circ \times 0.1^\circ$) and time scale (d) of the China Meteorological Forcing Data are still rough. The extraction accuracy of extreme precipitation events is insufficient. Third, the fixed window shape of the STSM limits the fine-grained extraction of spatio-temporal clusters; thus, it is easy to obtain false spatio-temporal clusters. Finally, the RR can reflect the spatio-temporal persistence of the clusters of extreme precipitation, but it cannot indicate the concentration degrees of different intensities or frequencies. It is necessary to improve the quantitative description index. If more refined data and more accurate models can be used, the detection of spatio-temporal clusters of extreme precipitation events will have more practical significance.

5. Conclusions

In this study, we coupled the spatial extent and the temporal range of extreme precipitation events to analyze the spatio-temporal aggregation characteristics by using the STSM (spatio-temporal scanning model) and LSAM (local spatial autocorrelation model), then applied this method to China. Through the STSM's dynamic scanning window, the spatio-temporal clusters break the limitation of subjective divisions, better synthesizing the temporal and spatial properties of extreme precipitation with an unbiased result. Combined with LSAM, we can detect the precise location of extreme precipitation in spatio-temporal clusters. The result showed that China's summer extreme precipitation events in 2016 are significantly aggregated. The clusters of extreme precipitation events are mainly distributed in eastern and northern China, such as cluster 1 located on Hebei, cluster 2 and cluster 3 located around Xinjiang, cluster 4 located on the middle basin of the Yangtze River and Xinjiang.

The LLR and RR in STSM are important quantitative evaluation indicators, which are not only helpful detect the location of extreme precipitation, but also for the quantitative evaluation of the aggregation degree. Although the clusters of extreme precipitation events with a larger RR also have a larger LLR, there is no obvious positive correlation among them. RR is more representable to catastrophic extreme precipitation.

Author Contributions: Conceptualization, S.Y. and C.C.; methodology, C.W. and T.Z.; software, C.W.; validation, C.W.; formal analysis, C.W., S.S. and T.Z.; investigation, S.S.; resources, S.Y.; data curation, S.Y.; writing—original draft preparation, C.W. and S.Y.; writing—review and editing, C.W. and S.Y.; visualization, C.W.; supervision, S.Y.; project administration, S.Y.; funding acquisition, C.C. All authors have read and agreed to the published version of the manuscript.

Funding: This work was supported by National Key Research and Development Plan of China [grant numbers No.2019YFA0606901]; National Natural Science Foundation of China [grant number 41801300, 41901316].

Acknowledgments: We would like to thank the high-performance computing support from the Center for Geodata and Analysis, Faculty of Geographical Science, Beijing Normal University [https://gda.bnu.edu.cn/]. We thank Chiyuan Miao for helpful discussion.

Conflicts of Interest: The authors declare no conflict of interest.

References

1. Tammets, T.; Jaagus, J. Climatology of precipitation extremes in Estonia using the method of moving precipitation totals. *Theor. Appl. Climatol.* **2013**, *111*, 623–639. [CrossRef]
2. Zhang, Q.; Zhou, Y.; Singh, V.P.; Li, J. Scaling and clustering effects of extreme precipitation distributions. *J. Hydrol.* **2012**, *454*, 187–194. [CrossRef]
3. Shukla, R.; Agarwal, A.; Sachdeva, K.; Kurths, J.; Joshi, P.K. Climate change perception: An analysis of climate change and risk perceptions among farmer types of Indian Western Himalayas. *Clim. Chang.* **2019**, *152*, 103–119. [CrossRef]
4. Alexander, L.V.; Zhang, X.; Peterson, T.C.; Caesar, J.; Gleason, B.; Klein Tank, A.M.G.; Tagipour, A. Global observed changes in daily climate extremes of temperature and precipitation. *J. Geophys. Res. Atmos.* **2006**, *111*, 5. [CrossRef]
5. Allan, R.P.; Soden, B.J. Atmospheric warming and the amplification of precipitation extremes. *Science* **2008**, *321*, 1481–1484. [CrossRef] [PubMed]
6. Gao, T.; Xie, L. Study on progress of the trends and physical causes of extreme precipitation in China during the last 50 years. *Adv. Earth Sci.* **2014**, *29*, 577–589.
7. IPCC Climate Change. *The Physical Science Basis. Working Group I Contribution to the Fifth Assessment Report of the Intergovernmental Panel on Climate Change*; Cambridge University Press: Cambridge, UK, 2013.
8. Dai, A.; Trenberth, K.E.; Karl, T.R. Global variations in droughts and wet spells: 1900–1995. *Geophys. Res. Lett.* **1998**, *25*, 3367–3370. [CrossRef]
9. Goswami, B.N.; Venugopal, V.; Sengupta, D.; Madhusoodanan, M.S.; Xavier, P.K. Increasing trend of extreme rain events over India in a warming environment. *Science* **2006**, *314*, 1442–1445. [CrossRef]
10. Liu, B.; Chen, J.; Chen, X.; Lian, Y.; Wu, L. Uncertainty in determining extreme precipitation thresholds. *J. Hydrol.* **2013**, *503*, 233–245. [CrossRef]
11. Min, S.; Qian, Y.F. Regionality and persistence of extreme precipitation events in China. *Adv. Water Sci.* **2008**, *19*, 763–771.
12. Yang, P. Research of Group-occurring Extreme Temperature and Precipitation Events during 1960–2005. Ph.D. Thesis, Lanzhou University, Lanzhou, China, 2009.
13. Cheung, K.K.; Ozturk, U. Synchronization of extreme rainfall during the Australian summer monsoon: Complex network perspectives. *Chaos Interdiscip. J. Nonlinear Sci.* **2020**, *30*, 063117. [CrossRef] [PubMed]
14. Kunkel, K.E.; Easterling, D.R.; Redmond, K.; Hubbard, K. Temporal variations of extreme precipitation events in the United States: 1895–2000. *Geophys. Res. Lett.* **2003**, *30*, 17. [CrossRef]
15. Groisman, P.Y.; Knight, R.W.; Easterling, D.R.; Karl, T.R.; Hegerl, G.C.; Razuvaev, V.N. Trends in intense precipitation in the climate record. *J. Clim.* **2005**, *18*, 1326–1350. [CrossRef]
16. Ren, F.; Cui, D.; Gong, Z.; Wang, Y.; Zou, X.; Li, Y.; Wang, X. An objective identification technique for regional extreme events. *J. Clim.* **2012**, *25*, 7015–7027. [CrossRef]
17. Easterling, D.R.; Evans, J.L.; Groisman, P.Y.; Karl, T.R.; Kunkel, K.E.; Ambenje, P. Observed variability and trends in extreme climate events: A brief review. *Bull. Am. Meteorol. Soc.* **2000**, *81*, 417–426. [CrossRef]
18. Biondi, F.; Kozubowski, T.J.; Panorska, A.K. Stochastic modeling of regime shifts. *Clim. Res.* **2002**, *23*, 23–30. [CrossRef]

19. Biondi, F.; Kozubowski, T.J.; Panorska, A.K.; Saito, L. A new stochastic model of episode peak and duration for eco-hydro-climatic applications. *Ecol. Model.* **2008**, *211*, 383–395. [CrossRef]
20. Blanchet, J.; Ceresetti, D.; Molinié, G.; Creutin, J.D. A regional GEV scale-invariant framework for Intensity–Duration–Frequency analysis. *J. Hydrol.* **2016**, *540*, 82–95. [CrossRef]
21. Gentilucci, M.; Barbieri, M.; Lee, H.S.; Zardi, D. Analysis of rainfall trends and extreme precipitation in the Middle Adriatic Side, Marche Region (Central Italy). *Water* **2019**, *11*, 1948. [CrossRef]
22. Jing, C.; Jiang, T.; Wang, Y.J.; Chen, J.; Jian, D.; Luo, L.; Buda, S.U. A study on regional extreme precipitation events and the exposure of population and economy in China. *Acta Meteor. Sin.* **2016**, *74*, 572–582.
23. Andreadis, K.M.; Clark, E.A.; Wood, A.W.; Hamlet, A.F.; Lettenmaier, D.P. Twentieth-century drought in the conterminous United States. *J. Hydrometeorol.* **2005**, *6*, 985–1001. [CrossRef]
24. Yuan, N.; Fu, Z.; Mao, J. Different scaling behaviors in daily temperature records over China. *Phys. A Stat. Mech. Appl.* **2010**, *389*, 4087–4095. [CrossRef]
25. Chen, Y.; Zhai, P. Persistent extreme precipitation events in China during 1951–2010. *Clim. Res.* **2013**, *57*, 143–155. [CrossRef]
26. Ghosh, S.; Mallick, B.K. A hierarchical Bayesian spatio-temporal model for extreme precipitation events. *Environmetrics* **2011**, *22*, 192–204. [CrossRef]
27. Zhang, Q.; Singh, V.P.; Li, J.; Jiang, F.; Bai, Y. Spatio-temporal variations of precipitation extremes in Xinjiang, China. *J. Hydrol.* **2012**, *434*, 7–18. [CrossRef]
28. Agarwal, A. Unraveling Spatio-Temporal Climatic Patterns via Multi-Scale Complex Networks. Ph.D. Thesis, Universität Potsdam, Potsdam, Germany, 2019.
29. Naus, J.I. The distribution of the size of the maximum cluster of points on a line. *J. Am. Stat. Assoc.* **1965**, *60*, 532–538. [CrossRef]
30. Kulldorff, M.; Nagarwalla, N. Spatial disease clusters: Detection and inference. *Stat. Med.* **1995**, *14*, 799–810. [CrossRef]
31. Kulldorff, M.; Athas, W.F.; Feurer, E.J.; Miller, B.A.; Key, C.R. Evaluating cluster alarms: A space-time scan statistic and brain cancer in Los Alamos, New Mexico. *Am. J. Public Health* **1998**, *88*, 1377–1380. [CrossRef]
32. Kim, Y.; O'Kelly, M. A bootstrap based space–time surveillance model with an application to crime occurrences. *J. Geogr. Syst.* **2008**, *10*, 141–165. [CrossRef]
33. Horn, L.M. Precipitation Associated with Increased Diarrheal Disease in Mozambique; A Time Series Analysis. Ph.D. Thesis, University of Washington, Washington, DC, USA, 2017.
34. Ye, S.J.; Lu, S.H.; Bai, X.S.; Gu, J.F. ResNet-Locust-BN network-based automatic identification of east asian migratory locust species and instars from RGB images. *Insects* **2020**, *11*, 458. [CrossRef] [PubMed]
35. Gong, D.Y.; Wang, S.W. Severe summer rainfall in China associated with enhanced global warming. *Clim. Res.* **2000**, *16*, 51–59. [CrossRef]
36. Yao, C.; Yang, S.; Qian, W.; Lin, Z.; Wen, M. Regional summer precipitation events in Asia and their changes in the past decades. *J. Geophys. Res. Atmos.* **2008**, *113*, 17. [CrossRef]
37. Sun, Q.H.; Miao, C.Y.; Duan, Q.Y. Changes in the Spatial Heterogeneity and Annual Distribution of Observed Precipitation across China. *J. Clim.* **2017**, *30*, 9399–9416. [CrossRef]
38. Miao, C.Y.; Duan, Q.Y.; Sun, Q.H.; Lei, X.H.; Li, H. Non-uniform changes in different categories of precipitation intensity across China and the associated large-scale circulations. *Environ. Res. Lett.* **2019**, *14*. [CrossRef]
39. Xie, P.; Janowiak, J.E.; Arkin, P.A.; Adler, R.; Gruber, A.; Ferraro, R.; Curtis, S. GPCP pentad precipitation analyses: An experimental dataset based on gauge observations and satellite estimates. *J. Clim.* **2003**, *16*, 2197–2214. [CrossRef]
40. Higgins, R.W.; Silva, V.B.S.; Shi, W.; Larson, J. Relationships between climate variability and fluctuations in daily precipitation over the United States. *J. Clim.* **2007**, *20*, 3561–3579. [CrossRef]
41. Ye, S.J.; Zhu, D.H.; Yao, X.C.; Zhang, N.; Fang, S.; Li, L. Development of a Highly Flexible Mobile GIS-Based System for Collecting Arable Land Quality Data. *IEEE J. Sel. Top. Appl. Earth Obs. Remote Sens.* **2014**, *5*, 14. [CrossRef]
42. Yatagai, A.; Arakawa, O.; Kamiguchi, K.; Kawamoto, H.; Nodzu, M.I.; Hamada, A. A 44-year daily gridded precipitation dataset for Asia based on a dense network of rain gauges. *Sola* **2009**, *5*, 137–140. [CrossRef]
43. Yang, K. China Meteorological Forcing Data (1979–2018). *Nat. Tibet. Plateau Data Cent.* **2018**. [CrossRef]
44. Yang, K.; He, J.; Tang, W.; Qin, J.; Cheng, C.C. On downward shortwave and longwave radiations over high altitude regions: Observation and modeling in the Tibetan Plateau. *Agric. Forest Meteorol.* **2010**, *150*, 38–46. [CrossRef]
45. He, J.; Yang, K.; Tang, W.; Lu, H.; Qin, J.; Chen, Y.; Li, X. The first high-resolution meteorological forcing dataset for land process studies over China. *Sci. Data* **2020**, *7*, 1–11. [CrossRef]
46. Ye, S.J.; Yan, T.L.; Yue, Y.-L.; Lin, W.-Y.; Li, L.; Yao, X.-C.; Mu, Q.-Y.; Li, Y.-Q.; Zhu, D. Developing a reversible rapid coordinate transformation model for the cylindrical projection. *Comput. Geosci.* **2016**, *89*, 44–56. [CrossRef]
47. Ye, S.; Liu, D.; Yao, X.; Tang, H.; Xiong, Q.; Zhuo, W.; Du, Z.; Huang, J.; Su, W.; Shen, S.; et al. RDCRMG: A Raster Dataset Clean & Reconstitution Multi-Grid Architecture for Remote Sensing Monitoring of Vegetation Dryness. *Remote Sens.* **2018**, *10*, 1376. [CrossRef]
48. Sokal, R.R.; Thomson, J.D. Applications of spatial autocorrelation in ecology. In *Develoments in Numerical Ecology*; Springer: Berlin/Heidelberg, Germany, 1987; pp. 431–466.
49. Ord, J.K.; Getis, A. Testing for local spatial autocorrelation in the presence of global autocorrelation. *J. Reg. Sci.* **2001**, *41*, 411–432. [CrossRef]

50. Wang, P.A.; Luo, W.H.; Bai, Y.P. Comparative analysis of aggregation detection based on spatial autocorrelation and spatial-temporal scan statistics. *Hum. Geogr.* **2012**, *27*, 119–127.
51. Ye, S.J.; Song, C.Q.; Shen, S.; Gao, P.C.; Cheng, C.X.; Cheng, F.; Wan, C.J.; Zhu, D.H. Spatial pattern of arable land-use intensity in China. *Land Use Policy* **2020**, *99*, 104845. [CrossRef]
52. Ye, S.J.; Cheng, C.X.; Song, C.Q.; Shen, S. Visualizing bivariate local spatial autocorrelation between commodity revealed comparative advantage index of China and USA from a new space perspective. *Environ. Plan. A Econ. Space* **2020**, *9*. [CrossRef]
53. Sherman, R.L.; Henry, K.A.; Tannenbaum, S.L.; Feaster, D.J.; Kobetz, E.; Lee, D.J. Peer reviewed: Applying spatial analysis tools in public health: An example using SaTScan to detect geographic targets for colorectal cancer screening interventions. *Prev. Chronic Dis.* **2014**, *11*, E41. [CrossRef] [PubMed]
54. BI, B.; Zhang, X.; Dai, K. Characteristics of 2016 severe convective weather and extreme rainfalls under the background of super El Niño. *Chin. Sci. Bull.* **2017**, *62*, 928–937. [CrossRef]
55. Gao, R.; Song, L.C.; Zhong, H.L. Characteristics of extreme precipitation in China during the 2016 flood season and comparison with the 1998 situation. *Meteor Mon.* **2018**, *44*, 699–703.
56. Zhou, G.L.; Gao, W.Q.; Huang, C.X. Analysis of the extreme rainstorm events in China in 2016. *China Flood Control Drought Relief* **2017**, *27*, 75. (In Chinese)

atmosphere

MDPI

Article

Inter-Comparison of Gauge-Based Gridded Data, Reanalysis and Satellite Precipitation Product with an Emphasis on Hydrological Modeling

Sridhara Setti [1,2], Rathinasamy Maheswaran [1], Venkataramana Sridhar [3,*], Kamal Kumar Barik [2], Bruno Merz [4] and Ankit Agarwal [4,5]

1 Department of Civil Engineering, MVGR College of Engineering, Vizianagaram, Andhra Pradesh 535005, India; sridhara.setti@gmail.com (S.S.); maheswaran27@yahoo.co.in (R.M.)
2 Department of Civil Engineering, Centurion University of Technology and Management, Bhubaneswar, Odisha 752050, India; kamal.barik@cutm.ac.in
3 Department of Biological Systems Engineering, Virginia Polytechnic Institute and State University, Blacksburg, VA 24061, USA
4 GFZ German Research Centre for Geosciences, Section 4.4: Hydrology, Telegrafenberg, 14473 Potsdam, Germany; bruno.merz@gfz-potsdam.de (B.M.); agarwal@pik-potsdam.de (A.A.)
5 Department of Hydrology, Indian Institute of Technology, Roorkee, Uttarakhand 247667, India
* Correspondence: vsri@vt.edu

Received: 13 October 2020; Accepted: 17 November 2020; Published: 20 November 2020

check for updates

Abstract: Precipitation is essential for modeling the hydrologic behavior of watersheds. There exist multiple precipitation products of different sources and precision. We evaluate the influence of different precipitation product on model parameters and streamflow predictive uncertainty using a soil water assessment tool (SWAT) model for a forest dominated catchment in India. We used IMD (gridded rainfall dataset), TRMM (satellite product), bias-corrected TRMM (corrected satellite product) and NCEP-CFSR (reanalysis dataset) over a period from 1998–2012 for simulating streamflow. The precipitation analysis using statistical measures revealed that the TRMM and CFSR data slightly overestimate rainfall compared to the ground-based IMD data. However, the TRMM estimates improved, applying a bias correction. The Nash–Sutcliffe (and R^2) values for TRMM, TRMMbias and CFSR, are 0.58 (0.62), 0.62 (0.63) and 0.52 (0.54), respectively at model calibrated with IMD data (Scenario A). The models of each precipitation product (Scenario B) yielded Nash–Sutcliffe (and R^2) values 0.71 (0.76), 0.74 (0.78) and 0.76 (0.77) for TRMM, TRMMbias and CFSR datasets, respectively. Thus, the hydrological model-based evaluation revealed that the model calibration with individual rainfall data as input showed increased accuracy in the streamflow simulation. IMD and TRMM forced models to perform better in capturing the streamflow simulations than the CFSR reanalysis-driven model. Overall, our results showed that TRMM data after proper correction could be a good alternative for ground observations for driving hydrological models.

Keywords: parameter and prediction uncertainty; IMD; TRMM; CFSR; Nagavali River Basin Region (NRB)

1. Introduction

Precipitation data are crucial to many applications related to human life and water management. Examples include estimating the hydrological water balance [1], improving management practices, hydropower planning, development projects and flood controls. Conventionally, rain gauges provide the direct measurement of precipitation. However, rain gauges are often not enough to correctly resolve precipitation and precipitation-related processes. Moreover, the spatial coverage of the rain gauge over a large area is low, and they require huge investments [2].

With the advancements in remote sensing and high computation facilities, several precipitation products from various sources are available. For instance, the National Center for Environmental Prediction Climate Forecast System Reanalysis (NCEP-CFSR) [3], Modern-Era Retrospective analysis for Research and Applications (MERRA) [4] and Global Land Data Assimilation System (GLDAS) [5] are different reanalysis data. Similarly, the Tropical Rainfall Measuring Mission (TRMM) [6], Precipitation Estimation from Remotely Sensed Information using Artificial Neural Networks (PERSIANN) [7], Climate Prediction Center MORPHing (CMORPH) [8], Climate Hazards Group Infrared Precipitation with Station data (CHIRPS) [9], Global Precipitation Measurement (GPM), Integrated Multi-Satellite Retrievals GPM (IMERG) [10,11] are satellite dataset commonly available. A few gauge-gridded datasets are the Asian Precipitation Highly Resolved Observational Data Integration towards Evaluation (APHRODITE) [12,13], Indian Meteorological Department (IMD) gridded datasets [14]. All these different precipitation products have been shown to be accurate and comparable with the ground observations in various circumstances. Moreover, remote sensing-based products [15] and reanalysis data [16] could be a viable alternative for ground-based observation [17] in complex terrain where observed gauge data are of low quality, sparse or non-existent for flood forecasting, drought monitoring [15,18–20] and water balance studies [21].

Among different reanalysis products, NCEP-CFSR is one of the commonly used for precipitation application [3,22], and various studies investigated the evaluation and intercomparison of the reanalysis data sets. Wang et al. [23] reported that CFSR has better precipitation distribution in comparison with the NCEP- NCAR data and ERA data sets. Similarly, Rienecker et al. [4] reported that the performance of CFSR and MERRA was similar in capturing the quantity of precipitation. Dile and Srinivasan [3] tested the applicability of CFSR data for hydrological modeling of the Nile River Basin and concluded that the CFSR could be a valuable alternative for data-scarce regions. Roth and Lehmann [24] compared the conventional weather data and CFSR data for simulating discharge and sediment volume for three catchments in Ethiopia and found that the CFSR simulation is not satisfactory. On similar lines, Tolera et al. [25] showed that the CFSR data set could be reliably used for the streamflow simulation with a caution that CFSR estimates are higher than the observed rainfall.

Moreover, high-resolution satellite remote-sensing products are vital in driving hydrological models, especially in flood-prone complex terrain [26–28]. Satellite data were heavily used for driving different hydrological models, including the SWAT model [15,21,29–32] and variable infiltration capacity (VIC) [33–35]. Furthermore, satellite data are useful in understanding the anthropogenic impacts on hydrology, evaluating the utility of watershed management practices, and predicting the future status of water quality and quantity. A multitude number of studies showed that the ability and reliability of the satellite-based precipitation products in driving the hydrological model fluctuate mainly due to variability in the catchment area, seasonality, topography, climatic characteristics, geographical location, satellite product type [15,36–42]. Studies mentioned above reported satisfactory model performance by using different TRMM precipitation products. Specifically, Zhu et al. [32] reported that performances were achieved on daily and monthly scales, respectively, for TRMM (3B42) data. Similarly, Li et al. [15] investigated the adequacy of TRMM satellite rainfall data in driving a hydrological model for Tiaoxi catchment, Taihu lake basin, China. Yuan et al. [43] used two TRMM products for statistical and hydrological assessment of these products at a sub-daily time scale in Myanmar using the SWAT model; they found all satellite data products showed acceptable results during the simulation of streamflow at the sub-daily scale. Apart from these studies, various other researchers also reported that the direct input of satellite-based precipitation products in hydrological models underperform in comparison with the observed ground-based measurements [29,44]. However, it is to be noted that, by applying a suitable correction, the performance of these datasets is drastically improved. Zhang et al. [41] found that the corrected TRMM multi-satellite precipitation analysis

(TMPA 3B42V7) showed a better understanding than the original 3b42v7 data. Similar inferences were reported by Bitew et al. [29] and Tuo et al. [2] in their respective analysis.

All the studies mentioned above strengthened the need for thorough validation of these precipitation products and bias correction before using them as an input for the hydrological model. In addition, errors and uncertainties associated with inputs (precipitation products) have a high probability of inducing it in hydrologic simulations [45]. A multitude of studies showed that the model recalibration using the precipitation product considerably increases the model performance [2,17,41,46–52]. However, the model parameter ranges obtained through the recalibration with the precipitation data may be improbable, thereby questioning the model's applicability for real-world applications. Further, in such recalibration of the models, it is essential to estimate the parameter uncertainty due to the precipitation input. Thus, validations and evaluations of these precipitation products are critical for any hydrologic modeling study [53,54]. Further, the calibration and output of the hydrological model is a function of the input precipitation characteristics. Therefore, it is imperative to evaluate the effect of the different precipitation products on the parameter estimation (calibration) and the streamflow simulations. It can also be stated that there is enough scope for advancement in understanding the effect of the different precipitation products on the hydrologic model- based streamflow simulations, parameter estimation (Calibration), predictive uncertainty and its applicability in Indian subcontinents.

Even though there has been a plethora of work done in terms of comparison of the reanalysis and satellite dataset individually, there has been a limited number of studies on intercomparison of different classes of precipitation data sets. Therefore, in this study, we compared three different types of precipitation products that are commonly used, (i) gauge-based rainfall product (IMD), (ii) reanalysis data (NCEP-CFSR) and (iii) satellite precipitation (TRMM and TRMM corrected). This study aimed to (1) compare the statistical characteristics of four different precipitation data sets, (2) investigate the adequacy of these precipitation products in driving the semi-distributed SWAT hydrological model and (3) evaluate the parameter uncertainty involved. These precipitation inputs are used to develop a semi-distributed SWAT model for a semi-arid, forest dominated coastal basin, "Nagavali River Basin in India".

2. Material and Methods

2.1. Study Area

We selected the Nagavali River Basin (*NRB*) since the literature on semi-arid and semi-humid climatic regions is limited. NRB lies between the Godavari and Mahanadi river basins in the eastern part of India, with a catchment area of about 9510 km^2 (Figure 1). The river begins from the Eastern Ghats at an elevation of 1600 m. It traverses about 160 km through Odisha and enters Andhra Pradesh near Kuneru village of Vizianagaram district at an elevation of about 152 m [55]. The extent of the basin is 18°17′ to 19°44′ north latitudes and 82°53′ to 83°54′ east longitudes. The stream length ranges around 255 km, out of which the main 62% lies in Odisha and the rest in Andhra Pradesh.

In this study, we considered the catchment area of 9056 km^2 up to the gauging and discharge measurement station at Srikakulam. NRB gets typical yearly precipitation of around 1140 mm, and the annual average temperature of the river basin is 28.2 °C in post-monsoon and 34.2 °C in pre-monsoon.

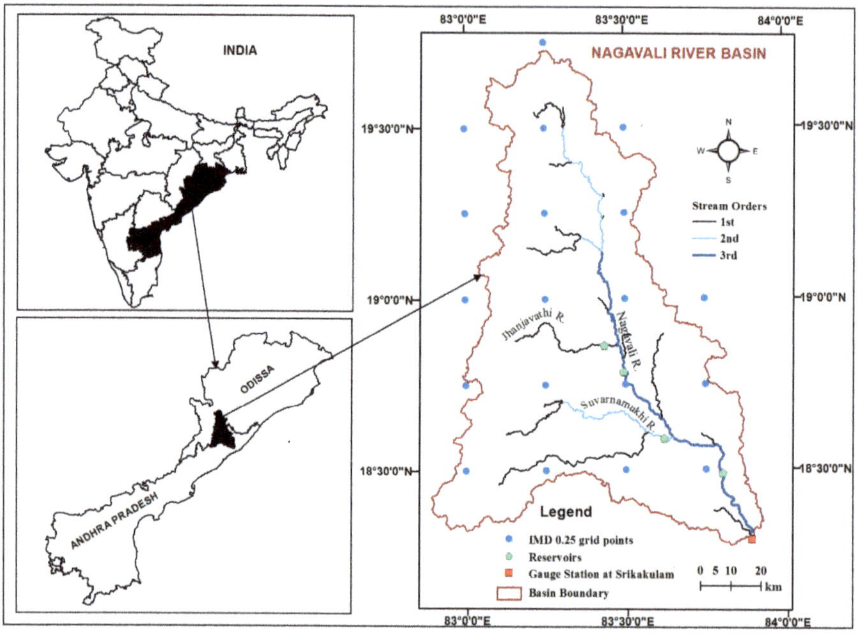

Figure 1. Index map showing the geographical location of Nagavali River Basin (NRB) along with the stream network, gauge location in red and precipitation grid points in blue. Total 19 grid points of the Indian Meteorological Department (IMD) were located within the NRB river basin.

2.2. Datasets

Four precipitation products from different sources and of different resolutions that are the most frequently used in the hydrological application are used. This include gridded IMD (0.25° × 0.25°), TRMM 3B42 (0.25° × 0.25°), bias-corrected TRMM (0.25° × 0.25°) and reanalysis NCEP-CFSR (0.31° × 0.31°). All the datasets are selected for the common observation period from 1998–2012. Table 1 and Supplementary Information Section S1 present detailed information about the precipitation data.

Table 1. Detailed information and sources of different rainfall datasets used in this study. IMD—Indian Meteorological Department; TRMM—Tropical Rainfall Measurement Mission; NCEP-CFSR National Center for Environmental Prediction Climate Forecast System Reanalysis.

Name and Type	Spatial Resolution/Coverage	Temporal Resolution/Period	Remarks, References and Data Sources
IMD (gridded rainfall dataset)	0.25°/Pan India	Daily/1901–2013	The gridded data were generated from the observed data of 6995 gauging stations across India using spatial interpolation for the period 1901–2013 [14] Details about these data can be obtained from http://imd.gov.in (homepage/rainfall information).
TRMM 3B42V7 (satellite product)	0.25°/Global (50 N-S)	Daily/1998–2014	Joint mission between the National Aeronautics and Space Administration (NASA) and the Japan Aerospace Exploration Agency (JAXA) to monitor tropical precipitation [6] one type of the TRMM multi-satellite precipitation analysis (TMPA) products) https://giovanni.gsfc.nasa.gov
Bias-corrected TRMM (satellite product)	0.25°/Global (50 N-S)	Daily/1998–2014)	The bias of TRMM rainfall was corrected by applying monthly multiplicative correction coefficients (MCC) of IMD rainfall [56] $MCC_{daily} = IMD_{total}/TRMM_{total}$ https://giovanni.gsfc.nasa.gov
CFSR (reanalysis dataset)	0.31°/Global (50 N-S)	Daily/1979–2014	Global, high resolution, coupled atmosphere–ocean–land surface-sea ice system started in 1979 by the National Centers for Environmental Prediction [22] https://globalweather.tamu.edu/

2.3. Verification Strategy

Various statistical metrics used to evaluate the satellite precipitation products with reference to gauge-based IMD gridded data both qualitatively and quantitatively. IMD dataset is a popular validated dataset used in several other works [57–61]. Pai et al. [14] and Yeggina et al. [62] evaluated the performance of the IMD gridded dataset against the gauge rainfall estimates and showed that the IMD gridded dataset could be used as a reliable alternative for gauge rainfall. Various qualitative methods were used to measure the correspondence between estimates by precipitation product and IMD data. Here, we use the false alarm ratio (FAR), probability of detection (POD) and critical success index (CSI)—to evaluate the TRMM, bias-corrected TRMM and CFSR data with the reference rainfall data (IMD). Detailed information on these statistical metrics is presented in Table 2 and Supplementary Information Section S2. Instead of point-to-point analysis, we analyzed the spatial average rainfall over the entire catchment considering the size and homogeneity of the catchment [63]. Based on the average rainfall for a given day, these statistical measures were computed by comparing rainfall and non-rainfall events for satellite (TRMM and CFSR) and reference (IMD) rainfall data.

Table 2. Definition of statistical metrics used to evaluate the TRMM, bias-corrected TRMM and CFSR data with the reference rainfall data (IMD). These statistical measures were computed by comparing rainfall and non-rainfall events for satellite (TRMM and CFSR) and reference (IMD) rainfall data. In general, when the values of probability of detection (POD) < 0.65, FAR > 0.35 and CSI < 0.45, it is assumed that the product performance is low in terms of detecting rainfall. The calculation of these three metrics is explained in [64].

S. No	Statistical Metrics	Equation	Description
1	Probability of detection (POD)	POD = H/(H + M)	It represents the detected rainfall events correctly by satellite out of rainfall events by reference data. Its ranges from 0 to 1 (1 is the best value)
2	False alarm ratio (FAR)	FAR = F/(F + H)	It represents the false, detected as rainfall event by satellite when reference data showed no rainfall. Its ranges from 0 to 1 (0 is the best value)
3	Critical success index (CSI)	CSI = H/(H + M + F)	It is more accurate when correct negative values are not considered. Its ranges from 0 to 1 (1 is the perfect value)

For the verification of the discharge simulation from the hydrological model, we adopted the popular performance evaluation measures such as Nash–Sutcliffe coefficient (NS) [65], percentage of bias (PBIAS) and the coefficient of determination (R2) in this study.

$$NS = 1 - \left\{ \frac{\sum_{i=1}^{n} \left(Y_i^{obs} - Y_i^{sim} \right)^2}{\sum_{i=1}^{n} \left(Y_i^{obs} - \overline{Y} \right)^2} \right\} \tag{1}$$

$$PBIAS = \left\{ \frac{\sum_{i=1}^{n} \left(Y_i^{obs} - Y_i^{sim} \right) \times 100}{\sum_{i=1}^{n} \left(Y_i^{obs} \right)} \right\} \tag{2}$$

$$\text{Correlation coefficient} = \frac{\sum_{i-1}^{n} (Y_i^{obs} - \overline{Y}^{obs}) \left(Y_i^{sim} - \overline{Y}^{sim} \right),}{\sqrt{\sum_{i-1}^{n} (Y_i^{obs} - \overline{Y}^{obs})^2} \sqrt{\sqrt{\sum_{i-1}^{n} (Y_i^{sim} - \overline{Y}^{sim})^2}}} \tag{3}$$

where Y_i^{sim} and Y_i^{obs} are the ith time step simulated and observed values, respectively. $\overline{Y}$ is the mean of 'n' observed values.

The values of NS ranges from $-\infty$ to 1, with $NS = 1$ indicating the best model. $PBIAS$ (Equation (2)) measures the average tendency of the simulated data (Gupta et al. [66], and the optimal value of $PBIAS$ is zero. If $PBIAS$ is a positive value, the model is underestimating the streamflow, and if $PBIAS$ is negative, the model overestimates the streamflow. The model is unsatisfactory if NSE ≤ 0.5, satisfactory performance ($0.5 < NS \leq 0.65$), good performance ($0.65 < NS \leq 0.75$), very good performance ($0.75 < NS \leq 1.00$) and $PBIAS >\pm$ 25% for the streamflow [67]. The R2 pattern similarity between observed and simulated data and its value ranges between 0 and 1. Higher the value of R2, the higher the similarity.

2.4. Hydrological Model

The precipitation products mentioned above were used to drive a semi-distributed Soil and Water Assessment Tool (SWAT) model. Here, we briefly describe the SWAT model and its governing equations, procedures to set up the model, and finally, the way to calibrate and evaluate the model.

SWAT is a comprehensive, physically based, continuous in time, semi-distributed hydrological model [68,69]. It was developed by the Agriculture Research Service of the United States Department of Agriculture and used widely to study large scale hydrological conditions in different regions [70–73]. It mimics various hydrological components of watersheds by solving process-based equations. The various processes in the SWAT model are simulated at daily time steps based on the water balance equation (Equation (1), [74])

$$SW_t = SW_O + \sum_{i=1}^{t}(R_{day} - Q_{surf} - E_a - W_{seep} - Q_{gw}) \tag{4}$$

where SW_o and SW_t represent initial and water content at any time 't' days (in mm), respectively. E_a, R_{day} *and* Q_{surf} represent the amount of evapotranspiration, precipitation and surface runoff (in mm) on any day. The amount of percolation and return flow on a day i (in mm) is denoted by W_{seep}&Q_{gw}, respectively. The Soil Conservation Service (SCS)-curve number (CN) method (Equations (5) and (6)) is used for estimating the surface runoff.

$$Q_{surf} = \frac{\left(R_{day} - \mathrm{I_a}\right)^2}{\left(R_{day} - \mathrm{I_a} + S\right)} \tag{5}$$

$$S = 254\left(\frac{CN}{100}\right) - 1 \tag{6}$$

Q_{surf} and R_{day} represent the accumulated runoff or rainfall excess (in mm) and rainfall received on that particular day (in mm), respectively [74]. $\mathrm{I_a}$ denotes an initial loss that includes detention, interception and percolation before the runoff. The retention factor is denoted by S marks retention parameter. The retention factor is subjected to spatial variation owing to land-use change, soil type, various management options, slope and temporal variation because of the changes in soil water.

The storage routing technique is used for calculating the percolation (Arnold et al. [75], assuming that the percolation occurs whenever water content exceeds the field capacity, and the layer below is unsaturated (Mosbahi et al. 2020 [76], Sridhar, 2013 [53]). In this study, the Penman–Monteith method is adopted for estimating potential evapotranspiration. Using DEM, SWAT divides the entire basin into different subbasins and subsequently into hydrological response units (HRUs) based on the unique land use, soil classification, slope and various management combinations.

Primary meteorological variables such as temperature, precipitation, solar radiation, relative humidity and wind speed are required in addition to land use, soil characteristics and land cover to set up a hydrological model using SWAT. A brief description of the input data and data sources are presented in Table 3. Moreover, the land use land cover and soil map of the study area are shown in Figure 2. The daily river discharge data were obtained from the water resources information system of

India (Central Water Commission, [77] for the period from 1985–2012 at Srikakulam station, as shown in Table 3. The reservoir details, inflow and outflow data were obtained from the Andhra Pradesh Water Resources Department and were incorporated in the model for implementing regulations.

Table 3. Detailed information on primary meteorological variables required to set up a hydrological model using the soil water assessment tool (SWAT).

Data Type	Scale	Source
Digital elevation model (DEM)	30 m × 30 m	Shuttle radar topography mission (SRTM) of USGS (https://earthexplorer.usgs.gov/)
Land use/land cover (LULC)	500 m	GIAM-IWMI (http://www.iwmi.cgiar.org)
Soil	1:1,500,000	http://www.fao.org/soils-portal/soil-survey/soil-maps-and-databases/faounesco-soil-map-of-the-world/en/ Food and Agriculture Organization (FAO)
Rainfall, temperature	0.25° (~32 km)	http://imdpune.gov.in/Clim_Pred_LRF_New/Grided_Data_Download.html Indian Metrological Department (IMD), PUNE
Solar radiation, relative humidity, wind speed	0.31° (~38 km)	https://rda.ucar.edu/datasets/ds093.1/ Climate Forecast System Reanalysis (CFSR)
Hydrological data	Daily	https://indiawris.gov.in/wris, (Central Water Commission (CWC)_

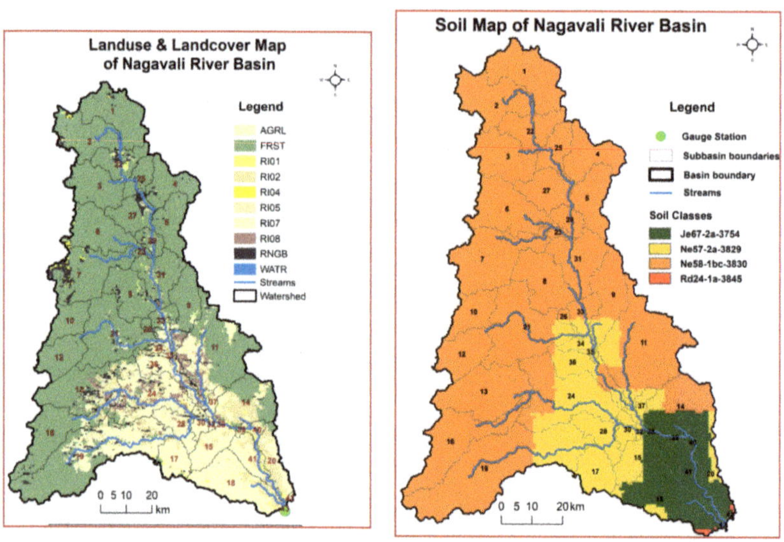

(**a**) (**b**)

Figure 2. Land use land cover classification (**a**) and soil map (**b**) with subbasins for Nagavali River Basin (NRB). QSWAT interfaced with a quantum geographical information system (QGIS) was used for developing the hydrological model.

QSWAT interfaced with a quantum geographical information system (QGIS) was used for developing the model. Preliminary analysis was done to re-project all the information to a common projection and was resampled to 30 m spatial resolution. TauDEM tool in QSWAT was used for stream network generation and watershed delineation. The entire basin was divided into 43 sub-basins using a threshold of 150 km^2. A total of 200 HRUs were formed by fixing a threshold of 200 HRUs with distinct land use, soil and slope. Here, the analysis was carried out at the hydrologic response unit (HRU) level on a monthly time step. Further, SWAT calibration and uncertainty procedures (CUP) 2012 with sequential uncertainty fitting (SUFI-2) was used for auto-calibration, sensitivity analysis and validation purposes (for more detail, please refer to Setti et al. [55].

3. Model Development and Evaluation

For simulating streamflow from different precipitation inputs, two different scenarios were considered, which are as follows:

Scenario A: calibrating the model with IMD gridded data and rerunning the model with other considered precipitation products; and

Scenario B: calibrating the model with each of the rainfall products.

Scenario A would allow investigating the impact of differences in rainfall products on streamflow simulation accuracy, whereas Scenario B (with each of rainfall products) will help in investigating the impact of input rainfall data on calibrated parameters, sensitivity analysis and streamflow simulation accuracy and in verifying whether these rainfall products can be used as an alternative source of input rainfall data for model calibration.

For both the scenarios, the sensitive parameters selection, automatic calibration and validation were performed by SUFI-2 optimization algorithms in the SWAT-CUP tool package. Sensitive parameters were identified using Latin-Hypercube One-factor-At-a-Time method with 2000 simulations. This procedure tests the model sensitivity by modifying only one parameter at a time and keeping the rest unchanged. In total, 18 parameters were identified from past literature (Setti et al. [63], and sensitivity analysis was performed.

In this study, we considered the first two years, i.e., 1998–2000, as the warm-up period to reduce the effect of initial conditions. The monthly simulated streamflow of the Nagavali River Basin was calibrated over a time period of 2001–2008 and then validated in the period of 2009–2012 with observation streamflow at Srikakulam station (as shown in Figure 1) using SWAT. For each iteration, 2000 simulations were run for the calibration period. Following each iteration, parameter ranges were modified (nearest to fitted value) with reference to the values suggested by the program and also by their reasonable physical limitations. Interested readers refer [78,79] for more details about the protocol to calibrate the SWAT model.

4. Results and Discussion

We present the results in three categories, first comparing satellite data and reanalysis data with ground-based IMD data, second, evaluating hydrological model based on the precipitation products for two different scenarios, and finally, comparison of the annual water balance components.

4.1. Comparison of Satellite Data and Reanalysis Data with Ground-Based IMD Data

First, we computed statistical parameters such as minimum rainfall, maximum rainfall, standard deviation, average and skewness coefficient of all precipitation products (Table 4) only for rainy days (rainfall magnitude ≥2.5 mm/day).

We observed that the maximum precipitation values were slightly overestimated by TRMM and CFSR datasets in comparison with the IMD product. Similarly, the CFSR is more positively skewed than the IMD, showing the general tendency for overestimation. The empirical cumulative density function (CDF) of the daily precipitation distribution during 1998–2012 was computed for the four datasets (Figure 3). There is a slight difference in the CDF obtained for IMD, TRMM and CFSR data at low values of rainfall. However, the deviation is significant in the precipitation range from 30 to 60 mm/day for TRMM and CFSR data. The CDF of the bias-corrected TRMM is closely matching with that of IMD.

Table 4. Statistical measures of four considered rainfall data sets estimated using only rainy days (rainfall magnitude ≥2.5 mm/day) for the period from 1998–2013. TRMM—Tropical Rainfall Measurement Mission; IMD—Indian Meteorological Department; NCEP-CFSR National Center for Environmental Prediction Climate Forecast System Reanalysis.

Statistical Measures	IMD	TRMM	Bias Corrected TRMM	CFSR
Minimum rainfall (mm)	2.5	2.5	2.5	2.5
Maximum rainfall (mm)	140.2	157.7	119.4	176.4
Mean rainfall (mm)	11.9	13.3	11.9	12.6
Standard deviation(mm)	12.7	13.7	11.9	13.1
Skewness	3.8	3.4	3.2	4.3

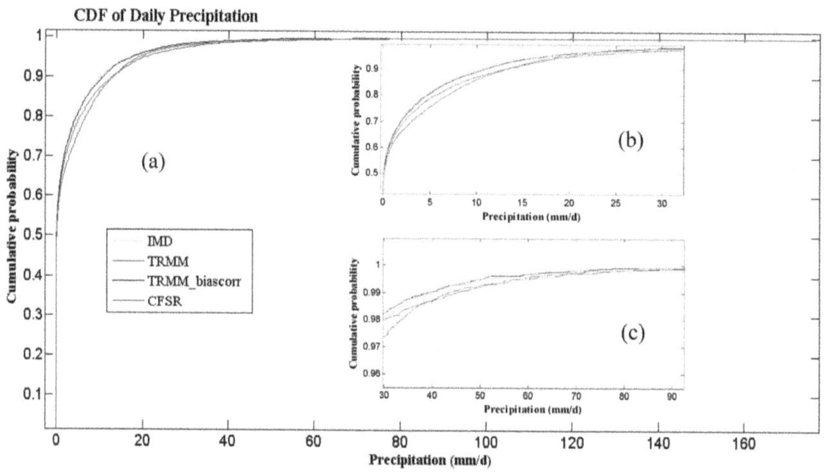

Figure 3. Cumulative probability of four daily precipitation datasets (IMD, TRMM, bias-corrected TRMM, CFSR) for NRB. Distribution of all precipitation values (**a**); distribution of precipitation below 30 mm (**b**); distribution of precipitation values between 30 mm to 60 mm (**c**).

IMD dataset for the time period from 1998–2013 revealed a total of 1591 rainy days and 4252 non-rainy days. On the other hand, the total rainy days estimated by the TRMM, bias-corrected TRMM and CFSR are 978, 932 and 1161, respectively (Table 5).

Table 5. Contingency table for category estimation of rainy days and non-rainy days between observed rainfall and satellite rainfall (TRMM and CFSR) with 2.5 mm/day threshold value during the period of 1998–2013.

Event	IMD	Observed Event	TRMM	Bias Corrected TRMM	CFSR
Rainy days	1591	Hit (H)	978	932	1161
		Miss (M)	613	659	430
Non-rainy days	4253	False alarm (F)	718	658	712
		Correct negative	3535	3595	3541

Using the threshold of 2.5 mm/day, the values of POD, FAR and CSI were estimated. The value of POD was found to be around 0.61, 0.69 and 0.73 (Table 6) for TRMM, bias-corrected TRMM and CFSR, respectively, indicating the CFSR product is close to the IMD dataset in terms of hits. Similarly, in terms of FAR and CSI, the CFSR data seems to perform better than the TRMM-based datasets. For example, the CFSR data have the lowest value of FAR (0.38) and the highest value of CSI (0.50). These analyses showed that the CFSR product has better performance than the other products like

TRMM and bias-corrected TRMM. However, it to be noted that this analysis does not consider the quantity of the rainfall; instead, it only finds hits and misses into consideration.

Table 6. Statistical metrics are calculated for daily data of Nagavali River Basin based on an average of 19 rainfall grids for the time series of 1998–2013.

Statistical Metrics		TRMM	Bias Corrected TRMM	CFSR
Probability of detection	POD	0.61	0.69	0.73
False alarm ratio	FAR	0.42	0.39	0.38
Critical success index	CSI	0.42	0.45	0.50

In order to compare the data products based on the rainfall amount, we choose the rainfall intensity classification of the World Meteorological Organization (WMO) standard Geneva, Switzerland, 2012 [80] (1) rain < 2.5 mm (no/tiny rain), (2) 2.5 mm ≤ rain < 5 mm (low moderate rain), (3) 5 mm ≤ rain < 10 mm (high moderate rain), (4) 10 mm ≤ rain < 20 mm (low heavy rain),(5) 20 mm ≤ rain < 50 mm (high heavy rain),and (6) rain ≥ 50 mm (violent rain). The four considered precipitation datasets display a similar probability of occurrence of dry days (rain < 2.5 mm/d), which are 49%, 44%, 44% and 37% for IMD, TRMM, bias-corrected TRMM and CFSR, respectively. However, within the wet days, the probability of occurrence of high, moderate rain (low heavy rain) was found to be 30% (22%), 26% (25%), 28% (24%) and 31% (27%) for IMD, TRMM, bias-corrected TRMM and CFSR, respectively. It is observed that the CFSR data have more probability of occurrence in the low heavy rain range than what is observed in IMD data.

To evaluate the overall pattern and rainfall quantity, rainfall datasets on three different time scales—daily, monthly and annual—were considered. On a daily scale, the linear Pearson correlation between the IMD and other data sets such as TRMM, bias-corrected TRMM and CFSR datasets was found to be 0.44, 0.83 and 0.82, respectively. The daily and average monthly time series of the rainfall events of four datasets are shown in Figure 4a,b. It is evident that there is a significant difference between precipitation products. The monthly variation of the precipitation values shows there is no systematic over or under prediction by CFSR or TRMM data. For example, during the 2003 monsoon, CFSR and TRMM overestimated the rainfall, whereas during the years 2004 and 2005, the rainfall was underestimated. On the annual time scale (Figure 4c), different products show that there is a strong similarity in terms of rainfall totals for the most part of the study period. However, in the latter part of the study period (2009–2013), TRMM and CFSR data overestimate the rainfall compared to the IMD-based values. The standard deviation of the four precipitation datasets computed at the annual time scale over the 16 years considered is 270 mm, 222 mm, 193 mm and 415 mm for the IMD, TRMM, bias-corrected TRMM and CFSR, respectively and the results of two-sample Kolmogorov–Smirnov (Setti et al. [63] tests at the 5% significance level indicate that the four precipitation datasets have different statistical distributions in the study area.

4.2. *Hydrological Model-Based Evaluation of the Precipitation Products*

As mentioned, two different scenarios were considered-calibrating the SWAT model with IMD gridded data and rerunning the model with other considered precipitation products.

4.2.1. Scenario A: SWAT Model Calibrated with IMD Data

This scenario attempts to unravel the effects of the precipitation products on streamflow simulation, specifically when the precipitation products driving the model is calibrated with IMD data.

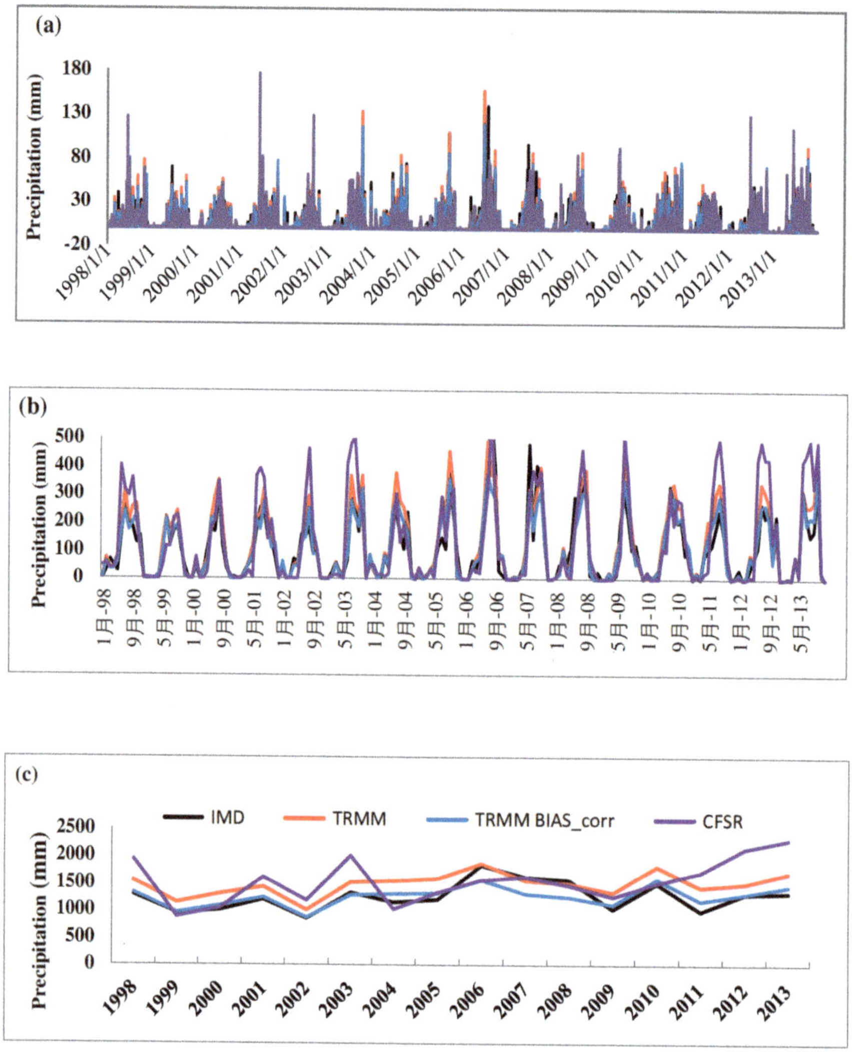

Figure 4. Comparison between observed rainfall product (IMD), satellite precipitation products (TRMM, bias-corrected) and reanalysis data (CFSR) at daily (**a**), monthly (**b**) and (**c**) annual scale.

4.2.2. Calibration of the SWAT Model with IMD Data

Initially, model calibration was done using the IMD rainfall data as input to the model after finding the sensitive parameters (using the method mentioned in Section 3). The calibration was done using the SWAT-CUP with the objective function of maximizing the *NS* between the observed and simulated flows. The performance statistics of the models' simulation showed that the calibration is good with reference to *NS* and correlation coefficient on the order of 0.85 and 0.79 and 0.91 and 0.88, respectively, during the calibration and validation periods.

In the second step, the other precipitation products are used for driving the calibrated model. Figure 5 shows the streamflow simulation results from each of the precipitation products, and the corresponding performance statistics are shown in Table 7. It is observed that even though the TRMM dataset-based model is performing satisfactorily in driving the model, according to Moriasi et al. [67].

However, streamflow estimates show systematic overestimation with *PBIAS* = −15.3%. Interestingly, the systematic difference has significantly reduced in the bias-corrected TRMM-based model with *PBIAS* = −7.4%. The streamflow simulation using the CFSR data seems to be overestimated with *PBIAS* = −16.2%. However, it is to be noted that in the case of the CFSR, the streamflow overestimations are not systematic; rather, the pattern streamflow closely follows the precipitation pattern. There is considerable streamflow overestimation in the year 2003, during which the rainfall estimates were higher than the IMD data. On the other hand, the streamflow simulation was underestimated during the years 2000, 2004 and 2005, where the rainfall estimates from CFSR were lower than IMD. Thus, there is no systematic under or overestimation in the streamflow simulation.

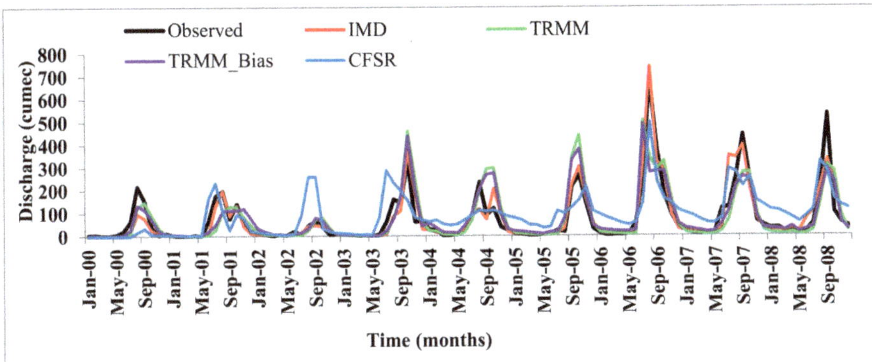

Figure 5. Comparison of streamflow simulation obtained from model calibrated with IMD data when driven using different precipitation datasets during the period (2000–2008).

Table 7. Comparison of the satellite streamflow simulation performance statistics for Scenario A (model calibrated with IMD data).

Statistical Metrics	Performance of Different Datasets under Scenario A			
	IMD	TRMM	Bias Corrected TRMM	CFSR
R^2	0.87	0.60	0.68	0.54
NS	0.85	0.55	0.62	0.52
PBIAS	−5.5	−15.3	−7.4	−16.2
P-factor	0.83	0.65	0.8	0.65
r-factor	0.67	0.48	0.63	0.91

These aforementioned results demonstrated that the bias-corrected TRMM data are a potential alternative to IMD data.

4.2.3. Scenario B: SWAT Model Calibrated Individually with Precipitation Products

This scenario evaluates the applicability of the precipitation products for calibrating the model and investigates the differences in model calibration, sensitive parameters and the ability to simulate streamflow accurately.

4.2.4. Sensitive Parameters

To evaluate the difference introduced at the stage of sensitivity analysis by using different precipitation products, we compared the ranking of the sensitive parameters obtained. Out of several parameters, we selected 18 commonly used parameters (Table 8) for streamflow simulation using four precipitation products.

Table 8. Description of parameters and their range values used for sensitivity analysis.

S. No	Parameter	Description	Min	Max
1	CN2.mgt *	SCS runoff curve number for moisture condition II, which controls the surface runoff	−0.1	0.1
2	ALPHA_BF.gw	Baseflow alpha factor for groundwater flow response to changes in recharge	0	1
3	GW_DELAY.gw	Groundwater delay represents the time taken by water move from the bottom soil layer to the upper shallow aquifer (days)	30	450
4	GWQMN.gw	Threshold depth of water in the shallow aquifer required for return flow to occur. If it is decreased, baseflow will be decreased (in mm)	0	5000
5	GW_REVAP.gw	Groundwater "revap" coefficient (Water movement from shallow aquifer to vegetation roots). If it is close to zero, the movement of water is restricted; if it approaches 1, the rate of water approaches the rate of evapotranspiration	0.02	0.2
6	RCHRG_DP.gw	Deep aquifer percolation fraction from the root zone which recharges in a deep aquifer	0	1
7	REVAPMN.gw	Threshold depth of water in the shallow aquifer for "revap" to occur (mm). Controls the water movement from a shallow aquifer to a deep aquifer	0	1000
8	SLSUBBSN.hru	Average slope length	10	150
9	HRU_SLP.hru	Average slope steepness (m/m)	0	1
10	OV_N.hru	Manning's "n" value for overland flow	0.01	30
11	LAT_TTIME.hru	Lateral flow travel time (days)	0	180
12	ESCO.hru	Soil evaporation compensation factor (represents the process of evapotranspiration from soil moisture to atmosphere, if the ESCO is reduced, the evaporation is increased)	0	1
13	EPCO.hru	Plant uptake compensation factor (The amount of water uptake by a plant for transpiration, If EPCO value is 1, the model allows more of the water uptake demand)	0	1
14	SURLAG.bsn	Surface runoff lags time (days). It controls the fraction of water to allow to the reach on the day	0.05	24
15	SOL_AWC().sol	Available water capacity of the soil layer	0	1
16	SOL_K().sol	Saturated hydraulic conductivity (mm/h). This relates to soil water flow rate to the hydraulic gradient.	0	2000
17	CH_K2.rte	Effective hydraulic conductivity in main channel alluvium (mm/h). If streams have continuous groundwater contribution, the effective conductivity will be zero.	0.1	500
18	ALPHA_BNK.rte	Baseflow alpha factor for bank storage (Contributes flow to the reach within the subbasin)	0	1

* Indicates that the CN2 values are reported as percentages of variation of the original value.

To identify the sensitivity parameters of each precipitation dataset, the model was run with 2000 simulations of eighteen selected parameters. Interestingly, the most sensitive parameters obtained for the four datasets were significantly different (Table 9) except for the deep percolation factor and curve number (CN2) that comes out to be significant for all products. The second most significant factor was ESCO using IMD and TRMM, whereas it was ALPHA_BF and GWQMN using bias-corrected TRMM and CFSR data. The most sensitive parameters from all the data sets include RCHRG_DP, ESCO and GWQMN, which show that the crucial processes governing the hydrology of the system are evapotranspiration and groundwater recharge, which is in line with the general understanding of the dominant processes in forest and agricultural catchment.

Table 9. Sensitivity parameter ranking for each dataset with default parameter values using sequential uncertainty fitting (SUFI)-2 algorithm. Interestingly, the most sensitive parameters obtained for the four datasets were significantly different except for the deep percolation factor and curve number (CN2) that comes out to be significant for all products. The second most significant factor was soil evaporation compensation factor (ESCO) using IMD and TRMM, whereas it was baseflow alpha factor (ALPHA_BF) and threshold depth of water in the shallow aquifer (GWQMN) using bias-corrected TRMM and CFSR data.

S. No	Parameter	IMD	TRMM	Bias Corrected TRMM	CFSR
1	r__CN2.mgt	6	1	2	1
2	v__ALPHA_BF.gw	18	18	17	14
3	v__GW_DELAY.gw	12	13	15	10
4	v__GWQMN.gw	3	6	8	3
5	v__GW_REVAP.gw	5	8	14	5
6	v__RCHRG_DP.gw	1	2	3	2
7	v__REVAPMN.gw	13	10	10	15
8	v__SLSUBBSN.hru	7	4	4	6
9	v__HRU_SLP.hru	8	5	6	7
10	v__OV_N.hru	16	11	12	11
11	v__LAT_TTIME.hru	17	7	7	9
12	v__ESCO.hru	2	3	1	4
13	v__EPCO.hru	11	12	11	17
14	v__SURLAG.bsn	15	16	18	18
15	r__SOL_AWC().sol	9	9	5	8
16	r__SOL_K().sol	14	14	9	13
17	v__CH_K2.rte	10	17	16	12
18	v__ALPHA_BNK.rte	4	15	13	16

4.2.5. Model Calibration

Here, out of the eighteen parameters, the ten most sensitive parameters for each dataset were further considered for calibration of the corresponding models.

The individual models were developed based on four different data sets and were calibrated at a monthly scale using the SUFI-2 algorithm. The results from the model calibration are enumerated for all the models in Table 10. Following the model performance classification of Moriasi et al. [67], the model based on IMD precipitation data performs well with *NS* (R^2) values on the range 0.85(0.87) and 0.79(0.82) during the calibration and validation period, respectively. On the other hand, the CFSR data-based model performance is satisfactory with the *NS* values 0.76 and 0.73 during calibration and validation, respectively. The models based on TRMM and bias-corrected TRMM dataset produced better results with *NS* equal to 0.71 and 0.74, respectively, as shown in Table 10. A similar pattern was observed in the *PBIAS* values during the calibration and validation period.

Figure 6 shows the streamflow simulation results during calibration and validation. Based on the simulation results obtained during calibration and validation, it can be seen that the IMD data produces the best results in comparison with the other datasets. However, closer simulation was obtained using the TRMM bias-corrected dataset. It is also to be noted that the bias-corrected TRMM datasets are yielding better results when compared to the TRMM data. Further, the results based on CFSR are also satisfactory when compared to its performance in Scenario A. However, there is an overestimation of streamflow for most of the simulations except for the years 2000, 2004 and 2005. This implies that the recalibration of the model with the individual dataset has increased the model performance to a greater extent. A similar observation was also reported by Tuo et al. [2] and Zhang et al. [41].

Table 10. Comparison of *NS* and R^2 values during the calibration (2000–2008) and validation (2009–2012) periods obtained when models were recalibrated with individual precipitation product as input.

Statistical Metrics	Calibration				Validation			
	IMD	TRMM	TRMM-Bias	CFSR	IMD	TRMM	Bias Corrected TRMM	CFSR
R^2	0.87	0.76	0.78	0.77	0.82	0.77	0.74	0.74
NS	0.85	0.71	0.74	0.76	0.79	0.72	0.72	0.73
PBIAS	−5.5	−13.4	−7.2	−14.3	−6.2	−15.2	−8.1	−17.2
P-factor	0.83	0.68	0.72	0.64	0.85	0.67	0.69	0.61
r-factor	0.67	0.46	0.59	0.40	0.79	0.44	0.61	0.41

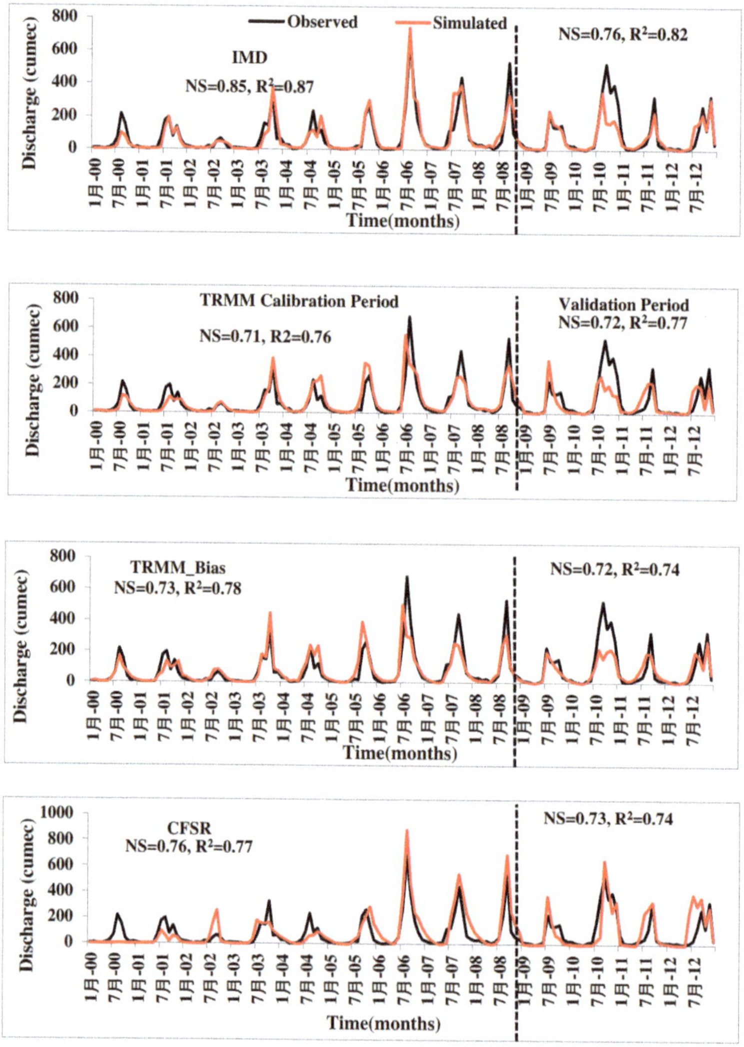

Figure 6. Simulated and observed hydrograph of the best simulation during the calibration (2000–2008) and validation periods (2009–2012) of four dataset products when the models were individually calibrated with the corresponding precipitation datasets.

Besides the evaluation of the results from the best simulation from each precipitation data set, as explained above, we also investigated the ensemble of all available simulations. We calculated the distribution of the NSC values obtained for the total of 2000 simulations. Figure 7 shows the empirical CDF of the NSC values obtained from the four models. All four models reached a satisfactory level of performance. However, interesting results were obtained using the bias-corrected TRMM data set, where most of the simulations (around 80% of the model simulations) were above NSC = 0.5, while IMD, TRMM, CFSR data sets had a lower fraction on the order of 70%, 20% and 65%, respectively. IMD data-based model displays a larger fraction (50%) of model simulations having NSC > 0.65, which represents excellent model performance. On comparing the bias-corrected TRMM dataset and TRMM-based models, there is a significant improvement in the bias-corrected TRMM-based results. This analysis indicates the uncertainty in terms of model performance when using different precipitation data sets.

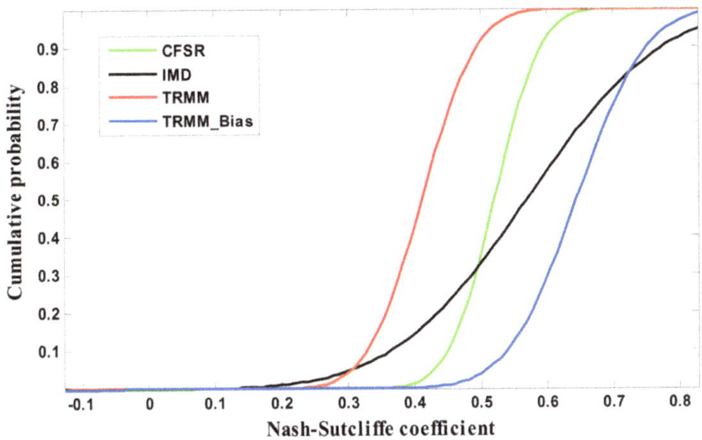

Figure 7. Distribution of Nash simulation values during calibration obtained using different precipitation datasets for the Nagavali River Basin.

4.2.6. Parameter Uncertainty

The parameter uncertainty when using different precipitation data sets was estimated by investigating the variations in the best-fit parameters, and the range of parameters obtained during the calibration. Table 11 and Figure 8 show the values and range of the best-fit parameters of the calibrated parameters of nine global sensitive hydrological parameters, respectively. Among the critical parameters shown in Figure 8, ESCO, an important parameter related to soil evaporation, have different ranges for the different datasets. It is also to be noted that the range of the parameters for ESCO obtained using IMD is different from the ones obtained from other precipitation data sets. TRMM and CFSR dataset-based models have a high ESCO value >0.5 when compared to IMD, and TRMM bias-corrected-based models indicate lower evapotranspiration compared to other datasets, which are also reflected in overestimation of streamflow.

Deep aquifer percolation fraction (RCHRG_DP) (the most dominant parameter in four precipitation datasets) displays variation in both best values and the ranges considering all the precipitation inputs. This indicates that the IMD-based model allows more water for deep percolation when compared to the CFSR-driven model. Similarly, the best parameter ranges for GWQMN (a parameter controlling the depth of water in the shallow aquifer) were higher for the IMD-based model when compared to the other models, which indicates the model-based in TRMM and CFSR data resulted in less shallow aquifer storage.

Interestingly, there is no significant difference in the CN2 parameter, which reflects the surface runoff in terms of both the best parameter and the ranges. However, for other parameters, such as SOL _AWC, HRU_SLP, CH_K2, the ranges and the best fit values are different.

Table 11. Values for best-fit parameters obtained using the four different datasets. Among the important parameters, ESCO, an important parameter related to soil evaporation, have different ranges for the different datasets. The range of the parameters for ESCO obtained using IMD is different from the ones obtained from other precipitation data sets.

S. No	Parameter	IMD	TRMM	Bias Corrected TRMM	CFSR
1	v__ESCO.hru	0.11	0.56	0.06	0.67
2	v__SLSUBBSN.hru	42.59	145.7	36.7	145.6
3	v__HRU_SLP.hru	0.28	0.06	1.0	0.11
4	r__SOL_AWC().sol	−0.03	0.037	0.054	0.03
5	r__CH_K2.rte	367.1	X	X	X
6	r__CN2.mgt	−0.152	−0.09	-0.15	−0.19
7	V__ALPHA_BNK.rte	0.33	X	X	X
8	V__RCHRG_DP.gw	0.073	0.047	0.06	0.011
9	V__GW_REVAP.gw	0.13	0.079	X	0.115
10	V__GWQMN.gw	4577	3196	3234	3942
11	V__OV_N.hru	X	13.47	X	21.1
12	V__LAT_TTIME.hru	X	85	47	28.8
13	V__GW_DELAY.gw	X	X	X	72

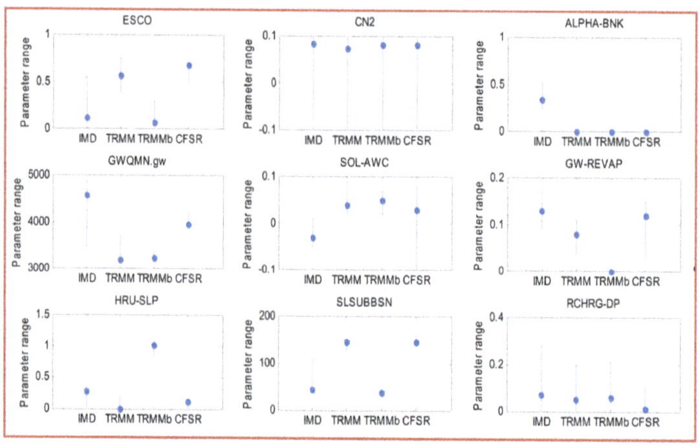

Figure 8. Calibrated parameter ranges (*y*-axis) of global sensitive hydrological parameters for the four precipitation datasets within the initial parameter range.

Overall, the analysis shows that different precipitation inputs affect both the best estimate of a parameter as well as its ranges. It is to be noted that the SWAT-CUP adjusts parameters in such a way that the observed streamflow values are matched. Even though the models capture the observed flow, the parameter values, and hence the distribution of water in the catchment, are entirely different. For example, the average water balances of the various components in the study area using different precipitation data set lead to different values of evapotranspiration, base flow, percolation, but more or less similar surface runoff values. It can be observed that the calibrated bias-corrected TRMM forced the model to have low ESCO when compared to the other models resulting in higher values of evapotranspiration.

4.2.7. Uncertainty in Streamflow Simulation

Even though the parameter uncertainties contribute to uncertainty in all outputs, variables were obtained from the model. In this section, we limit our discussion only to the uncertainty in the streamflow, as shown in Figure 9. We evaluated the uncertainty using the p-factor, and R-factor obtained during the calibration. The p-factor represents the fraction of observed streamflow falling within the 95PPU band and varies from 0 to 1, where $p = 1$ indicates that 100% of the observed flow falls within the 95% confidence band (i.e., a perfect model simulation considering the uncertainty). The R-factor is estimated as the ratio of the average width of the 95% confidence band and the standard deviation of the observed streamflow. A value of $R < 1.5$, again depending on the situation, would be desirable for this index. According to Abbaspour et al. [81], R-value less than 1.5 would be desirable for satisfactory model simulation.

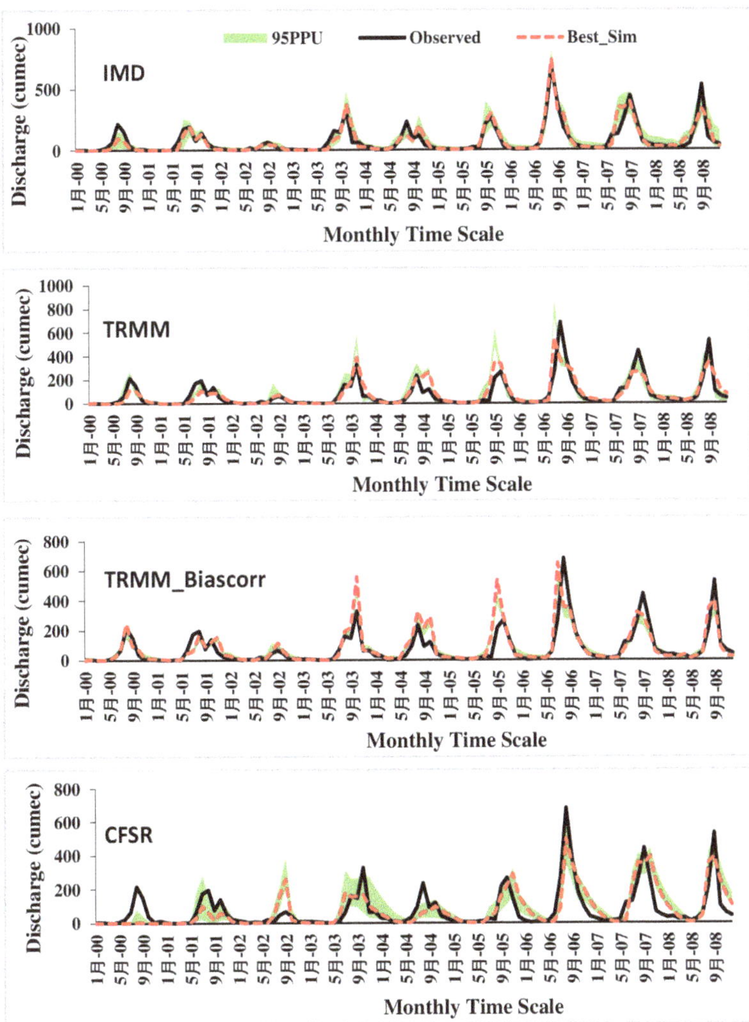

Figure 9. 95PPU plots along with the observed and simulated flows using best parameter values for the four datasets during the calibration period (2000–2008) at the monthly time scale.

In this study, the percentage of observed data bracketed by 95PPU uncertainty for IMD, TRMM, bias-corrected TRMM and CFSR is 83%, 68%, 72% and 64%, respectively during the calibration and 85%, 67%, 69% and 61%, respectively during the validation period as shown in Figure 9 and Table 10. However, comparing the r-factors (which measures the width of the 95PPU), it can be observed that the CFSR and bias-corrected TRMM-based models have low values indicating the higher levels of uncertainty. Further, the observed peak values do not fall within the 95PPU obtained from these data sets. Thus, it is clear from these results different precipitation inputs generate discrete prediction uncertainties for estimation of streamflow.

4.3. Comparison of the Annual Water Balance

Figure 10 represents the annual average of the water balance components based on the different rainfall inputs. For comparison, the observed streamflow and evapotraspiration (ET) are provided from the literature. Figure 10a presents the results of the models calibrated with IMD data (Scenario A), and Figure 10b shows the results of the model obtained when calibrated with individual data rainfall type.

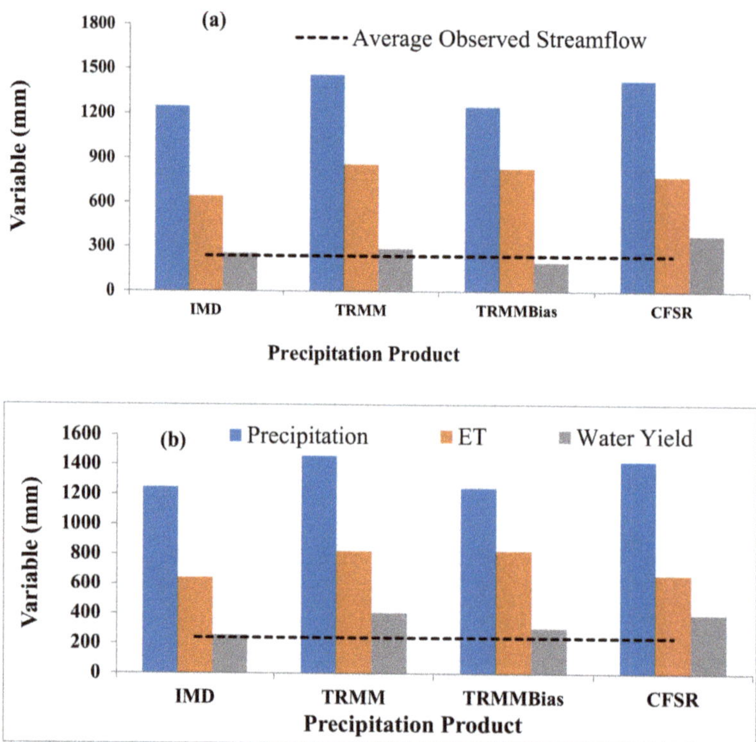

Figure 10. The annual average of hydrologic components such as precipitation, streamflow and evapotranspiration obtained using different rainfall inputs for a model calibrated with (**a**) IMD data and (**b**) individual rainfall data.

4.3.1. Precipitation

The catchment averaged rainfall estimates from the TRMM and CFSR indicate that these data sets have a general tendency for overestimating the rainfall when compared to the IMD datasets. This is consistent with the results reported in [41].

4.3.2. Streamflow

On comparing the observed and the simulated streamflow volume, we noted that the model based on IMD data produced annual streamflow accurately; nevertheless, the performance of other data sets depends on how the model was calibrated.

When the model was calibrated with IMD data (Scenario A), the CFSR-based model overestimated the average annual streamflow. In particular, it can be seen from Figure 5 that the non-monsoon streamflow was higher. However, TRMM and bias-corrected TRMM models result in slight over and underestimation, respectively. The results closely follow the pattern in the rainfall estimates showing that rainfall is the primary driving factor in streamflow simulation. Under Scenario B, all the models show comparatively better ability in capturing the streamflow volume; however, bias-corrected TRMM showed the ability to reproduce the streamflow accurately. Moreover, significant improvement in the CFSR data-based model was reported.

4.3.3. Evapotranspiration

The IMD-based model simulation of ET (641.5 mm) was well within the range of 600–700 mm in the study area [53]. Under Scenario A, TRMM, CFSR, and bias-corrected TRMM resulted in an overestimation of the ET, but the amount of overestimation is different. Even though TRMM and CFSR have similar precipitation amounts, but the distribution of the water balance components is different. This difference may be attributed to the difference in the rainfall pattern and intensity, as discussed in Section 4.1. On calibrating the model with the satellite data sets (the objective function used here is the performance of streamflow volume), TRMM and bias-corrected TRMM significantly overestimated the ET, whereas the CFSR data resulted in lesser value and close to IMD-based model estimation. Overall, we notice a considerable increase in both ET and streamflow estimates for all the data sets due to the overestimation of the rainfall.

It was observed that the precipitation dataset is primary data in driving hydrological model and also the main source of uncertainty as different precipitation dataset leads to different best parameter ranges. Further, uncertainty in the input precipitation propagates to the uncertainty to the model estimates of the water balance components and, therefore, on the water resources management and policy decisions made based on the model results.

5. Conclusions

We compared four different precipitation datasets of different categories, source and resolution, namely IMD, TRMM, bias-corrected TRMM and CFSR using a hydrological model. We investigated the impact of precipitation products on hydrological model calibration, model predictions for a forest-dominated river basin in India. The following are salient features obtained from the study.

All the considered precipitation products showed a good correlation with each other as well as the IMD gridded data on a daily scale. The CC of CFSR and bias-corrected TRMM with the IMD data were approximately 0.8. However, a quantitative comparison of the precipitation showed that CFSR and TRMM data have a slight tendency to overestimate precipitation. We also observed that bias-corrected TRMM data are comparable with the IMD data. The deviation is significant in the precipitation range from 30 mm/day to 60 mm/day.

We created two scenarios were for evaluating the precipitation using the SWAT model. In Scenario A, the model was calibrated using IMD precipitation data as input, and later this model was run with other data sets. In Scenario B, individual models were developed with each of the rainfall products as input. From these scenarios, we observed that the performance of streamflow modeling increased when the model was calibrated individually for each of the rainfall products.

The results from Scenario B showed that the different precipitation datasets resulted in multiple sets of sensitive parameters, parameter ranges and water balance components. The IMD data-based model yielded the best results in terms of streamflow simulation. The model based on the bias-corrected

TRMM dataset produced closer results and thus can be used as an alternate for gauge precipitation. In summary, the choice of precipitation products has a vital role in model performance, prediction uncertainties and parameter uncertainties in streamflow simulations. Water balance estimation based on different precipitation datasets can lead to different conclusions, and therefore uncertainty generated by the use of different precipitation inputs must be taken into consideration.

Supplementary Materials: The following are available online at http://www.mdpi.com/2073-4433/11/11/1252/s1, Section S1: Detailed Information on Precipitation Products; Section S2: Statistical Metrics.

Author Contributions: Data curation, Formal analysis, S.S.; Funding acquisition, Investigation, R.M.; Methodology-S.S., R.M., V.S., A.A.; Resources-R.M.; Supervision, R.M., K.K.B., A.A., B.M., V.S.; Writing—original draft, S.S., R.M.; Writing—review & editing, R.M., V.S., A.A., K.K.B., B.M. All authors have read and agreed to the published version of the manuscript.

Funding: This research was supported by the Early Career Research Award under SERB, India under grant No. ECRA/2016/01721. The authors are also grateful for the partial support from the Inspire Faculty grant held by Maheswaran from DST. AA and BM acknowledge the joint funding support from the University Grant Commission (UGC) and DAAD under the framework of the Indo-German Partnership in Higher Education (IGP). We also acknowledge the partial support received by the corresponding author from the Virginia Agricultural Experiment Station (Blacksburg) and the Hatch Program of the National Institute of Food and Agriculture, the USS Department of Agriculture (Washington, DC, USA).

Conflicts of Interest: The authors have no conflicts of interest.

References

1. Tuo, Y.; Marcolini, G.; Disse, M.; Chiogna, G. A multi-objective approach to improve SWAT model calibration in alpine catchments. *J. Hydrol.* **2018**, *559*, 347–360. [CrossRef]
2. Agarwal, A.; Marwan, N.; Maheswaran, R.; Ozturk, U.; Kurths, J.; Merz, B. Optimal Design of Hydrometric Station Networks Based on Complex Network Analysis. *J. Hydrol. Earth Syst. Sci.* **2020**, *24*, 2235–2251. [CrossRef]
3. Dile, Y.T.; Srinivasan, R. Evaluation of CFSR climate data for hydrologic prediction in data-scarce watersheds: An application in the Blue Nile River Basin. *J. Am. Water Resour. Assoc. (JAWRA)* **2014**, *50*, 1226–1241. [CrossRef]
4. Rienecker, M.M.; Suarez, M.J.; Gelaro, R.; Todling, R.; Bacmeister, J.; Liu, E.; Woollen, J. MERRA: NASA's Modern-Era Retrospective Analysis for Research and Applications. *J. Clim.* **2011**, *24*, 3624–3648. [CrossRef]
5. Rodell, M.; Houser, P.R.; Jambor, U.E.A.; Gottschalck, J.; Mitchell, K.; Meng, C.J.; Arsenault, K.; Cosgrove, B.; Radakovich, J.; Bosilovich, M.; et al. The global land data assimilation system. *Bull. Am. Meteorol. Soc.* **2004**, *85*, 381–394. [CrossRef]
6. Huffman, G.J.; Adler, R.F.; Bolvin, D.T.T.; Gu, G.; Nelkin, E.J.; Bowman, K.P.; Hong, Y.; Stocker, E.F.; Wolff, D.B. The TRMM Multisatellite Precipitation Analysis (TMPA): Quasi-global, multiyear, combined-sensor precipitation estimates at fine scales. *J. Hydrometeorol.* **2007**, *8*, 38–55. [CrossRef]
7. Hong, Y.; Hsu, K.L.; Sorooshian, S.; Gao, X.G. Precipitation estimation from remotely sensed imagery using an artificial neural network cloud classification system. *J. Appl. Meteorol.* **2004**, *43*, 1834–1852. [CrossRef]
8. Joyce, R.J.; Janowiak, J.E.; Arkin, P.A.; Xie, P. CMORPH: A method that produces global precipitation estimates from passive microwave and infrared data at high spatial and temporal resolution. *J. Hydrometeorol.* **2004**, *5*, 487–503. [CrossRef]
9. Funk, C.; Peterson, P.; Peterson, S.; Shukla, S.; Davenport, F.; Michaelsen, J.; Knapp, K.R.; Landsfeld, M.; Husak, G.; Harrison, L.; et al. A High-Resolution 1983–2016 T max Climate Data Record Based on Infrared Temperatures and Stations by the Climate Hazard Center. *J. Clim.* **2019**, *32*, 5639–5658. [CrossRef]
10. Hou, A.; Jackson, G.S.; Kummerow, C.; Shepherd, J.M. Global precipitation measurement. In *Precipitation: Advances in Measurement, Estimation, and Prediction*; Michaelides, S., Ed.; Springer: Berlin, Germany, 2008; pp. 131–170.
11. Hou, A.Y.; Kakar, R.K.; Neeck, S.; Azarbarzin, A.A.; Kummerow, C.D.; Kojima, M.; Oki, R.; Nakamura, K.; Iguchi, T. The global precipitation measurement mission. *Bull. Am. Meteorol. Soc.* **2014**, *95*, 701–722. [CrossRef]
12. Hamada, A.; Arakawa, O.; Yatagai, A. An automated quality control method for daily rain-gauge data. *Glob. Environ. Res.* **2011**, *15*, 165–172.

13. Yatagai, A.; Kamiguchi, K.; Arakawa, O.; Hamada, A.; Yasutomi, N.; Kitoh, A. APHRODITE: Constructing a long-term daily gridded precipitation dataset for Asia based on a dense network of rain gauges. *Bull. Am. Meteorol. Soc.* **2012**, *93*, 1401–1415. [CrossRef]
14. Pai, D.S.; Sridhar, L.; Rajeevan, M.; Sreejith, O.P.; Satbhai, N.S.; Mukhopadhyay, B. (1901–2010) daily gridded rainfall data set over India and its comparison with existing data sets over the region. *Mausam* **2014**, *1*, 1–18.
15. Li, D.; Christakos, G.; Ding, X.; Wu, J. Adequacy of TRMM satellite rainfall data in driving the SWAT modeling of Tiaoxi catchment (Taihu lake basin, China). *J. Hydrol.* **2018**, *556*, 1139–1152. [CrossRef]
16. Wang, A.; Zeng, X. Evaluation of multireanalysis products with in situ observations over the Tibetan Plateau. *J. Geophys. Res. Atmos.* **2012**, *117*, 1–12. [CrossRef]
17. Stisen, S.; Sandholt, I. Evaluation of remote-sensing-based rainfall products through predictive capability in hydrological runoff modelling. *Hydrol. Process.* **2010**, *24*, 879–891. [CrossRef]
18. Skinner, C.J.; Coulthard, T.J.; Parsons, D.R.; Ramirez, J.A.; Mullen, L.; Manson, S. Simulating tidal and storm surge hydraulics with a simple 2D inertia based model, in the Humber Estuary, U.K. *Estuar. Coast. Shelf Sci.* **2015**, *155*, 126–136. [CrossRef]
19. Qi, J.; Zhang, X.; McCarty, G.W.; Sadeghi, A.M.; Cosh, M.H.; Zeng, X.; Gao, F.; Daughtry, C.S.; Huang, C.; Lang, M.W.; et al. Assessing the performance of a physically-based soil moisture module integrated within the Soil and Water Assessment Tool. *J. Environ. Model. Softw.* **2018**, *109*, 329–341. [CrossRef]
20. Yu, Y.; Wang, J.; Cheng, F.; Deng, H.; Chen, S. Drought monitoring in Yunnan Province based on a TRMM precipitation product. *J. Nat. Hazards* **2020**, *104*, 2369–2387. [CrossRef]
21. Tuo, Y.; Duan, Z.; Disse, M.; Chiogna, G. Evaluation of precipitation input for SWAT modeling in Alpine catchment: A case study in the Adige river basin (Italy). *Sci. Total Environ.* **2016**, *573*, 66–82. [CrossRef]
22. Fuka, D.R.; Walter, M.T.; MacAlister, C.; Degaetano, A.T.; Steenhuis, T.S.; Easton, Z.M. Using the climate forecast system reanalysis as weather input data for watershed models. *Hydrol. Process.* **2013**, *28*, 5613–5623. [CrossRef]
23. Wang, J.; Hong, Y.; Li, L.; Gourley, J.J.; Khan, S.I.; Yilmaz, K.K.; Limaye, A.S. The coupled routing and excess storage (CREST) distributed hydrological model. *Hydrol. Sci. J.* **2011**, *56*, 84–98. [CrossRef]
24. Roth, V.; Lemann, T. Comparing CFSR and conventional weather data for discharge and soil loss modelling with SWAT in small catchments in the Ethiopian highlands. *Hydrol. Earth Syst. Sci.* **2016**, *20*, 921–934. [CrossRef]
25. Tolera, M.B.; Chung, I.M.; Chang, S.W. Evaluation of the climate forecast system reanalysis weather data for watershed modeling in Upper Awash basin, Ethiopia. *Water* **2018**, *10*, 725. [CrossRef]
26. Zhu, D.; Ren, Q. Hydrological evaluation of hourly merged satellite—Station precipitation product in the mountainous basin of China using a distributed hydrological model. *Meteorol. Appl.* **2020**, *27*, e1909. [CrossRef]
27. Ma, M.; Wang, H.; Jia, P.; Tang, G.; Wang, D.; Ma, Z.; Yan, H. Application of the GPM-IMERG Products in Flash Flood Warning: A Case Study in Yunnan, China. *Remote Sens.* **2020**, *12*, 1954. [CrossRef]
28. Yeditha, P.K.; Venkatesh, K.; Maheswaran, R.; Ankit, A. Forecasting of extreme flood events using different satellite precipitation products and wavelet-based machine learning methods. *Chaos Interdiscip. J. Nonlinear Sci.* **2020**, *30*, 063115. [CrossRef]
29. Bitew, M.M.; Gebremichael, M. Evaluation of satellite rainfall products through hydrologic simulation in a fully distributed hydrologic model. *Water Resour. Res.* **2011**, *47*, 1–11. [CrossRef]
30. Vu, M.T.T.; Raghavan, S.V.; Liong, S.Y.Y. SWAT use of gridded observations for simulating runoff—A Vietnam river basin study. *Hydrol. Earth Syst. Sci.* **2012**, *16*, 2801–2811. [CrossRef]
31. Tang, G.; Zeng, Z.; Long, D.; Guo, X.; Yong, B.; Zhang, W.; Hong, Y. Statistical and hydrological comparisons between TRMM and GPM Level-3 products over a midlatitude Basin: Is day-1 IMERG a good successor for TMPA 3B42V7? *J. Hydrometeorol.* **2016**, *17*, 121–137. [CrossRef]
32. Zhu, H.; Li, Y.; Huang, Y.; Li, Y.; Hou, C.; Shi, X. Evaluation and hydrological application of satellite-based precipitation datasets in driving hydrological models over the Huifa river basin in Northeast China. *J. Atmos. Res.* **2018**, *207*, 28–41. [CrossRef]
33. Huang, S.; Chang, J.; Leng, G.; Huang, Q. Integrated index for drought assessment based on variable fuzzy set theory: A case study in the Yellow River basin, China. *J. Hydrol.* **2015**, *527*, 608–618. [CrossRef]
34. Sun, R.; Yuan, H.; Liu, X.; Jiang, X. Evaluation of the latest satellite-gauge precipitation products and their hydrologic applications over the Huaihe River basin. *J. Hydrol.* **2016**, *536*, 302–319. [CrossRef]

35. Wang, Z.; Zhong, R.; Lai, C.; Chen, J. Evaluation of the GPM IMERG satellite-based precipitation products and the hydrological utility. *Atmos. Res.* **2017**, *196*, 151–163. [CrossRef]
36. Yilmaz, K.K.; Hogue, T.S.; Hsu, K.L.; Sorooshian, S.; Gupta, H.V.; Wagener, T. Intercomparison of rain gauge, radar, and satellite-based precipitation estimates with emphasis on hydrologic forecasting. *J. Hydrometeorol.* **2005**, *6*, 497–517. [CrossRef]
37. Su, F.; Hong, Y.; Lettenmaier, D.P. Evaluation of TRMM multisatellite precipitation analysis (TMPA) and its utility in hydrologic prediction in the La Plata Basin. *J. Hydrometeorol.* **2008**, *9*, 622–640. [CrossRef]
38. Gao, Z.; Long, D.; Tang, G.; Zeng, C.; Huang, J.; Hong, Y. Assessing the potential of satellite-based precipitation estimates for flood frequency analysis in ungauged or poorly gauged tributaries of China's Yangtze River basin. *J. Hydrol.* **2017**, *550*, 478–496. [CrossRef]
39. Maggioni, V.; Massari, C. On the performance of satellite precipitation products in riverine flood modeling: A review. *J. Hydrol.* **2018**, *558*, 214–224. [CrossRef]
40. Jiang, L.; Bauer-Gottwein, P. How do GPM IMERG precipitation estimates perform as hydrological model forcing? Evaluation for 300 catchments across Mainland China. *J. Hydrol.* **2019**, *572*, 486–500. [CrossRef]
41. Zhang, Z.; Tian, J.; Huang, Y.; Chen, X.; Chen, S.; Duan, Z. Hydrologic evaluation of TRMM and GPM IMERG satellite-based precipitation in a humid basin of China. *Remote Sens.* **2019**, *11*, 431. [CrossRef]
42. Jiang, D.; Wang, K. The role of satellite-based remote sensing in improving simulated streamflow: A review. *Water* **2019**, *11*, 1615. [CrossRef]
43. Yuan, F.; Zhang, L.; Soe, K.M.W.; Ren, L.; Zhao, C.; Zhu, Y.; Jiang, S.; Liu, Y. Applications of TRMM- and GPM-era multiple- satellite precipitation products for flood simulations at sub-daily scales in a sparsely gauged watershed in Myanmar. *Remote Sens.* **2019**, *11*, 140. [CrossRef]
44. Grassotti, C.; Hoffman, R.N.; Vivoni, E.R.; Entekhabi, D. Multiple-timescale intercomparison of two radar products and rain gauge observations over the Arkansas—Red River basin. *J. Weather Forecast.* **2003**, *18*, 1207–1229. [CrossRef]
45. Yang, F.; Zhang, G.L.; Yang, J.L.; Li, D.C.; Zhao, Y.G.; Liu, F.; Yang, R.M.; Yang, F. Organic matter controls of soil water retention in an alpine grassland and its significance for hydrological processes. *J. Hydrol.* **2014**, *519*, 3010–3027. [CrossRef]
46. Le, M.H.; Lakshmi, V.; Bolten, J.; Du Bui, D. Adequacy of Satellite-derived Precipitation Estimate for Hydrological modeling in Vietnam Basins. *J. Hydrol.* **2020**, *586*, 124820. [CrossRef]
47. Guo, D.; Wang, H.; Zhang, X.; Liu, G. Evaluation and analysis of grid precipitation fusion products in Jinsha river basin based on China meteorological assimilation datasets for the SWAT model. *Water* **2019**, *11*, 253. [CrossRef]
48. Jiang, L.; Madsen, H.; Bauer-Gottwein, P. Simultaneous calibration of multiple hydrodynamic model parameters using satellite altimetry observations of water surface elevation in the Songhua River. *Remote Sens. Environ.* **2019**, *225*, 229–247. [CrossRef]
49. Harris, A.; Rahman, S.; Hossain, F.; Yarborough, L.; Bagtzoglou, A.C.; Easson, G. Satellite-based flood modeling using TRMM-based rainfall products. *J. Sens.* **2007**, *7*, 3416–3427. [CrossRef]
50. Su, J.; Lü, H.; Zhu, Y.; Cui, Y.; Wang, X. Evaluating the hydrological utility of latest IMERG products over the Upper Huaihe River Basin, China. *J. Atmos. Res.* **2019**, *225*, 17–29. [CrossRef]
51. Jiang, Q.; Li, W.; Wen, J.; Qiu, C.; Sun, W.; Fang, Q.; Xu, M.; Tan, J. Accuracy evaluation of two high-resolution satellite-based rainfall products: TRMM 3B42V7 and CMORPH in Shanghai. *Water* **2018**, *10*, 40. [CrossRef]
52. Shukla, R.; Agarwal, A.; Gornott, C.; Sachdeva, K.; Joshi, P.K. Farmer typology to understand differentiated climate change adaptation in Himalaya. *Sci. Rep.* **2019**, *9*, 20375. [CrossRef] [PubMed]
53. Abbaspour, K.C.; Yang, J.; Maximov, I.; Siber, R.; Bogner, K.; Mieleitner, J.; Zobrist, J.; Srinivasan, R. Modelling hydrology and water quality in the pre-alpine/alpine Thur watershed using SWAT. *J. Hydrol.* **2007**, *333*, 413–430. [CrossRef]
54. Sridhar, V. Tracking the influence of irrigation on land surface fluxes and boundary layer climatology. *J. Contemp. Water Res. Educ.* **2013**, *152*, 79–93. [CrossRef]
55. Setti, S.; Rathinasamy, M.; Chandramouli, S. Assessment of water balance for a forest dominated coastal river basin in India using a semi distributed hydrological model. *Model. Earth Syst. Environ.* **2017**, *4*, 127–140. [CrossRef]
56. Haerter, J.O.; Hagemann, S.; Moseley, C.; Piani, C. Climate model bias correction and the role of timescales. *Hydrol. Earth Syst. Sci.* **2011**, *15*, 1065–1079. [CrossRef]

57. Sridhar, V.; Wedin, D.A. Hydrological behavior of Grasslands of the Sandhills: Water and Energy Balance Assessment from Measurements, Treatments and Modeling. *Ecohydrology* **2009**, *2*, 195–212. [CrossRef]
58. Guntu, R.K.; Rathinasamy, M.; Agarwal, A.; Sivakumar, B. Spatiotemporal variability of Indian rainfall using multiscale entropy. *J. Hydrol.* **2020**, *587*, 124916. [CrossRef]
59. Guntu, R.K.; Maheswaran, R.; Agarwal, A.; Singh, V.P. Accounting for temporal variability for improved precipitation regionalization based on self-organizing map coupled with information theory. *J. Hydrol.* **2020**, *590*, 125236. [CrossRef]
60. Agarwal, A. Unraveling Spatio-Temporal Climatic Patterns via Multi-Scale Complex Networks. Ph.D. Thesis, Universität Potsdam, Potsdam, Germany, 2019.
61. Kurths, J.; Agarwal, A.; Marwan, N.; Rathinasamy, M.; Caesar, L.; Krishnan, R.; Merz, B. Unraveling the spatial diversity of Indian precipitation teleconnections via nonlinear multi-scale approach. *Nonlinear Process. Geophys.* **2019**, *26*, 251–266. [CrossRef]
62. Yeggina, S.; Teegavarapu, R.S.; Muddu, S. Evaluation and bias corrections of gridded precipitation data for hydrologic modelling support in Kabini River basin, India. *Theor. Appl. Climatol.* **2020**, *140*, 1495–1513. [CrossRef]
63. Setti, S.; Maheswaran, R.; Radha, D.; Sridhar, V.; Asce, M.; Barik, K.K.; Narasimham, M.L. Attribution of Hydrologic Changes in a Tropical River Basin to Rainfall Variability and Land-Use Change: Case Study from India. *J. Hydrol. Eng.* **2020**, *25*, 1–15. [CrossRef]
64. AghaKouchak, A.; Mehran, A. Extended contingency table: Performance metrics for satellite observations and climate model simulations. *Water Resour. Res.* **2013**, *49*, 7144–7149. [CrossRef]
65. Nash, J.E.; Sutcliffe, J.V. River flow forecasting through conceptual models part I- A discussion of principles. *J. Hydrol.* **1970**, *10*, 282–290. [CrossRef]
66. Gupta, H.V.; Sorooshian, S.; Yapo, P.O. Status of automatic Calibration for hydrologic models: Comparison with multilevel expert calibration. *J. Hydrol. Eng.* **1999**, *4*, 135–143. [CrossRef]
67. Moriasi, D.N.; Arnold, J.G.; Van Liew, M.W.; Bingner, R.L.; Harmel, R.D.; Veith, T.L. Model evaluation guidelines for systematic quantification of accuracy in watershed simulations. *J. Trans. ASABE* **2007**, *50*, 885–900. [CrossRef]
68. Arnold, J.G.; Moriasi, D.N.; Gassman, P.W.; Abbaspour, K.C.; White, M.J.; Griensven, V.; Liew, V. DigitalCommons@University of Nebraska-Lincoln SWAT: Model use, calibration, and validation. *Biol. Syst. Eng. Pap. Publ.* **2012**. [CrossRef]
69. Arnold, J.G.; Srinivasan, R.; Muttiah, R.S.; Williams, J.R. Large area hydrologic modeling and assessment part I: Model development. *J. Am. Water Resour. Assoc.* **1998**, *34*, 73–89. [CrossRef]
70. Kang, H.; Sridhar, V. Combined statistical and spatially distributed hydrological model for evaluating future drought indices in Virginia. *J. Hydrol. Reg. Stud.* **2017**, *12*, 253–272. [CrossRef]
71. Kang, H.; Sridhar, V. Improved drought prediction using near real-time climate forecasts and simulated hydrologic conditions. *Sustainability* **2018**, *10*, 1799. [CrossRef]
72. Sehgal, V.; Sridhar, V.; Juran, L.; Ogejo, J. Integrating Climate Forecasts with the Soil and Water Assessment Tool (SWAT) for High-Resolution Hydrologic Simulation and Forecasts in the Southeastern USS. *Sustainability* **2018**, *10*, 3079. [CrossRef]
73. Sridhar, V.; Kang, H.; Ali, S.A. Human-induced alterations to land use and climate and their responses on hydrology and water management in the Mekong River basin. *Water* **2019**, *11*, 1307. [CrossRef]
74. Neitsch, S.L.; Arnold, J.G.; Kiniry, J.R.; Srinivasan, R.; Williams, J.R. *Soil & Water Assessment Tool: Input/Output Documentation*; Version 2012. TR-439; Texas Water Resources Institute: College Station, TX, USA, 2012.
75. Arnold, J.G.; Williams, J.R.; Maidment, D.R. Continuous-time water and sediment-routing model for large basins. *J. Hydraul. Eng.* **1995**, *121*, 171–183. [CrossRef]
76. Mosbahi, M.; Benabdallah, S. Assessment of land management practices on soil erosion using SWAT model in a Tunisian semi-arid catchment. *J. Soils Sediments* **2020**, 1129–1139. [CrossRef]
77. CWC (Central Water Commission). Stream Flow Discharge Data for the Period 1985–2012. 2018. Available online: http://cwc.gov.in/water-resources-information-system-wris (accessed on 22 March 2018).
78. Abbaspour, K.C. SWAT-CUP: SWAT Calibration and Uncertainty Programs—A User Manual. 2015. Available online: http://swat.tamu.edu/software/swatcup/ (accessed on 10 August 2016).

79. Abbaspour, K.C.; Rouholahnejad, E.; Vaghefi, S.; Srinivasan, R.; Yang, H.; Kløve, B. A continental-scale hydrology and water quality model for Europe: Calibration and uncertainty of a high-resolution large-scale SWAT model. *J. Hydrol.* **2015**, *524*, 733–752. [CrossRef]
80. World Meteorological Organization. Observation of present and past weather state of the ground. In *Guide to Meteorological Instruments and Methods of Observation*; Chapter 14; WMO: Geneva, Switzerland, 2012; pp. 14–19.
81. Abbaspour, K.C.; Johnson, C.A.; van Genuchten, M.T. Estimating uncertain flow and transport parameters using a sequential uncertainty fitting procedure. *Vadose Zone J.* **2004**, *3*, 1340–1352. [CrossRef]

atmosphere

Article

Train Performance Analysis Using Heterogeneous Statistical Models

Jianfeng Wang * **and Jun Yu**

Department of Mathematics and Mathematical Statistics, Umeå University, SE 901 87 Umeå, Sweden; jun.yu@umu.se
* Correspondence: jianfeng.wang@umu.se

Abstract: This study investigated the effect of a harsh winter climate on the performance of high-speed passenger trains in northern Sweden. Novel approaches based on heterogeneous statistical models were introduced to analyse the train performance to take time-varying risks of train delays into consideration. Specifically, the stratified Cox model and heterogeneous Markov chain model were used to model primary delays and arrival delays, respectively. Our results showed that weather variables including temperature, humidity, snow depth, and ice/snow precipitation have a significant impact on train performance.

Keywords: stratified Cox model; heterogeneous Markov chain model; likelihood ratio test; primary delay; arrival delay

check for updates

Citation: Wang, J.; Yu, J. Train Performance Analysis Using Heterogeneous Statistical Models. *Atmosphere* **2021**, *12*, 1115. https://doi.org/10.3390/atmos12091115

Academic Editors: Ankit Agarwal, Naiming Yuan, Kevin K. W. Cheung and Roopam Shukla

Received: 5 August 2021
Accepted: 26 August 2021
Published: 30 August 2021

1. Introduction

Coldness, heavy snow and ice/snow precipitation are well-known winter phenomena in the northern region of Sweden. Such a climate can cause severe problems to railway transportation as well as the people who rely on it, which leads to ineluctable impacts on the normal operation of the whole society. It has become an especially prominent problem recently, as the railway network has become more complicated, the trains run faster, and more people choose rail as their travel mode. The aim of this study is thus to analyse the harsh winter effects on railway operation in northern Sweden. Regarding railway operation, punctuality is one key criteria to minimise societal costs and increase the reliability of railway operation. Therefore, the aim of the study is to investigate and figure out how train delays are affected by the winter climate.

Primary delay and arrival delay are two commonly used measurements in train operation. Primary delay measures the increment in delay within two consecutive measuring spots in terms of running time, and arrival delay is the delay in terms of arrival time at a measuring spot. The time limits to define primary delays and arrival delays vary from country to country [1]. According to the Swedish Transport Administration (STA), a train arriving at one measuring spot within five minutes is not considered to be arrival delay, and a delay of three minutes or more in terms of running time within two consecutive measuring spots is considered to be primary delay. One of the main interests from the STA is to investigate how the two kinds of train delays are affected by winter weather. Therefore, we apply the STA criteria throughout the study.

Several studies of train performance analysis have been conducted. Yuan [1] used probability models based on blocking time theory to estimate the knock-on delays of trains caused by route conflicts and late transfer connections in stations. In Murali et al. [2], the authors modelled travel-time delay as a function of the train mix and the network topology. Lessan et al. [3] proposed a hybrid Bayesian network model to predict arrival and departure delays in China. Huang et al. [4] pointed out in their paper that arrival delay was highly correlated with capacity use of the train line. In a more recent study, Huang et al. [5] applied a Bayesian network to predict disruptions and disturbances during train operations in China. In addition to those, a few earlier studies of the relationships between train performance and weather have also been investigated. Thornes and Davis [6] investigated the effects of temperature, snow, ice, humidity and other weather variables on railway delays in the UK. Four case studies in Ludvigsen and Klæboe [7] showed that harsh winter weather in 2010, for example low temperatures, heavy snowfall and strong winds, affected freight train delays in Norway, Sweden, Switzerland and Poland. Xia et al. [8] fitted a linear model and showed that weather variables such as snow, temperature, precipitation and wind had significant effects on the punctuality of trains in the Netherlands. The effect of snowfall and rainfall in winter on delays of passenger trains in Hungary was identified in Nagy and Csiszár [9]. Brazil et al. [10] used a simple multiple linear regression model and demonstrated that weather variables, such as wind speed and rainfall, can have a significantly negative impact on arrival delays in the Dublin-area rapid-transit rail system. A machine-learning approach was used to create a predictive model to predict the arrival delay at each station for a train line in China with the help of weather observations in Wang and Zhang [11]. Ottosson [12] used negative binomial regression and a zero-inflated model and showed that weather variables, such as snow depth, temperature and wind direction, had significant effects on the train performance. A recent study by Wang et al. [13] applied a non-stratified Cox model and homogeneous Markov chain model to analyse the weather effects on primary delay and arrival delay, respectively. The authors treated primary delay as recurrent time-to-event data, and the transitions between states (arrival) delay and punctuality in a train trip as a Markov chain. One limitation was that the hazard function in the Cox model was assumed to be constant over events, and the transition intensity in the Markov chain model cannot change at any specified time. However, these assumptions are often not realistic.

In this study, we relax the restrictions in Wang et al. [13] by assuming heterogeneity in the models, i.e., hazard functions vary among events, and transition intensity may change at any specified time point. The main contribution is that we prove that the heterogeneous models outperform the homogeneous counterparts. In addition, to the best of our knowledge, this is the first study to apply heterogeneous models to investigate the weather impacts on the train-delay issues, i.e., a stratified Cox model is used to investigate how winter climate affects the occurrence of primary delays, and a heterogeneous Markov chain model is applied to study the effect of winter climate on the transitions between delayed and punctual states.

The paper is organized as follows. In Section 2, we introduce the statistical models in detail. Data processing and analysis methods are described in Section 3. Section 4 is reserved for results. Section 5 is devoted to the conclusion and discussion.

2. Statistical Modelling

In this section, the two statistical models, i.e., stratified Cox model and heterogeneous Markov chain model, are introduced in detail.

2.1. Stratified Cox Model with Time-Dependent Covariates for Recurrent Event

As an extension of original Cox models in Cox [14], Andersen and Gill [15], Prentice et al. [16] proposed a stratified Cox model, which is commonly used for modelling recurrent events in survival analysis. It will be used in this study to analyse the relationship between hazards of trains with recurrent events (primary delay) and weather covariates by assuming that the hazard function of a train is correlated with its preceding events through an event-specific baseline hazard function. Formally, the stratified Cox model with time-dependent covariates for recurrent events is an expression of the hazard function and covariates

$$h_{ij}(t) = h_{0j}(t) \exp\left(\boldsymbol{\beta}^T \mathbf{x}_{ij}(t)\right), \tag{1}$$

where

- $h_{ij}(t)$ represents the hazard function for the jth event of the ith train at time t.

- $h_{0j}(t)$ is an event-specific baseline hazard and the order number j is the stratification variable, e.g., $h_{01}(t)$ is a common baseline hazard of the first event for each train.
- $\mathbf{x}_{ij}(t)$ represents the weather covariate vector for the ith train and the jth event at time t.
- $\boldsymbol{\beta}$ is an unknown coefficient vector to be estimated, the exponential of which indicates how the hazard ratios are affected by the covariate vector.

The coefficients can be estimated by maximising the partial likelihood, given by

$$L(\boldsymbol{\beta}) = \prod_{i=1}^{n}\prod_{j=1}^{k_i}\left(\frac{\exp\left(\boldsymbol{\beta}^T\mathbf{x}_i(t_{ij})\right))}{\sum_{l\in R(t_{ij})}\exp\left(\boldsymbol{\beta}^T\mathbf{x}_l(t_{ij})\right)}\right)^{\delta_{ij}}, \tag{2}$$

where j is the event index with k_i being the train-specific maximum number of events, $\mathbf{x}_i(t_{ij})$ denotes the covariate vector for the ith train at the jth event time t_{ij}, δ_{ij} is an event indicator that equals 1 for the jth event of the ith train and 0 for censoring, $R(t_{ij}) = \{l, l = 1, \cdots, n : t_{l(j-1)} < t_{ij} \le t_{lj}\}$ is a group of trains that is at risk of the jth event at time t_{ij}. Please note that the partial likelihood takes into account the conditional probabilities for the events that occur for trains.

The fitted model can then be used to predict the hazard function, $\hat{h}_{ij}(t)$, for the jth event of train i of interest given the values of covariates, as well as corresponding survival function, $\hat{S}_{ij}(t)$, which gives the probability that the train i does not suffer the jth event up to time t. The survival function is an exponential function of the hazards function, i.e., $\hat{S}_{ij}(t) = \exp\left(-\int_0^t \hat{h}_{ij}(x)\,\mathrm{d}x\right)$.

2.2. Heterogeneous Markov Chain Model with Time-Dependent Covariates

Let $\{Y(t), t \ge 0\}$ denote a continuous time Markov chain. At each time point t, $Y(t)$ takes a value over a countable state space. The probability of chain $Y(t)$ being in state s at time t is $P(Y(t) = s)$. The conditional probability $p_{rs}(t, t+u) = P(Y(t+u) = s|Y(t) = r)$ represents the transition probability of moving from the state r at time t to the state s at time $t+u$. The instantaneous movement from state r to state s at time t is governed by transition intensity, $q_{rs}(t)$, through the transition probabilities

$$q_{rs}(t) = \lim_{\Delta t\to 0} P(Y(t+\Delta t) = s|Y(t) = r)/\Delta t. \tag{3}$$

With these definitions, a Markov chain can be used to describe train running states (delay/punctuality) on a train line, where the time t refers to running distance of a train from the starting point throughout the study instead of time, since the running distance is more meaningful in practice. The $q_{rs}(t)$ of a q states process forms a $q \times q$ transition intensity matrix $Q(t)$, whose rows sum to zero, so that the diagonal entries are defined by $q_{rr}(t) = -\sum_{s\neq r} q_{rs}(t)$. An example of transition intensity matrix $Q(t)$ with two states can be seen below

$$Q(t) = \begin{bmatrix} q_{11}(t) & q_{12}(t) \\ q_{21}(t) & q_{22}(t) \end{bmatrix}, \tag{4}$$

where $q_{11}(t) = -q_{12}(t)$ and $q_{22}(t) = -q_{21}(t)$ at time t.

A homogeneous Markov chain in time means that the transition intensity $Q(t)$ is independent of t, and the transition probability from one state to another depends solely on the time difference between two time points, i.e.,

$$P(Y(t+u) = s|Y(t) = r) = P(Y(u) = s|Y(0) = r). \tag{5}$$

Corresponding to the transition intensity matrix Q, the entry in a transition probability matrix $P(t, t+u)$ is the transition probability $p_{rs}(t, t+u)$. The relationship between transition intensity matrix and transition probability matrix is specified through the Kolmogorov differential equations [17]. In particular, when a process is homogeneous, the transition probability matrix can be calculated by taking the matrix exponential of the transition intensity matrix

$$P(t, t+u) = P(u) = \mathrm{Exp}(uQ). \tag{6}$$

In a homogeneous Markov chain model, to take account of the effect of covariates, a Cox-like model was proposed by Marshall and Jones [18]

$$q_{rs} = q_{rs}^{(0)} \exp\left(\boldsymbol{\beta}_{rs}^T \mathbf{x}_{rs}\right), \tag{7}$$

where $q_{rs}^{(0)}$ is a baseline transition intensity from state r to state s when all covariates are zero and $\mathbf{x}_{rs}$ is a covariate vector under the corresponding transition. The value $\exp(\beta_{rs})$, where β_{rs} is one element of the vector $\boldsymbol{\beta}_{rs}$, reflects how the corresponding covariate affects the hazard ratio given that all other covariates are held constant. More specifically, $\exp(\beta_{rs}) > 1$ indicates the transition intensity from r to s increases as the value of the covariate increases, $\exp(\beta_{rs}) < 1$ indicates the transition intensity decreases as the value of the covariate increases, while $\exp(\beta_{rs}) = 1$ implies the covariate has no effect on the transition intensity.

The coefficient vectors $\boldsymbol{\beta}_{rs}$ as well as the transition intensity matrix Q and the transition probability matrix $P(t)$ can be estimated by maximising the likelihood

$$L(Q) = \prod_{i=1}^{n} \prod_{j=1}^{c_i} p_{Y(t_{i,j}),Y(t_{i,j+1})}(t_{i,j+1} - t_{i,j}), \tag{8}$$

where j is a sequence index of observed states with c_i being number of measuring spots for train i on the train line, $Y(t_{i,j})$ represents the jth observed state of the ith train at time $t_{i,j}$ and the transition probability is evaluated at the time difference $t_{i,j+1} - t_{i,j}$.

Contrary to the homogeneous Markov chain model, a heterogeneous Markov chain model assumes that the transition intensity may change continuously at any time. However, the transition probability matrix as well as the likelihood (8) are analytically intractable under this situation [19]. An exception is that the transition intensity changes at countable time points. For example, the transition intensity is assumed to change at time point t_0 for each train. To achieve it, one can introduce an indicator covariate in the model to represent the two time periods

$$q_{rs}(t) = q_{rs}^{(0)} \exp\left(\boldsymbol{\beta}_{rs}^T \mathbf{x}_{rs}^{(\mathbb{1}_{\{t \geq t_0\}})} + z_{rs} \mathbb{1}_{\{t \geq t_0\}}\right), \tag{9}$$

where $\mathbb{1}$ is an indicator function taking value 1 if $t \geq t_0$, otherwise, 0, and z_{rs} is the coefficient. Please note that the covariate vector under the same transition is separated into two at t_0 through the indicator function, since (9) can be formulated as two homogeneous models and each model has its own covariate vector, i.e., $\mathbf{x}_{rs}^{(0)}$ for the first model when $t < t_0$ and $\mathbf{x}_{rs}^{(1)}$ for the second model when $t \geq t_0$. Similar to $\exp(\beta_{rs})$, the value $\exp(z_{rs})$ is the hazard ratio of intensities between $t \geq t_0$ and $t < t_0$ for the transition from r to s.

After fitting the heterogeneous Markov chain model, one can calculate the predicted transition probability matrix for any operational interval of interest on the train line using (6) provided that values of covariates for the interval are given.

3. Data and Method

This section describes the train data and weather variables used for the analysis as well as an imputation method for the missing train records. Moreover, a model comparison method, likelihood ratio test, is presented which is used to compare the performance between the heterogeneous models and homogeneous models.

3.1. Train Data

Our investigation focuses on high-speed passenger trains, which is a type of train with a top speed of between 200 and 250 km/h, between Umeå and Stockholm in the northern region. The high-speed passenger train is chosen because this type of train has higher priority on the train line and often travels longer distances, which can minimise non-natural effects on the train line so that it is easier to detect the pure weather impacts. The data window chosen is December 2016–February 2017, which is typical wintertime in Sweden.

A train line comprises of several measuring spots where the operational times are recorded such as departure and arrival times. The train line between Umeå and Stockholm includes 116 measuring spots in total. The total length of the train line is 711 km and the planned drive time for a high-speed passenger train is approximately 6.5 h. The lengths of any two consecutive measuring spots vary from 0.3 km to 15 km. The key variables are listed in Table 1.

Table 1. List of variables in the train operation data.

Variables	Description
Train Number	An identification number for train used in the trip
Arrival location	Name of arrival measuring spot
Departure location	Name of departure measuring spot
Departure date	The departure date (yyyy-mm-dd) for a train at a location
Arrival date	The arrival date (yyyy-mm-dd) for a train at a location
Train type	Type of train, for example: high-speed, commuter train and regional
Section Length	Length (km) between two consecutive measuring spots
Planned departure time	The planned departure time (hh:mm) at a measuring spot
Planned arrival time	The planned arrival time (hh:mm) at a measuring spot
Actual departure time	The Actual departure time (hh:mm) at a measuring spot
Actual arrival time	The Actual arrival time (hh:mm) at a measuring spot

To fit the two statistical models to the train data, the data should be organized to include the following variables, e.g., each record has one departure spot of the train run (it is not necessary in the Markov chain model), its subsequent arrival spot, distances of these two measuring spots from starting station, and indicators of primary delay and arrival delay, 0/1, for this running section, as well as corresponding weather covariates and train identification number. To obtain the indicator variables for primary delay and arrival delay, one needs to calculate the running time difference and arrival time difference compared to the schedule, which are (actual arrival time—actual departure time)—(planned arrival time—planned departure time) and (actual arrival time—planned arrival time), respectively. Afterwards, the values for the two indicator variables can be assigned, i.e., 1 stands for a primary/arrival delay, 0 otherwise. An example of how to derive the indicator variables along a train line is illustrated in Figure 1.

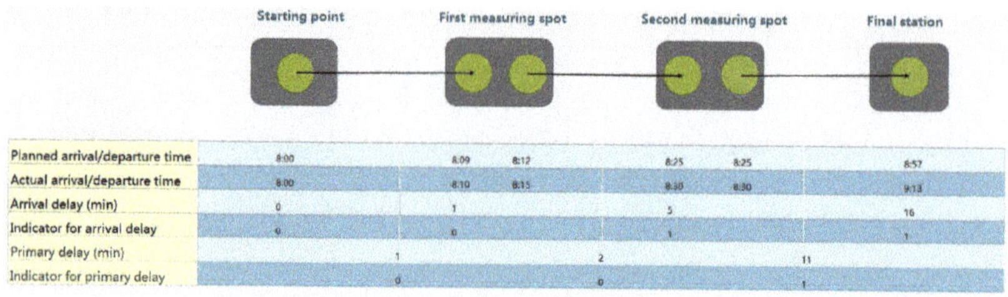

Figure 1. Illustration of a train run with derived indicators for primary and arrival delays.

3.2. Weather Data

The weather data from December 2016 to February 2017 is simulated from the Weather Research and Forecasting (WRF) model instead of using real meteorological observations, since the distances between the nearest meteorological station and measuring spot along the train line range from 17 to 24 km [12]. Thus, using meteorological data is not an ideal choice in the analysis. However, a WRF model is a numerical weather prediction system that is used for research and operational purposes. Its reliable performance has been assessed in several studies [20–23]. The WRF model simulates the desired weather variable estimations over grids. Higher spatial resolution implies smaller grids over a region of interest. Temporal resolution decides the time interval between each simulation. Therefore, a WRF with high spatio-temporal resolution is a good alternative under this situation. In this study, the spatial resolution is set as 3×3 km and the temporal resolution is set as 1 h. The simulation region as well as the train line of interest are shown in Figure 2.

The weather variables of interest are shown in Table 2. These variables are chosen because they are believed to have impacts on the train operation in winter and have been used in Ottosson [12], Wang et al. [13].

Table 2. The weather variables of interest.

Variables	Description
Temperature	The temperature (°C) at 2 m above the ground
Humidity	Relative Humidity (%) at 2-m
Snow depth	The snow depth in centimetres (cm)
Ice/snow precipitation	Hourly accumulated ice/snow in millimetre (mm)

The measuring time in train operation data must be rounded to the closest hour, so that every measuring spot on the train line can be matched with the closest grid point by date and time.

The average of the weather variables within any two consecutive spots are calculated and used in the analysis. Since a large number of the ice/snow precipitation values are zero along the train line, a categorical variable is used instead of a continuous variable, i.e., 0 if ice/snow precipitation is zero, 1 otherwise.

Figure 2. Train line in the region with simulated WRF data.

3.3. Missing Values in the Train Operation Data

A section between two consecutive measuring spots for a train trip often has missing departure/arrival times that can be classified into three different classes, which are defined in Table 3.

Table 3. Classes of missing times.

Class	Departure Time Missing	Arrival Time Missing
1	True	False
2	False	True
3	True	True

A common method to impute missing values in such longitudinal data is called last observation carried forward (LOCF), i.e., the latest recorded value is used to impute the missing value. The advantages of using LOCF are that the number of observations removed from the study decreases and makes it possible to study all subjects over the whole time period. A disadvantage of the method is the introduction of bias of the estimates if the values change considerably with time, or the time period between the most recent value and the missing value is long. Because the intervals with missing values are short in the dataset, which decreases the risk of bias, it is reasonable to apply this approach. Based on the LOCF, the imputation procedure is explained further below.

1. Start from the beginning of the trip and save the latest arrival and departure time until a missing time is occurring.
2. If the missing time is arrival time, then (a); if departure time is missing then (b):
    (a) Replace the missing arrival time with the latest departure time + the planned driving time for the previous section
    (b) Replace the missing departure time with the latest arrival time + the planned dwell time.
3. Save the imputed time as the new latest time.
4. If the section is not the last section of the trip, go back to step 1.

3.4. Likelihood Ratio Test

The likelihood ratio test is a hypothesis test that helps to determine whether adding complexity to a simple model makes the complex model significantly better compared to the simple model. Under the study context, comparisons occur between the two (complex) heterogeneous models against the two (simple) homogeneous models, respectively. The likelihood ratio test statistic is given by

$$\lambda = -2\ln\left(\frac{\mathcal{L}_{homo}(\hat{\boldsymbol{\theta}})}{\mathcal{L}_{heter}(\hat{\boldsymbol{\theta}})}\right), \tag{10}$$

where the numerator in the bracket is the likelihood value for a homogeneous model with estimated parameter vector $\hat{\boldsymbol{\theta}}$, while the denominator represents the likelihood for the corresponding heterogeneous model. The null hypothesis in the simple model is better and a low p value leads to the rejection of the null hypothesis and favouring of the complex model.

For the purposes of clarity, Figure 3 summarises the related statistical methodologies used for the analysis of the two types of delays as well as the model comparison method, respectively.

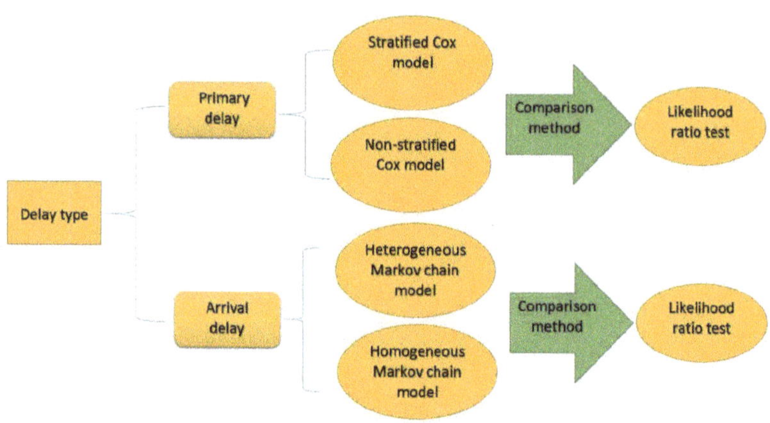

Figure 3. Statistical methodologies in the analysis.

3.5. Analysis Tool

R is the software used for data processing and modelling. Specifically, the package survival is used for the stratified Cox model and the package msm is used for the heterogeneous Markov chain model.

4. Results

4.1. Stratified Cox Model

The estimates from the fitted stratified Cox model with 95% confidence intervals (CIs) and p-values can be found in Table 4. Temperature and humidity are two variables that have significant effects on the occurrence of the primary delay. To be specific, as temperature increases with 1 °C, the hazard decreases 3.6%, and as humidity increases 1%, the hazard increases 1.7%. Comparison between the stratified Cox model and the non-stratified Cox model in Wang et al. [13] using a likelihood ratio test shows that the stratified model is significantly better than the non-stratified model ($p < 0.0001$).

Table 4. Estimates from the fitted stratified Cox model.

Predictor	Hazard Ratio	CI: Lower	CI: Upper	p-Value
Temperature	0.964	0.933	0.997	0.0338
Humidity	1.017	1.003	1.030	0.0107
Snow depth	1.017	0.990	1.044	0.2197
Ice/snow precipitation	1.055	0.841	1.323	0.6465

Besides hazard ratios, a survival plot is also produced to show how survival probabilities vary between the first and second occurrence of primary delays in Figure 4. The survival curves for the higher orders of primary delays are not shown due to the data deficiency. The curves are plotted under the condition with the average of temperature, humidity and snow depth among the whole data together with ice/snow precipitation, i.e., temperature is −1.2 °C, humidity is 85%, snow depth is 3 cm and ice/snow precipitation is 1.

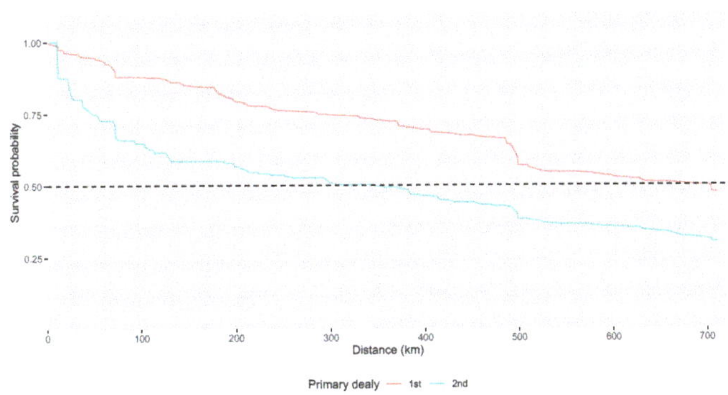

Figure 4. Survival probabilities for the first two primary delays.

The figure clearly indicates that slightly less than 50% of trains do not experience any primary delay during the trip, and 50% of trains that have experienced the first primary delay suffer second primary delays after running 330 km from the starting point. It is interesting to note that there is a substantial reduction in survival probability right before running 500 km from the starting point for the first primary delay. The reason for this might be related to some mechanical problems of trains in winter.

4.2. Heterogeneous Markov Chain Model

As indicated in Figure 4, under the average weather conditions, half of the number of the first two primary delays occur in the second part of the trip partitioned at 330 km, thus it is reasonable to assume the transition intensities are different before and after running 330 km. Therefore, $t_0 = 330$ is chosen when modelling heterogeneous Markov chain (9).

Tables 5 and 6 present the hazard ratios from the heterogeneous Markov chain model with 95% CIs and p-values. The ice/snow precipitation has a significant impact on the transition from punctual to delayed states in Table 5, which means that the transition intensity from punctuality to delay increases 23% with ice/snow precipitation. In contrast, temperature, humidity and snow depth have significant impacts on the transition from delayed to punctual states. It indicates in Table 6 that as the temperature increases 1 °C, the transition intensity from delayed to punctual states increases 4.4%, as the humidity increases 1%, the transition intensity decreases 1.6%, and as the snow depth increases 1 cm, the transition intensity decreases 4.8%. Likelihood ratio test between the heterogeneous Markov chain model (9) and the homogeneous Markov chain model in Wang et al. [13] is also performed, which shows that our new model (9) fits significantly better with $p < 0.0001$ than the homogeneous one.

Table 5. Hazard ratios from punctual to delayed states.

Predictor	Hazard Ratio	CI: Lower	CI: Upper	p-Value
Temperature	0.988	0.962	1.015	0.3881
Humidity	0.990	0.980	1.001	0.0683
Snow depth	1.000	0.977	1.026	0.9442
Ice/snow precipitation	1.230	1.001	1.512	0.0489

Table 6. Hazard ratios from delayed to punctual states.

Predictor	Hazard Ratio	CI: Lower	CI: Upper	p-Value
Temperature	1.044	1.016	1.073	0.0022
Humidity	0.984	0.973	0.995	0.0036
Snow depth	0.952	0.924	0.980	0.0012
Ice/snow precipitation	0.890	0.719	1.103	0.2921

Using the average of temperature, humidity and snow depth together with ice/snow precipitation, Table 7 and Figure 5 show the estimated transition intensities and probabilities of evolution of delayed status for the two segments divided at 330 km, respectively. In Table 7, the transition intensity of the second segment of the trip from punctual to delayed states is 58.6% higher than the first; however, transition intensity of the second segment of the trip from delayed to punctual states is 31.9% lower than the first segment. In other words, the second segment is much more likely to suffer delay and more difficult to recover from a delay. It is also verified in Figure 5 that the first segment has a higher probability of being punctual. Since the arrival delay analysis is conducted over all the measuring spots after the starting point, and the state of the train at the initial station is not considered, and for the purpose of illustration of the difference of transition intensities between the two segments shown in Table 7, Figure 5 only presents the trip that begins with a punctual state. Under such an assumption, the probability of a train arriving at the final station on time is about 81% $(0.91 \times 0.81 + 0.09 \times 0.80 = 0.809)$.

Table 7. Estimated hazard ratios between segments $[330, \text{end})$ and $[0, 330)$.

Predictor	Hazard Ratio	CI: Lower	CI: Upper	p-Value
Punctuality—delay	1.586	1.337	1.883	<0.0001
Delay—punctuality	0.681	0.564	0.822	<0.0001

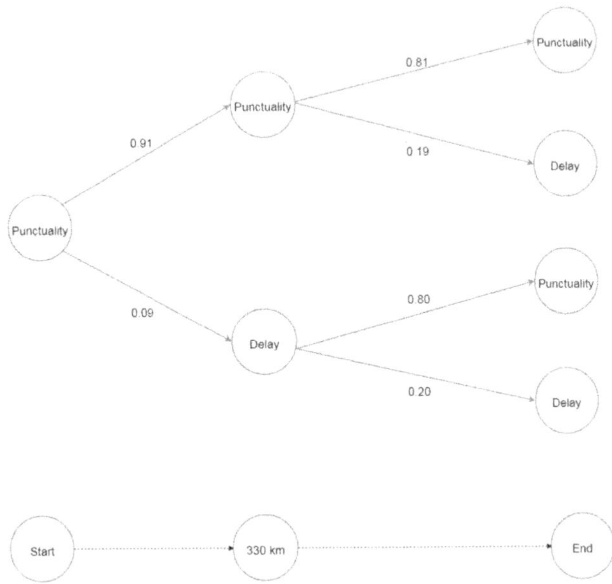

Figure 5. Probabilities of evolution of delayed status in the trip.

5. Discussion

In the study, we considered the heterogeneity within each train in both statistical models; however, the heterogeneity among trains is not touched on yet and could be considered for further investigation, for example a frailty Cox model and/or fitting the two models from a Bayesian perspective with random effects among trains [24]. In addition, choosing the changing point and the number of changing points in a heterogeneous Markov chain process become critical problems, since the estimated transition intensity matrix may be sensitive to the choices, which are very subjective. In this study, only one changing point at 330 km was used, which was decided by the fact that half of the number of the first two primary delays occurred in the second part of the trip partitioned at 330 km under the average weather condition in Figure 4. Continuously changing transition intensities, which are a smooth function of time, e.g., a Weibull-distributed time function, may be more plausible with the help of numerical approximation methods [19]. Moreover, more could be done in terms of statistical modelling. For instance, (1) a more-than-two-states Markov chain model can be used to acquire a deeper understanding of the climate effects; (2) more than one changing point of the transition intensity can be investigated in the model; and (3) interactions between weather variables and the indicator variable could be considered to account for the weather effects in each segment in the heterogeneous Markov chain model. Moreover, train operation data from more than one winter could to be included in the model-fitting procedure to acquire more robust inference.

6. Conclusions

This study investigated the effects of a harsh winter climate on the performance of high-speed passenger trains in northern Sweden, with respect to the occurrence of primary delays and the transition intensities between delayed and punctual states. Novel approaches based on heterogeneous statistical models were introduced to analyse the train performance to take the time-varying risks of train delays into consideration. Specifically, a stratified Cox model and a heterogeneous Markov chain model were used to modelling primary delays and arrival delays, respectively. We conclude that (1) the two heterogeneous models outperform the homogeneous counterparts; (2) the weather variables, including temperature, humidity, snow depth, and ice/snow precipitation, have significant impacts on train delays.

Author Contributions: All authors made significant contributions to the manuscript. J.W. conducted the statistical analysis and wrote the paper; J.Y. supervised the paper including writing, reviewing and editing. All authors have read and agreed to the published version of the manuscript.

Funding: This research was funded by European Regional Development Fund, Region Västerbotten, and Regional Council of Ostrobothnia. It is a part of the NoICE project in Interreg Botnia-Atlantica Programme with grant number 20201611.

Institutional Review Board Statement: Not applicable.

Informed Consent Statement: Not applicable.

Data Availability Statement: Not applicable.

Acknowledgments: We would like to thank the Swedish Transport Administration for providing the train operation data, the Atmospheric Science Group at Luleå University of Technology for providing the WRF data, and the High-Performance Computing Center North (HPC2N) and the Swedish National Infrastructure for Computing (SNIC) for providing the computing resources needed to generate the WRF data.

Conflicts of Interest: The authors declare no conflict of interest.

References

1. Yuan, J. Dealing with Stochastic Dependence in the Modeling of Train Delays and Delay Propagation. Ph.D. Thesis, Delft University of Technology, TRAIL Research School, Delft, The Netherlands, 2006, doi:10.1061/40932(246)641.
2. Murali, P.; Dessouky, M.; Ordóñez, F.; Palmer, K. A delay estimation technique for single and double-track railroads. *Transp. Res. Part E Logist. Transp. Rev.* **2010**, *46*, 483–495, https://doi.org/10.1016/j.tre.2009.04.016.
3. Lessan, J.; Fu, L.; Wen, C. A hybrid Bayesian network model for predicting delays in train operations. *Comput. Ind. Eng.* **2019**, *127*, 1214–1222, https://doi.org/10.1016/j.cie.2018.03.017.
4. Huang, P.; Wen, C.; Li, J.; Peng, Q.; Li, Z.; Fu, Z. Statistical Analysis of Train Delay and Delay Propagation Patterns in a High-Speed Railway System. In Proceedings of the 2019 5th International Conference on Transportation Information and Safety (ICTIS), Liverpool, UK, 14–17 July 2019; pp. 664–669, doi:10.1109/ICTIS.2019.8883805.
5. Huang, P.; Lessan, J.; Wen, C.; Peng, Q.; Fu, L.; Li, L.; Xu, X. A Bayesian network model to predict the effects of interruptions on train operations. *Transp. Res. Part C Emerg. Technol.* **2020**, *114*, 338–358, https://doi.org/10.1016/j.trc.2020.02.021.
6. Thornes, J.; Davis, B. Mitigating the impact of weather and climate on railway operations in the UK. In Proceedings of the ASME/IEEE Joint Railroad Conference, Washington, DC, USA, 23–25 April 2002; pp. 29–38, doi:10.1109/RRCON.2002.1000089.
7. Ludvigsen, J.; Klæboe, R. Extreme weather impacts on freight railways in Europe. *Nat. Hazards* **2013**, *70*, 767–787, doi:10.1007/s11069-013-0851-3.
8. Xia, Y.; Van Ommeren, J.N.; Rietveld, P.; Verhagen, W. Railway infrastructure disturbances and train operator performance: The role of weather. *Transp. Res. Part D Transp. Environ.* **2013**, *18*, 97–102, https://doi.org/10.1016/j.trd.2012.09.008.
9. Nagy, E.; Csiszár, C. Analysis of Delay Causes in Railway Passenger Transportation. *Periodica Polytech. Trans. Eng.* **2015**, doi:10.3311/PPtr.7539.
10. Brazil, W.; White, A.; Nogal, M.; Caulfield, B.; O'Connor, A.; Morton, C. Weather and rail delays: Analysis of metropolitan rail in Dublin. *J. Transp. Geogr.* **2017**, *59*, 69–76, https://doi.org/10.1016/j.jtrangeo.2017.01.008.
11. Wang, P.; Zhang, Q.P. Train delay analysis and prediction based on big data fusion. *Transp. Saf. Environ.* **2019**, *1*, 79–88, doi:10.1093/tse/tdy001.
12. Ottosson, L. Analysis of High-Speed Passenger Trains and the Influence of Winter Climate and Atmospheric Icing. Master's Thesis, Umeå University, Department of Mathematics and Mathematical Statistics, Umeå, Sweden, 2019.
13. Wang, J.; Granlöf, M.; Yu, J. Effects of winter climate on delays of high speed passenger trains in Botnia-Atlantica region. *J. Rail Transp. Plan. Manag.* **2021**, *18*, 100251, https://doi.org/10.1016/j.jrtpm.2021.100251.
14. Cox, D.R. Regression models and life-tables. *J. R. Stat. Soc. Ser. B (Methodol.)* **1972**, *34*, 187–220.
15. Andersen, P.K.; Gill, R.D. Cox's regression model for counting processes: A large sample study. *Ann. Stat.* **1982**, *10*, 1100–1120, doi:10.1214/aos/1176345976.
16. Prentice, R.L.; Williams, B.J.; Peterson, A.V. On the regression analysis of multivariate failure time data. *Biometrika* **1981**, *68*, 373–379, doi:10.1093/biomet/68.2.373.
17. Cox, D.R.; Miller, H.D. *The Theory of Stochastic Processes*; CRC Press: Boca Raton, FL, USA, 1977.
18. Marshall, G.; Jones, R.H. Multi-state models and diabetic retinopathy. *Stat. Med.* **1995**, *14*, 1975–1983, doi:10.1002/sim.4780141804.
19. Titman, A. Flexible nonhomogeneous Markov models for panel observed data. *Biometrics* **2011**, *67*, 780–787, doi:10.1111/j.1541-0420.2010.01550.x.

20. Wang, J.; Fonseca, R.M.; Rutledge, K.; Martín-Torres, J.; Yu, J. Weather simulation uncertainty estimation using Bayesian hierarchical models. *J. Appl. Meteorol. Climatol.* **2019**, *58*, 585–603, doi:10.1175/JAMC-D-18-0018.1.
21. Wang, J.; Fonseca, R.M.; Rutledge, K.; Martín-Torres, J.; Yu, J. A hybrid statistical-dynamical downscaling of air temperature over Scandinavia using the WRF model. *Adv. Atmos. Sci.* **2020**, *37*, 57–74, https://doi.org/10.1007/s00376-019-9091-0.
22. Mohan, M.; Bhati, S. Analysis of WRF model performance over subtropical region of Delhi, India. *Adv. Meteorol.* **2011**, *2011*, 621235, https://doi.org/10.1155/2011/621235.
23. Cassano, J.J.; Higgins, M.E.; Seefeldt, M.W. Performance of the weather research and forecasting model for month-long pan-arctic simulations. *Mon. Weather Rev.* **2011**, *139*, 3469–3488, doi:10.1175/MWR-D-10-05065.1.
24. van Niekerk, J.; Bakka, H.; Rue, H.; Schenk, O. New frontiers in Bayesian modeling using the INLA package in R. *arXiv* **2019**, arXiv:1907.10426.

atmosphere

MDPI

Article

Sensitivity of Microphysical Schemes on the Simulation of Post-Monsoon Tropical Cyclones over the North Indian Ocean

Gundapuneni Venkata Rao [1], Keesara Venkata Reddy [1] and Venkataramana Sridhar [2,*]

1 Department of Civil Engineering, National Institute of Technology Warangal, Telangana 506004, India; gvenkat@student.nitw.ac.in (G.V.R.); kvreddy@nitw.ac.in (K.V.R.)

2 Department of Biological Systems Engineering, Virginia Polytechnic Institute and State University, Blacksburg, VA 24061, USA

* Correspondence: vsri@vt.edu

Received: 11 November 2020; Accepted: 30 November 2020; Published: 30 November 2020

check for updates

Abstract: Tropical Cyclones (TCs) are the most disastrous natural weather phenomenon, that have a significant impact on the socioeconomic development of the country. In the past two decades, Numerical Weather Prediction (NWP) models (e.g., Advanced Research WRF (ARW)) have been used for the prediction of TCs. Extensive studies were carried out on the prediction of TCs using the ARW model. However, these studies are limited to a single cyclone with varying physics schemes, or single physics schemes to more than one cyclone. Hence, there is a need to compare different physics schemes on multiple TCs to understand their effectiveness. In the present study, a total of 56 sensitivity experiments are conducted to investigate the impact of seven microphysical parameterization schemes on eight post-monsoon TCs formed over the North Indian Ocean (NIO) using the ARW model. The performance of the Ferrier, Lin, Morrison, Thompson, WSM3, WSM5, and WSM6 are evaluated using error metrics, namely Mean Absolute Error (MAE), Mean Square Error (MSE), Skill Score (SS), and average track error. The results are compared with Indian Meteorological Department (IMD) observations. From the sensitivity experiments, it is observed that the WSM3 scheme simulated the cyclones Nilofar, Kyant, Daye, and Phethai well, whereas the cyclones Hudhud, Titli, and Ockhi are best simulated by WSM6. The present study suggests that the WSM3 scheme can be used as the first best scheme for the prediction of post-monsoon tropical cyclones over the NIO.

Keywords: ARW model; microphysical schemes; North Indian Ocean; tropical cyclones; average track error; skill score; WSM3

1. Introduction

Among the world's oceans, the North Indian Ocean (NIO) is a highly active region for the formation of tropical cyclones (TCs). The NIO accounts for 7% of the TCs that are formed over the globe [1]. TCs are one of the most dangerous natural weather calamities, which have a significant impact on the socio-economic aspects of the countries along the rim of the NIO [2]. TCs will cause substantial loss to humans, physical property, ecology and the environment at different levels when they make landfall [3]. The damage caused by TCs along the rim of the NIO may be attributed to its shallow bathymetry, low-lying flood prone areas, dense population along the coastlines, and poor socio-economic conditions [4].

The TCs over the NIO are highly seasonal and occur during the pre- and post-monsoon seasons with very few in the rest of the year. The seasonal occurrence of the TCs over the NIO may be attributed to the presence of nearby equatorial troughs over the open ocean in the pre- and post-monsoon seasons [5,6]. The post-monsoon season is a highly active period for the formation of TCs over the NIO. The TCs that are formed in the post-monsoon season are approximately twice the TCs that are formed in the pre-monsoon season [3]. As per the annual and primary reports of the Regional Specialized Meteorological Centre, Indian Meteorological Department (RSMC-IMD) for the TCs over the NIO, a total of 20 cyclones were formed in the last five years (i.e., from 2014 to 2018). Of these, 60% of the TCs were formed in the post-monsoon season and the rest in the pre-monsoon [7–11]. By using the data from atlas maps published by Indian Meteorological Department (IMD) and their records for a period of 122 years (1877–1998), Singh et al. (2000) [12] evaluated the trends in post-monsoon TCs over the NIO and found an increasing trend of 20% in both the intensity and frequency of TCs. Mohanty et al. (2012) [13] found a large increase in the number of severe cyclonic storms (SCS) in the Bay of Bengal (BoB) by evaluating the trends in cyclone data for a period of 120 years (1891–2010). The increasing number of tropical cyclones in the post-monsoon season may be attributed to the rise in sea surface temperature, weak vertical wind shear, El Niño Southern Oscillation (ENSO) and the Indian Ocean Dipole (IOD) [13–16]. ENSO is an ocean-atmospheric coupled phenomenon which affects the intensity of the TCs [17]. The track of the TC is influenced by large-scale wind dynamics which include vertical wind shear, low-level rotational wind, low-level relative vorticity, and mid-troposphere [18,19].

With the advancements in computational power, significant improvements are seen in the field of TC prediction over the NIO using the Numerical Weather Prediction (NWP) models. Due to inconsistencies (initial and boundary conditions, grid resolution, representation of physics schemes, and geographical location) in the NWP models, there is need for further improvement in the prediction of TCs. Accurate representation of cloud processes in NWP models is crucial for the track and intensity prediction of TCs. Representation of cloud processes plays an important role in the production and distribution of heat, mass, and momentum in the atmosphere in both the horizontal and vertical directions with the help of precipitation, winds, and turbulence. The representation of physics schemes in the NWP model are important when the cloud processes and their effects are unresolved by the model [20,21]. In the last two decades, based on various assumptions, researchers have developed a number of parameterization schemes for track and intensity prediction and they are being used for both operational and research purposes [22]. Among all the physics schemes, Cloud Microphysics (CMP), Cumulus Parameterization Scheme (CPS), Planetary Boundary Layer (PBL), radiation (longwave and shortwave), and land-surface schemes are being used for weather predictions [23]. The cloud process in the model can be implicitly treated by CPS and explicitly treated by CMP schemes. CPS reduces the convective instability in a model through the redistribution of temperature and moisture in a grid column [24]. CMP schemes represent cloud and precipitation processes (condensation, nucleation, coalescence, phase changes, etc.) according to the atmospheric conditions in terms of temperature, wind, and moisture. Both CPS and CMP schemes control the spatio-temporal variations in precipitation and yield different profiles of moistening and heating in the atmosphere. Without double counting the thermo-dynamical impacts, both types of schemes represent the convective activity [20,21].

Numerous studies have been conducted to assess the impact of physics schemes on the prediction of track and intensity of the TCs using the Advanced Research Weather Research and Forecasting (ARW) model. Among all the schemes in the ARW model, convective processes play an important role in the development of TCs and boundary layer dynamics in their intensification [4,20,25,26], whereas microphysical schemes have significant impacts on the track prediction of TCs [25]. Pattanayak et al. (2012) [4] found that the track and intensity of the TC Nargis is well simulated by the Ferrier CMP scheme along with Yonsei University (YSU) PBL, Simplified Arakawa Schubert (SAS) CPS schemes. With the same CPS and PBL schemes, the Kessler CMP scheme provided better results for the TC Vardah [21]. Kanase and Salvekar (2015) [27] showed that the WSM6 scheme, in combination with Bettes-Miller-Janjic (BMJ) CPS and YSU PBL schemes, simulated better results for the TC Laila. Based on the sensitivity experiments conducted by Srinivas et al. (2013) and Lakshmi and Annapurnaiah

(2016) [28,29], the Lin scheme improved the results for the TCs Sidr, Nisha, Tane, Jal, Nargis, and Hudhud along with the combination of Kain–Fritsch (KF) CPS and YSU PBL schemes. With the same combination of CMP and PBL schemes, Choudhury and Das (2017) [30] suggested the Goddard scheme, and the Ferrier scheme was suggested by Raju et al. (2011) [26] and Reddy et al. (2014) [31] for the prediction of TCs. Osuri et al. (2012) [32] and Mahala et al. (2015) [33] reported that the TCs over the NIO, were better simulated by the WRF Single-Moment-3 (WSM3) scheme with the same CPS and PBL schemes.

Based on previous studies, it was found that Kain–Fritsch (KF) and Yonsei University schemes can be used as convective and boundary layer schemes [26,28,30,32], and various microphysical schemes such as Ferrier [26,31] WSM3 [32], Lin [28] and Goddard [30] can be used in the prediction of the track and intensity of the TCs over NIO.

From these studies, it may be difficult to identify a suitable microphysical scheme for the prediction of the TCs over the NIO region. In this study, numerical experiments were conducted to revalidate the suggested microphysical schemes for the simulation of TCs. This study mainly concentrated on the prediction of post-monsoon TCs, because 71% of the TCs that are formed in the post-monsoon season make landfall over the Indian coast, whereas, in the pre-monsoon season, the majority of the TCs over the BoB make landfall near Myanmar, and the TCs that are formed over the Arabian Sea (AS) make landfall near the Gulf countries [3,12].

Tropical Cyclone Case Studies

In the recent years, RSMC-IMD has adopted various techniques in the analysis and forecasting of the TCs over the NIO. RSMC-IMD uses a blending technique based on conceptual models, dynamical and statistical models, meteorological datasets, technology and expertise for TC analysis, prediction and decision-making process. For the purpose of TCs analysis and prediction, RSMC-IMD uses data from conventional observational networks, automatic weather stations, buoy and ship observations, cyclone detection radars and satellite imagery. RSMC-IMD provides a bulletin on tropical weather look, tropical cyclone advisories, storm surge guidance, maritime forecast bulletins, tropical cyclone advisories for aviation, national bulletins, cone of uncertainty forecasts, and wind forecasts for different quadrants. RSMC-IMD issues a national bulletin to the public on the formation of cyclones from the stage of depression (D) onwards. During the stages of depression or deep depression, RSMC issues the bulletins based on 00, 03, 06, 12, and 18 UTC observations. When the system intensifies into a cyclonic storm over the NIO, these bulletins are issued at 3-hour intervals based on previous observations, which gives the information on present status of the system, expected damage and action suggested. These bulletins are completely made for national users and disseminated through the various modes of communication (i.e., All India Radio, National TV, Telephone, SMS, print electronic media) and are made available at the RSMC-IMD website [11].

The eight TCs formed over NIO from 2014 to 2018 were considered in the present study in order to assess the impact of CMP on their track and intensity predictions. Among these TCs, two had formed over AS and the remaining were in BoB. The best tracks provided by the study cyclones are presented in Figure 1. The details about these cyclones are given in Table 1 and a brief summary is given below.

1. Hudhud On the morning of 6 October 2014, cyclone Hudhud formed as a low-pressure area (LPA) over BoB. It gradually intensified into a Very Severe Cyclonic Storm (VSCS) in the afternoon of 10 October 2014. It made landfall near Visakhapatnam with a northwestward movement in the morning of 12 October 2014 as VSCS and moved in the same direction. It then gradually weakened into a Well-Marked Low-Pressure Area (WMLA) on the evening of 14th October 2014 over eastern Uttar Pradesh [7];
2. Nilofar A VSCS Nilofar formed as an LPA over the southeast AS on the morning of 21 October 2014. The cyclone initially moved northwestward on the day of formation and then recurved northeastwards. It exhibited rapid intensification as well as rapid weakening and weakened into a WMLA near the North Gujarat coast on the morning of 31 October 2014 [7];

3. Kyant Cyclone Kyant formed as a depression (D) over east central BoB on 21 October 2016. The track followed by this system is rare in nature as it experienced two re-curvatures during its life period. The rate of intensification was very slow and steady, taking about 4 days to become a cyclonic storm (CS) from the stage of D, and the rate of weakening was rapid as it reduced to a WMLA from the CS stage within 30 hours on the morning of 28 October 2016 [9];
4. Ockhi Cyclone Ockhi formed as an LPA over Andaman Sea on 22 November 2017. There was a rapid intensification during its genesis stage, as it intensified into a CS within 6 hours from the stage of deep depression (DD). While moving west–northwestwards, Ockhi further intensified into a Severe Cyclonic Storm (SCS) over Lakshadweep area in the early morning of 1 December 2017 and VSCS over the southeast AS in the afternoon of the same day. It then moved northwestwards and reached its peak intensity in the afternoon of 2 December 2017. It moved north–northwestwards and then northeastwards, and crossed the south coast of Gujarat between Surat and Dahanu as a WMLA in the early morning of 06 December 2017 [10];
5. Daye Daye is the first cyclonic storm formed over the NIO in the month of September after 2005. It formed as a D over the east central of BoB in the afternoon of 19 September 2018. Moving nearly west–northwestwards, it intensified into a DD on the morning of 20 September 2018 and further into a CS in the same day. It made landfall close to Gopalpur as a CS from 1900–2000 h UTC of 20 September 2018. It continued to move west–northwestwards, and weakened into an LPA over south Haryana on the morning of 24 September 2018 [11];
6. Titli Titli cyclone formed as an LPA over the southeast BoB on the morning of 7 October 2018. Moving nearly west–northwestwards, it intensified into a DD on the morning of 8 October 2018, and further into a CS around noon on 9 October 2018. It then moved northwestwards, and in the early morning of 10 October 2018, it intensified into an SCS. It then moved north–northwestwards and further intensified into a VSCS around noon of 10 October 2018 and crossed the northern Andhra Pradesh and south Odisha coasts near Palasa from 2300 to 0000 h UTC as a VSCS. Moving further west–northwestwards, it weakened into an SCS around the noon of 11 October 2018 and a CS in the same evening. Under the influence of southwesterly winds, the system recurved northeastwards from 11 p.m., and gradually weakened into an LPA over Gangetic West Bengal and adjoined Bangladesh in the morning of 13 October 2018 [11];
7. Gaja VSCS Gaja originated from an LPA which formed over the Gulf of Thailand and adjoining Malay Peninsula on the morning of 8 November 2018. Under the favorable conditions, it concentrated into a D over southeast BoB in the morning of 10 November. Moving west–northwestwards, it intensified into a DD in the same evening and further intensified into a CS in the early morning of 11 November 2018. It then moved nearly westwards till early morning of 12 November 2018. Thereafter, it recurved south–southwestwards and followed an anticlockwise looping track till the morning of 13 November 2018. It then moved west–southwestwards and intensified into an SCS southwest BoB on the morning of 15 November 2018 and into a VSCS in the same night. Moving further west–southwestwards, it crossed Tamil Nadu and Puducherry coast between Nagapattinam and Vedaranniyam from 1900 to 2100 h UTC of 16 Nov 2018. Thereafter, it moved nearly westwards, and weakened rapidly into an SCS, CS, and DD over interior Tamil Nadu on 16 November 2018. It then moved west–southwestwards and weakened into a D in the same evening over central Kerala. Moving nearly westwards, it intensified into a DD over southeast AS in the early morning of 17 November 2018. Thereafter, it moved nearly west–northwestwards and crossed Lakshadweep Islands on the afternoon of 17 November 2018, as a DD. It continued to move west-northwestwards and weakened into a D over the same region around noon of 19 November 2018, WMLA in the same night, and an LPA on 21 November 2018 [11];
8. Phethai A SCS Phethai formed as an LPA over the Equatorial Indian Ocean and adjoinined the central parts of south BoB on the evening of 9 December 2018. It was laid as a WMLA over the same area on the morning of 11 December 2018. It continued to be WMLA till the morning

of 13 December 2018, and, under favourable conditions, it concentrated into a D over southeast BOB. Moving north–northwestwards, it intensified into a DD over the same area on the same midnight. Continuing to move in the same direction, it intensified into a CS in the evening of the 15th and into an SCS in the afternoon of the 16th. It maintained its intensity of SCS till the early morning of the 17th and weakened into a CS in the same morning. Continuing to move north–northwestwards and then northwards, it crossed the Andhra Pradesh (south of and close to Yanam and 40 km south of Kakinada) coast during the 17th afternoon as a CS. After landfall, the cyclone moved north–northeastwards and weakened rapidly into a DD near the Kakinada coast in the same evening. Continuing to move in the same direction, it again crossed Andhra Pradesh coast near Tuni and weakened into a D over the coastal Andhra Pradesh during the same midnight. It further weakened into a WMLA over the northwest and adjoining west–central BoB and coastal Odisha in the early morning of the 18th, and into an LPA northwest BoB and adjoining Odisha in the same morning [11].

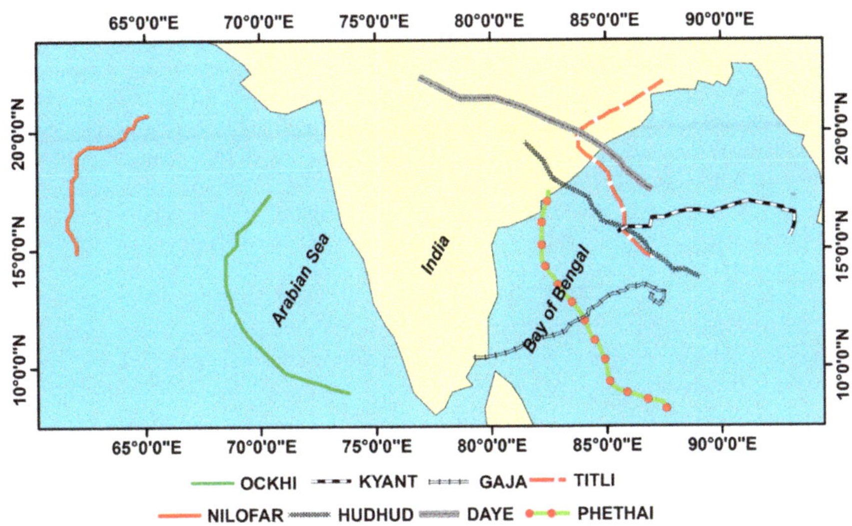

Figure 1. Best tracks of tropical cyclones (TCs) considered in the present study.

Table 1. Details of the study of tropical cyclones (TCs).

S.No	Period	Cyclone Name	Landfall	Category
1	7–17 October 2014	Hudhud	Visakhapatnam	Very Severe Cyclonic Storm
2	25–31 October 2014	Nilofar	No Landfall	Very Severe Cyclonic Storm
3	21–28 October 2016	Kyant	No Landfall	Cyclonic Storm
4	29 November–05 December 2017	Ockhi	South Gujarat Coast	Very Severe Cyclonic Storm
5	19–22 September 2018	Daye	Gopalpur	Cyclonic Storm
6	8–13 October 2018	Titli	Palasa	Very Severe Cyclonic Storm
7	10–19 November 2018	Gaja	Puducherry	Very Severe Cyclonic Storm
8	13–18 December 2018	Phethai	Yanam	Severe Cyclonic Storm

2. ARW Model and Sensitivity Experiments

Advanced Research Weather Research and Forecasting (ARW version 4.0) model is an open-source atmospheric modelling system designed for broad use in research and operational studies at the National

Centre for Atmospheric Research (NCAR) in collaboration with various universities [34]. The model supports atmospheric simulations across scales from large-eddy to globe, with a wide range of applications. The model integrates the compressible non-hydrostatic Eulerian equations using a terrain following vertical coordinates. The model uses the 3rd-order Runge–Kutta time integration scheme for low-frequency modes, whereas smaller time steps are used to integrate the high-frequency acoustic waves to maintain numerical stability. The horizontal propagating acoustic models and gravity waves are integrated using a forward–backward time integration scheme, and vertically propagating acoustic models and buoyancy oscillations are integrated using a vertical implicit scheme. This model and its earlier versions have the versatility to choose the region of interest, grid spacing in horizontal and vertical directions, interactive nested domains with various physics parameterization schemes for convection, boundary layer, microphysics, radiation, soil, and surface processes [35,36].

In this study, ARW v4.0 was used to conduct sensitivity experiments to predict the track and intensity of TCs. The initial and boundary conditions for the prediction of TCs were considered from the 0.5° × 0.5° resolutions of NCEP-GFS model forecasts with 6-hour intervals. The model was designed with two two-way nested domains with 27 and 9 km grid spacing. The terrain data of 10m resolution from the United States Geological Survey (USGS) were used for both domains. The model utilized a total of seven microphysical schemes, namely Lin, Thompson, Ferrier, Morrison, WSM3, WSM5 and WSM6, for both the domains. Yonsei University (YSU) PBL scheme, Rapid Radiative Transfer Model (RRTM) for long-wave radiation and Dudhia scheme for short-wave radiation were used for both the inner and outer domains, whereas the Kain–Fritsch (KF) convective scheme was used for the outer domain.

The various microphysics schemes deal with the mixing ratios of the prognostic variables with different approaches under different assumptions. The mixing ratios of the prognostic variables in all the microphysical schemes considered in the study are presented in Table 2. The Lin scheme includes all the prognostic variables. It is the most sophisticated scheme of the ARW model and suitable for research studies [37]. The new Eta Ferrier scheme has the ability to predict the changes in water vapor and estimates the precipitation ice density along with the mixing ratios [38]. The Thompson scheme used in the present study is a double-moment scheme which includes the prediction of ice concentration [39]. The scheme assumes that the snow size distribution depends on both the water and ice content and temperature [40]. Morrison scheme is also a double-moment scheme which predicts the mixing ratios and concentrations of all the prognostic variables. The scheme uses Kohler's theory to calculate the homogeneity and heterogeneity in the nucleation process and quasi-stationary saturation adjustment algorithm for droplet concentration [41]. WSM3 is a simple ice scheme, which predicts only the liquid hydrometers (i.e., Q_v, Q_c, and Q_r). The expressions considered by WSM3 to predict liquid hydrometers are assumed to be above the freezing point. Further, the scheme considers Q_c as Q_i and Q_r as Q_s when the temperature is less than or equal to the freezing point. The WSM5 scheme predicts the mixing ratios of all the prognostic variables except graupel [42]. The WSM6 scheme is similar to that of WSM3, but includes more complex processes for predicting the mixing ratios of all the prognostic variables [43]. Compared to the double-moment schemes, single-moment schemes have the capability of simulating TCs with a smaller eye, stronger tangential wind, high positive temperature and closer latent heating area to the cyclone center, and smaller radius of maximum wind [44].

For the sensitivity experiments, a total of eight TCs were selected to study the sensitivity of microphysical schemes to the prediction of track and intensity of TCs over the NIO region that occurred from 2014 to 2018. The model initiation time and simulation period for the TCs are presented in Table 3. The simulated results of the TCs are validated against the best track given by IMD.

Table 2. The mixing ratios of the prognostic variables in the Microphysical Schemes.

Microphysical Schemes	Mixed Phase Variable Processes
Ferrier	Water Vapor (Q_v), Cloud Water (Q_c), Rain (Q_r), Ice (Q_i)
Lin	Water Vapor, Cloud Water, Rain, Ice, Snow (Q_s), and Graupel (Q_g)
Morrison	Water Vapor, Cloud Water, Rain, Ice, Snow, and Graupel
Thompson	Water Vapor, Cloud Water, Rain, Ice, Snow, and Graupel
WSM3	Water Vapor, Cloud Water/Ice and rain/snow
WSM5	Water Vapor, Cloud Water, Rain, Ice, and Snow
WSM6	Water Vapor, Cloud Water, Rain, Ice, Snow, and Graupel

Table 3. Model initiation dates and simulation times considered for the study.

Tropical Cyclone	Initial Date: Time (DD-MM-Year: hh)	End Date: Time (DD-MM-Year: hh)	Simulation Period (h)	Maximum Sustained Wind (MSW) (Kt)	Mean Sea Level Pressure (MSLP)(hPa)	Intensity Stages of TCs for Model Initialization
Hudhud	09-10-2014: 00	13-10-2014: 00 Z	96	45	990	Severe Cyclonic Storm (SCS)
Nilofar	27-10-2014: 00	31-10-2014: 00 Z	96	55	990	Severe Cyclonic Storm (SCS)
Kyant	23-10-2016: 00	27-10-2016: 00 Z	96	20	996	Depression (D)
Ockhi	01-12-2017: 00	05-12-2017: 00 Z	96	50	992	Severe Cyclonic Storm (SCS)
Daye	20-09-2018: 00	22-09-2018: 00 Z	48	25	996	Depression (D)
Titli	09-10-2018: 00	13-10-2018: 00 Z	96	30	1000	Deep Depression (DD)
Gaja	12-11-2018: 00	16-11-2018: 00Z	96	40	998	Cyclonic Storm (CS)
Phethai	14-12-2018: 00	18-12-2018: 00 Z	96	30	1002	Deep Depression (DD)

3. Evaluation Method

Statistical analysis is the most common method used to find the uncertainty in the model forecasts with respect to the observations. The Direct Positional Error (DPE) was calculated by using the Haversine formula, which gives the geographical distance between the two points on a sphere. Mean Sea Level Pressure (MSLP) and Maximum Sustained Wind (MSW) were measured at each time step and evaluated against the IMD observations. The Mean Absolute Error (MAE), and Mean Square Error (MSE) were calculated with respect to IMD observations. MAE is an average prediction error and used to measure the forecast accuracy. MSE is a measure to determine the quality of a forecast with a positive value. If the value of MAE and MSE are close to zero, the quality of the forecast is better. The skill score (SS) of DPE, MSLP and MSW were calculated with respect to the reference forecast. SS is the relative accuracy score of a forecast over a reference forecast. The reference forecast was chosen based on the numerical experiments conducted by Srinivas et al. (2013) [29]. By conducting 65 numerical experiments, Srinivas et al. (2013) suggested that the Lin scheme provided better results for the prediction of the track and intensity of 21 TCs over the BoB. Hence, the sensitivity experiments with the Lin scheme considered as a reference forecast and the skill score for all the other microphysical schemes were calculated. Positive (negative) values of SS indicate that the model is more (less) skilled than the configuration using the Lin scheme.

The SS of MSLP and MSW was calculated by using Equation (3).

$$\text{Mean Absolute Error (MAE)} = \frac{1}{n}\sum_{i=1}^{n}|P_s - P_o| \quad (1)$$

$$\text{Mean Square Error (MSE)} = \frac{1}{n}\sum_{i=1}^{n}(P_s - P_o)^2 \quad (2)$$

$$\text{Skill Score } (SS_i) = 1 - \frac{MSE_{Simulation}}{MSE_{Reference}} \quad (3)$$

In the above equations, P_s = Simulated Value of the Parameter, P_o = Observed Value of the Parameter, and n = Number of Observations. The SS for DPE is calculated using Equation (4).

$$\text{Skill Score } (SS_{DPE}) = 1 - \frac{DPE_{Simulation}}{DPE_{Reference}} \quad (4)$$

4. Results

The sensitivity of seven CMP schemes was analyzed to find the optimum combination physics schemes for the prediction of post-monsoon TC over the NIO using the ARW model. Except for the Daye cyclone, the simulation period for the selected TCs is 96 h. For the Daye cyclone, the simulation period is 48 h, as the lifespan of the cyclone itself is 48 h. The model errors for MSW, MSLP, and track position are calculated with respect to the observed value.

4.1. Track and Intensity Errors

The predicted tracks of the selected cyclones for all the combinations of CMP schemes and the best track provided by IMD are depicted in Figures 2–5 along with their respective direct positional errors (DPE). From the results, it is observed that the intensity stages (Depression, Deep Depression, Cyclonic Storm, Severe Cyclonic Storm, and Very Severe Cyclonic Storm) of TCs during the model initialization have a significant impact on its track prediction. The cyclones initiated at deep depression or higher stages showed an increasing trend in average track error from the model initiation to the end of the simulation, whereas the cyclones initiated at the depression stage showed a decreasing trend in their average track error until 48 h of model simulation, and then gradually increased to the end of the simulation. Except for the cyclones Daye and Kyant, the predicted track was near to the observed track during the initial stages of the model simulation, with an average error of 64 km for all the schemes. The results are in good agreement with previous studies [27,32]. Subsequently, as the time of simulation increased, the predicted track also started moving away from the best track. At the end of the simulation period, the average track error was found to be 247 km. For cyclone Kyant, an average track error of 88 km was found during the initial stages and gradually reduced to 67 km up to 48 h of the model simulation, and then gradually increased to 347 km at the end. Similarly, for cyclone Daye, the average track error was gradually reduced from 162 to 78 km from the model initiation to the end of the simulation.

The model performance was evaluated by calculating the MAE, MSE, and average track errors at every 24 h interval (i.e., 4, 48, 72 h, and 96 h). The 24-hourly average track error for the TCs is presented in Figure 6. The WSM3 scheme simulated the cyclones Nilofar, Kyant, Ockhi, Daye, and Phethai with an average track error ranging from 83 to 190 km, 45 to 195 km, 42 to 75 km, 102 to 47 km, and 113 to 115 km at 24h to end of the simulation time, respectively. Hudhud cyclone is well simulated by all the CMP schemes, with a maximum average track error of 63 km at a 24 h simulation time. From then, the average track error gradually increased to 555 km at the end of the simulation with a least error of 219 km by the WSM6 scheme. Cyclone Gaja is well simulated by Ferrier, with least average track errors of 110, 264, 231, and 139 km at 24, 48, 72, and 96h of the model simulation time. In the case of Titli, the average track was considered for the overall simulation period because the Morrison scheme gave the lowest average track error of 64 and 39 km during the initial stages of model simulation, whereas the WSM6 scheme produced the least error of 111 and 37 km at the end of the simulation. Hence, the WSM6 scheme was considered to provide superior results for Titli cyclone. The single-moment or double-moment schemes did not shown any significant variations in the TCs track prediction. The deviations in the predicted tracks may be attributed to the variations in the intensification process during the model simulation. The schemes which showed rapid intensification process during the model simulation showed minimum deviation from the observed track, whereas the schemes which showed a slower intensification process during the model simulation showed maximum deviation from the observed track [45].

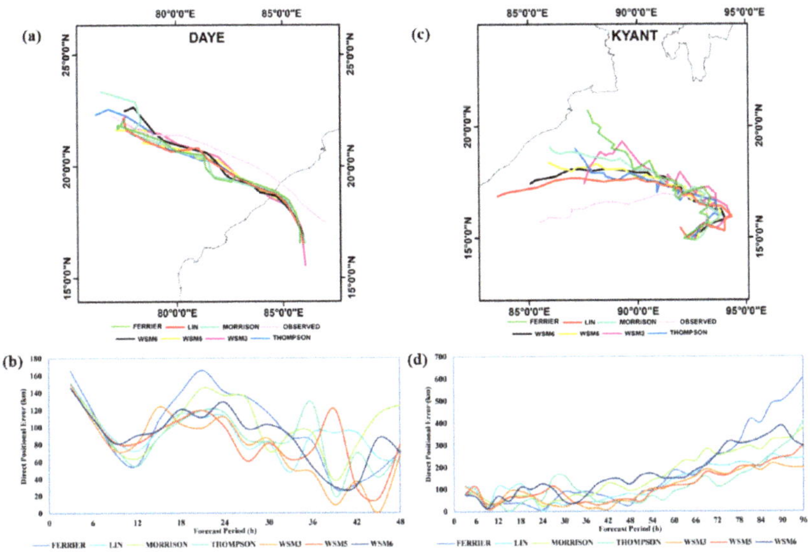

Figure 2. Observed and predicted tracks of tropical cyclones initiated at depression stage along with DPE (**a**) tracks of Daye Cyclone, (**b**) DEP of Daye Cyclone, (**c**) track of Kyant Cyclone, (**d**) DEP of Kyant Cyclone.

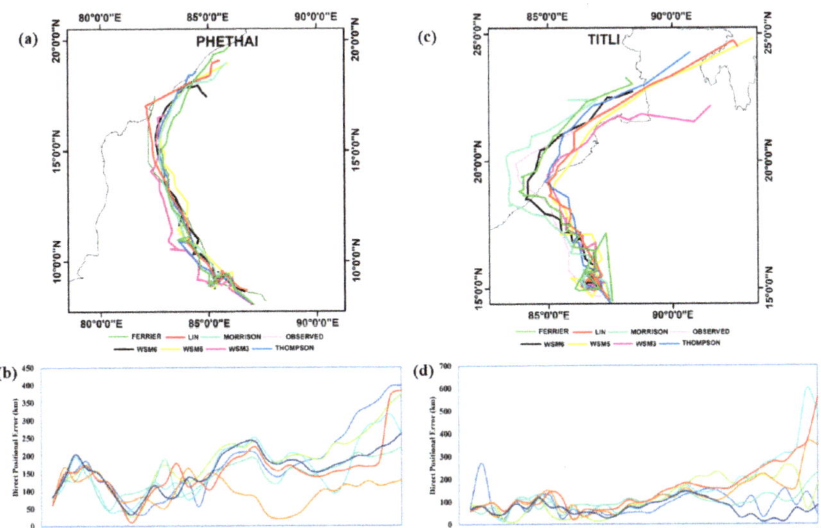

Figure 3. Observed and predicted tracks of tropical cyclones initiated at deep depression stage along with DPE (**a**) track of Phethai Cyclone, (**b**) DPE of Phethai Cyclone, (**c**) track of Titli Cyclone, (**d**) DPE of Titli Cyclone.

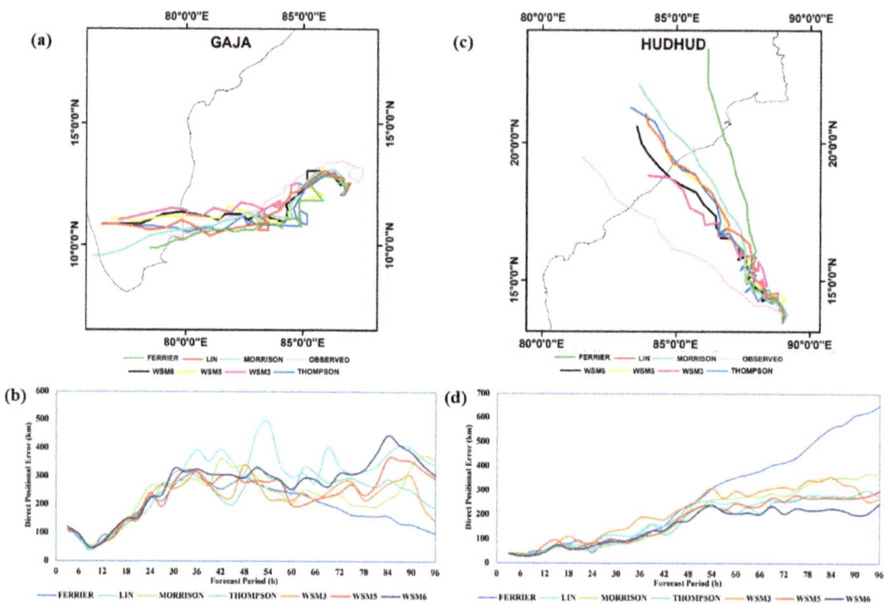

Figure 4. Observed and predicted tracks of tropical cyclones initiated at cyclonic stage (Gaja) and severe cyclonic stage (Hudhud) along with DPE (**a**) track of Gaja Cyclone, (**b**) DPE of Gaja Cyclone, (**c**) track of Hudhud Cyclone, (**d**) DPE of Hudhud Cyclone.

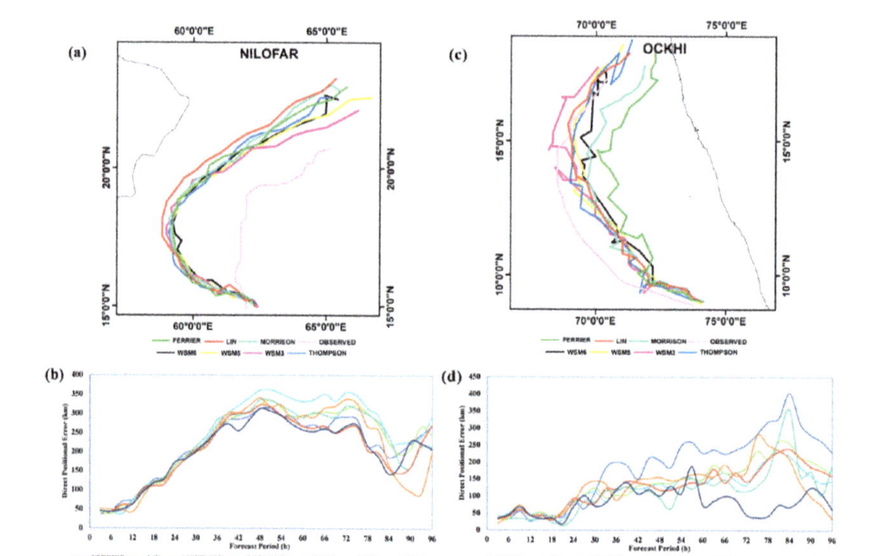

Figure 5. Observed and predicted tracks of tropical cyclones initiated at severe cyclonic stage along with DPE (**a**) track of Nilofar Cyclone, (**b**) DPE of Nilofar Cyclone, (**c**) track of Ockhi Cyclone, (**d**) DPE of Ockhi Cyclone.

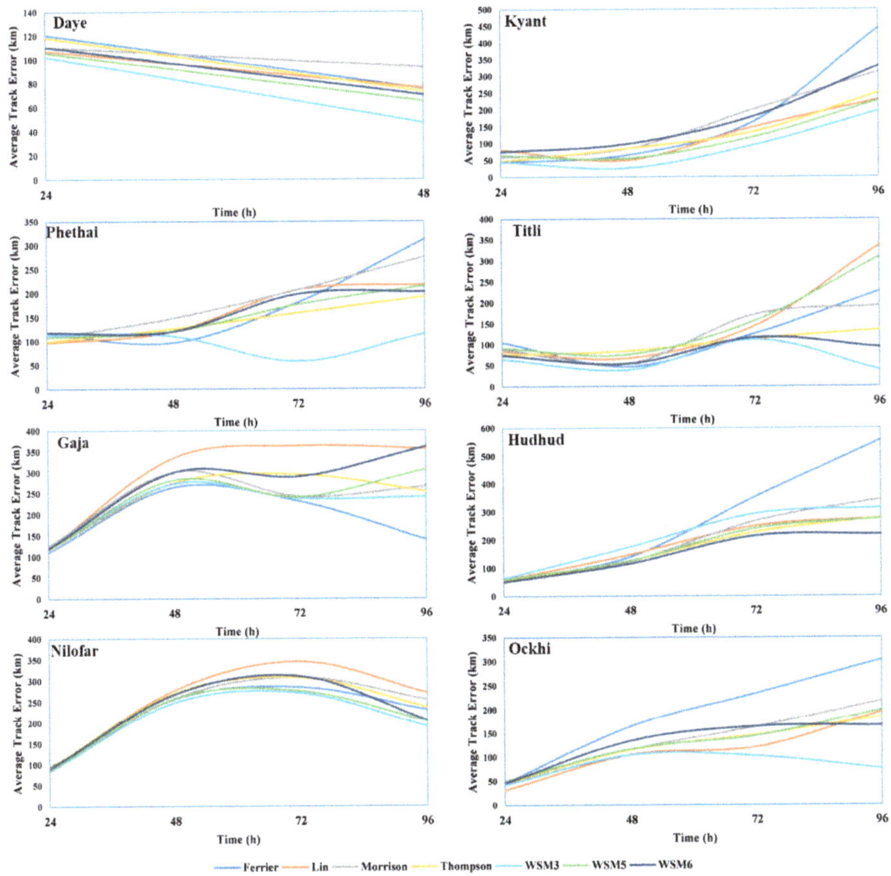

Figure 6. Average track error at every 24-hour interval of the tropical cyclones considered.

The Mean Absolute Error (MAE) and Mean Square Error (MSE) of MSW for all CMPs were calculated and the results of the TC Hudhud are presented in Figure 7. In the case of Nilofar, Kyant, Daye, and Phethai cyclones, WSM3 indicates the lowest MAE and MSE. The lowest MAE for the TCs ranged from 4.73 to 17.08 m/s, 6.84 to 4.63 m/s, 8.60 to 1.45 m/s, and 7.16 to 3.66 m/s for MSW from 24 h to the end of the simulation. The lowest MSE for the TCs ranged from 10.14 to 11.31 m^2/s^2, 2.89 to 4.97 m^2/s^2, 6.56 to 0.88 m^2/s^2, and 2.29 to 6.29 m^2/s^2. The lowest MAE and MSE for the Gaja cyclone were obtained from the Ferrier scheme and are ranged from 4.19 to 9.94 m/s and 5.77 to 13.46 m^2/s^2, respectively. However, for the Hudhud, Titli and Ockhi cyclones, the WSM6 and Lin scheme provided the lowest MAE, ranging from 2.44 to 9.49 m/s, 5.08 to 3.91 m/s and 4.85 to 3.85 m/s, and the MSE ranged from 1.60 to 1.48 m^2/s^2, 5.39 to 8.63 m^2/s^2 and 6.44 to 12.45 m^2/s^2. The schemes which predicted the MSW well also predicted MSLP for the respective TCs.

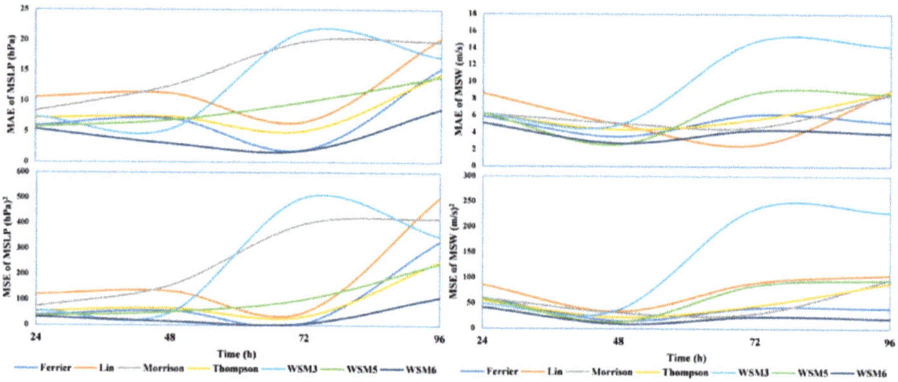

Figure 7. Mean Absolute Error (MAE), Mean Square Error (MSE) of Mean Sea Level Pressure (MSLP) and Maximum Sustained Wind (MSW) of the tropical cyclone Hudhud.

The intensity of the TC cyclone is influenced by the auto conversion process between the hydrometers and the amount of latent heat released during the conversion process [46,47]. To assess the impact of various microphysical schemes on the intensity of TCs, the vertical profile of the area averaged mixing ratios was calculated at every 3-hour interval over the entire numerical domain. An average value of the prognostic variables over all the time steps was taken for the analysis. The averaged values of the prognostic variables for the cyclone Hudhud are presented in Figure 8 and other cyclones are presented in Supplementary Figures S1–S7. From the results, it can be observed that the WSM3 scheme only predicted the liquid hydrometers for all the cyclones. This indicates that the WSM3 scheme assumed that the temperature of the clouds was above the freezing point. Compared to the other microphysical schemes, the WSM3 scheme produced significant amounts of cloud water and rain in the lower troposphere for the cyclones Nilofar, Kyant, Daye, and Phethai. For the cyclones Kyant, Daye, and Phethai, all the microphysical schemes showed a significant decrease in the frozen hydrometers in the middle troposphere results in slowing down the vertical acceleration of the intense updrafts in the eye wall of the storm, which might be reason for inhibiting the storm intensification [48]. For the cyclone Nilofar, the frozen hydrometers predicted by all the microphysical schemes in the middle troposphere are negligible in quantity and the liquid hydrometers predicted by Lin, WSM5 and WSM6 in the lower troposphere are also negligible in quantity. Compared to the Ferrier, Morrison, and Thompson schemes, which produced cloud water and rain in the lower troposphere, the WSM3 scheme produced a higher amount. The presence of cloud water and rain in the lower troposphere helps in its intensification. Therefore, the WSM3 scheme produced a higher intensity for the cyclones Nilofar, Daye, Kyant, and Phethai.

The WSM6 scheme produced a significantly large amount of cloud water and rain in the lower troposphere compared to WSM3 for the cyclones Hudhud, Ockhi, and Titli. Due to the presence of cloud water and rain in the lower troposphere helping in the release of latent heat and cyclone intensification, the WSM6 scheme has given a higher intensity for the cyclones Hudhud, Ockhi and Titli, whereas the other microphysical schemes showed a significant decrease in the frozen hydrometers in the middle troposphere, which prevents the intensification of cyclones. In the case of Gaja cyclone, the Ferrier, Lin, Thompson, and WSM6 schemes produced cloud water in the lower troposphere, and all the schemes produced negligible quantities of all the other prognostic variables. Compared to other schemes, due to the presence of a large amount of cloud water in the lower troposphere, the Ferrier scheme predicted the Gaja cyclone with greater intensity. The uncertainties in the predictions by various microphysical schemes can be attributed to the presence of a graupel hydrometer [49].

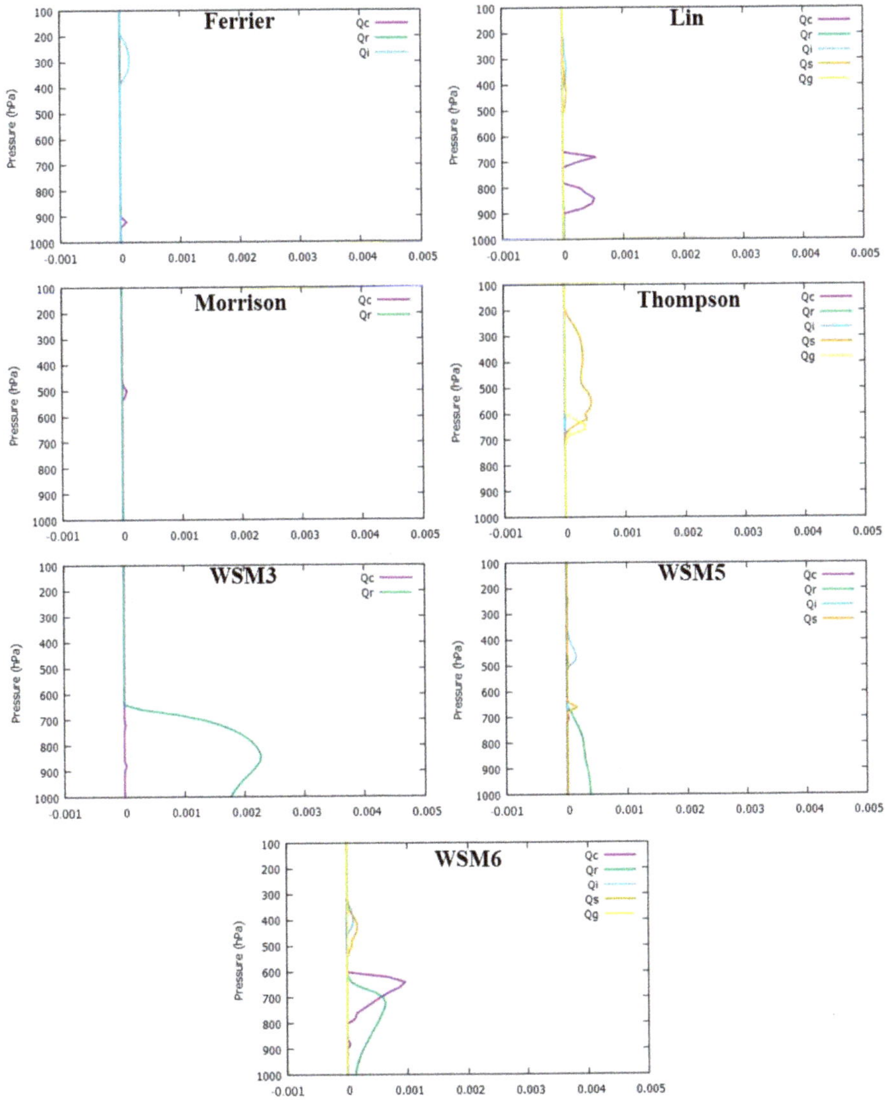

Figure 8. Time evolution of area averaged mixing ratios (g/kg) for the cyclone Hudhud (Qv-Water Vapor, Qc-Cloud Water, Qr-Rain Water, Qi-Ice, Qs-Snow, Qg-Graupel).

4.2. Skill Score

Skill Score (SS) was calculated to obtain information about the improvement in the model forecast over the reference forecast. It is easy to identify the improvement in the model performance as the SS provides a single value. In the present study, sensitivity experiments with the Lin scheme were considered as a reference forecast and the skill score for all the other microphysical schemes was calculated. The skill scores for the DPE, MSW and MSLP at the end of the simulation are provided in Tables 4–6.

Table 4. Skill score for Direct Positional Error (DPE).

Cyclone/CMP	Ferrier	Morrison	Thompson	WSM3	WSM5	WSM6
Hudhud	−0.52	−0.07	0.07	−0.16	0.04	**0.17**
Nilofar	0.13	0.08	0.08	**0.18**	0.17	0.14
Kyant	−0.44	−0.31	0.06	**0.17**	0.08	−0.39
Ockhi	−0.54	−0.12	0.07	−0.04	−0.04	**0.35**
Daye	−0.09	−0.16	0.04	**0.19**	0.08	0.01
Titli	0.49	0.28	0.37	0.21	0.01	**0.58**
Gaja	**0.35**	0.20	0.21	0.26	0.19	0.08
Phethai	−0.11	−0.17	0.10	**0.41**	0.03	0.01

Table 5. Skill score for maximum sustained wind (MSW).

Cyclone/CMP	Ferrier	Morrison	Thompson	WSM3	WSM5	WSM6
Hudhud	0.33	0.18	0.10	−1.33	0.00	**0.43**
Nilofar	0.22	0.01	0.23	**0.47**	0.21	0.11
Kyant	0.15	0.26	0.66	**0.70**	0.61	0.53
Ockhi	−0.24	−0.56	−0.38	−0.90	−0.33	**0.01**
Daye	0.01	−0.08	−0.02	**0.17**	−0.10	−0.07
Titli	0.13	0.15	0.05	0.09	−0.17	**0.26**
Gaja	**0.62**	0.61	0.33	−0.16	0.18	0.14
Phethai	0.27	0.12	0.20	**0.47**	0.33	0.26

Table 6. Skill score for mean sea level pressure (MSLP).

Cyclone/CMP	Ferrier	Morrison	Thompson	WSM3	WSM5	WSM6
Hudhud	0.45	−0.30	0.50	−0.16	0.51	**0.74**
Nilofar	0.41	0.05	0.30	**0.70**	0.44	0.25
Kyant	0.24	0.37	0.38	**0.68**	0.62	0.07
Ockhi	−0.85	−0.25	−0.94	−1.68	−0.92	**0.15**
Daye	−0.02	0.21	0.18	**0.29**	0.11	0.10
Titli	0.11	0.22	0.03	0.35	−0.15	**0.35**
Gaja	**0.47**	0.41	0.43	−0.23	0.03	0.33
Phethai	0.23	0.10	0.31	**0.40**	0.28	0.20

The WSM3 scheme showed an improvement of 18%, 17%, 19%, and 41% in direct positional error (DPE) for the cyclones Nilofar, Kyant, Daye, and Phethai, respectively, whereas the WSM6 scheme showed an improvement of 17%, 35% and 58% in DPE for the cyclones Hudhud, Ockhi, and Titli. For the Gaja cyclone, the Ferrier scheme showed an improvement of 35% over the reference forecast. Similar results were obtained for MSW and MLSP.

5. Conclusions

In the present study, a total of 56 sensitivity experiments were conducted to assess the impact of seven microphysical schemes on the track and intensity of eight tropical cyclones over the NIO region that occurred from 2014 to 2018. To assess the model performance, DPE, MAE and MSE errors were calculated based on the observations provided by IMD and skill score was calculated over the reference forecast. The WSM3 microphysical scheme showed the lowest DPE, MAE and MSE for the tropical cyclones Nilofar, Kyant, Daye and Phethai, and WSM6 scheme showed the lowest values for the tropical cyclones Hudhud, Ockhi and Titli. Due to the presence of liquid hydrometers (i.e., cloud water and rain) in the lower atmosphere, the WSM3 has predicted more intensity for Nilofar, Kyant, Daye and Phethai cyclones, and WSM6 for Hudhud, Ockhi and Phethai. Compared to the model performance with the Lin scheme, the WSM3 scheme provided a significant improvement in the prediction of track and intensity for the tropical cyclones Nilofar, Kyant, Daye and Phethai and WSM6 for the tropical cyclones Hudhud, Ockhi and Titli. From the results, the WSM3 scheme can be suggested as a first best scheme for the prediction of the track and intensity of tropical cyclones over the NIO region. WSM3 appears to have better skills than other, more complex parameterizations, maybe due to the presence of fewer prognostic variables. The presence of more prognostic variables in other schemes may add uncertainty to the model simulations.

A study by Choudhary and Das (2017) on the impact of grid resolution on the track and intensity indicated that the domains with less than 5 km resolution have produced accurate results. When the size of the domain increases, the model results will be more dependent on the parameterization schemes. However, for operational purposes, the disaster management agencies need real-time forecasts within a short time with reasonable accuracy. Therefore, the methodology and the information about the microphysical applied in the study can be used in real-time forecasting of TCs over the NIO region.

Supplementary Materials: The following are available online at http://www.mdpi.com/2073-4433/11/12/1297/s1, Figure S1. Time evolution of area averaged mixing ratios (g/kg) for the cyclone Daye (Qv-Water Vapor, Qc-Cloud Water, Qr-Rain Water, Qi-Ice, Qs-Snow, Qg-Graupel). Figure S2. Time evolution of area averaged mixing ratios (g/kg) for the cyclone Gaja (Qv-Water Vapor, Qc-Cloud Water, Qr-Rain Water, Qi-Ice, Qs-Snow, Qg-Graupel). Figure S3. Time evolution of area averaged mixing ratios (g/kg) for the cyclone Kyant (Qv-Water Vapor, Qc-Cloud Water, Qr-Rain Water, Qi-Ice, Qs-Snow, Qg-Graupel). Figure S4. Time evolution of area averaged mixing ratios (g/kg) for the cyclone Nilofar (Qv-Water Vapor, Qc-Cloud Water, Qr-Rain Water, Qi-Ice, Qs-Snow, Qg-Graupel). Figure S5. Time evolution of area averaged mixing ratios (g/kg) for the cyclone Ockhi (Qv-Water Vapor, Qc-Cloud Water, Qr-Rain Water, Qi-Ice, Qs-Snow, Qg-Graupel). Figure S6. Time evolution of area averaged mixing ratios (g/kg) for the cyclone Phethai (Qv-Water Vapor, Qc-Cloud Water, Qr-Rain Water, Qi-Ice, Qs-Snow, Qg-Graupel). Figure S7. Time evolution of area averaged mixing ratios (g/kg) for the cyclone Titli (Qv-Water Vapor, Qc-Cloud Water, Qr-Rain Water, Qi-Ice, Qs-Snow, Qg-Graupel).

Author Contributions: Conceptualization, G.V.R. and K.V.R.; methodology, V.S., K.V.R., G.V.R.; formal analysis, G.V.R.; investigation, G.V.R.; resources, K.V.R.; data curation, G.V.R.; writing—original draft preparation, G.V.R.; writing—review and editing, V.S., K.V.R.; supervision, V.S., K.V.R.; funding acquisition, K.V.R. All authors have read and agreed to the published version of the manuscript.

Funding: The research described in this paper, in part, is funded by the Ministry of Human Resource Development (MHRD), Government of India under Scheme for Promotion of Academic and Research Collabo-ration (SPARC) through project number P270.

Acknowledgments: The first author is thankful to Satya Prakash Ojha, Scientist, SAC-ISRO and Sathiyamoorthy, Scientist, SAC – ISRO, Ahmedabad, for providing me the opportunity to work as a research trainee in the Satellite Meteorology and OceAnography Research and Training (SMART) programme. The public access to the IMD best track, ARW model and GFS data, provided by NCAR, is gratefully acknowledged. We also acknowledge the partial support received by the corresponding author from the Virginia Agricultural Experiment Station (Blacksburg) and the Hatch Program of the National Institute of Food and Agriculture, U.S Department of Agriculture (Washington, D.C.).

Conflicts of Interest: The authors declare no conflict of interest.

References

1. Balaji, M.; Chakraborty, A.; Mandal, M. Changes in tropical cyclone activity in north Indian Ocean during satellite era (1981–2014). *Int. J. Climatol.* **2018**, *38*, 2819–2837. [CrossRef]
2. Kreimer, A. Social and economic impacts of natural disasters. *Int. Geol. Rev.* **2001**, *43*, 401–405. [CrossRef]
3. Wahiduzzaman, M.; Oliver, E.C.J.; Wotherspoon, S.J.; Luo, J.J. Seasonal forecasting of tropical cyclones in the North Indian Ocean region: The role of El Niño-Southern Oscillation. *Clim. Dyn.* **2020**, *54*, 1571–1589. [CrossRef]
4. Pattanayak, S.; Mohanty, U.C.; Osuri, K.K. Impact of Parameterization of Physical Processes on Simulation of Track and Intensity of Tropical Cyclone Nargis (2008) with WRF-NMM Model. *Sci. World J.* **2012**, *2012*, 1–18. [CrossRef]

5. Ramage Monsoonal influences on the annual variation of tropical cyclone development over the Indian and Pacific Oceans. *Mon. Weather Rev.* **1974**, *102*, 745–753. [CrossRef]
6. Vishnu, S.; Sanjay, J.; Krishnan, R. Assessment of climatological tropical cyclone activity over the north Indian Ocean in the CORDEX-South Asia regional climate models. *Clim. Dyn.* **2019**, *53*, 5101–5118. [CrossRef]
7. RSMC—Tropical Cyclone. In *Report on Cyclonic Disturbances over the North Indian Ocean During 2014*; Government of India: New Delhi, India, 2015.
8. RSMC—Tropical Cyclone. In *Report on Cyclonic Disturbances over the North Indian Ocean During 2015*; Government of India: New Delhi, India, 2016.
9. RSMC—Tropical Cyclone. In *Report on Cyclonic Disturbances over the North Indian Ocean During 2016*; Government of India: New Delhi, India, 2017.
10. RSMC—Tropical Cyclone. In *Report on Cyclonic Disturbances over the North Indian Ocean During 2017*; Government of India: New Delhi, India, 2018.
11. RSMC—Tropical Cyclone. In *Report on Cyclonic Disturbances over the North Indian Ocean During 2018*; Government of India: New Delhi, India, 2019.
12. Singh, O.P.; Ali Khan, T.M.; Rahman, M.S. Changes in the frequency of tropical cyclones over the North Indian Ocean. *Meteorol. Atmos. Phys.* **2000**, *75*, 11–20. [CrossRef]
13. Mohanty, U.C.; Osuri, K.K.; Pattanayak, S.; Sinha, P. An observational perspective on tropical cyclone activity over Indian seas in a warming environment. *Nat. Hazards* **2012**, *63*, 1319–1335. [CrossRef]
14. Zehr, R.M. Tropical Cyclogenesis in the Western North Pacific. In *NOAA Technical Report NESDIS 61*; National Oceanic and Atmospheric Administration: Washington, DC, USA, 1992.
15. Balaguru, K.; Taraphdar, S.; Leung, L.R.; Foltz, G.R. Increase in the intensity of postmonsoon Bay of Bengal tropical cyclones. *Geophys. Res. Lett.* **2014**, *41*, 3594–3601. [CrossRef]
16. Girishkumar, M.S.; Ravichandran, M. The influences of ENSO on tropical cyclone activity in the Bay of Bengal during October-December. *J. Geophys. Res. Ocean.* **2012**, *117*, 1–13. [CrossRef]
17. Mei, W.; Xie, S.P.; Primeau, F.; McWilliams, J.C.; Pasquero, C. Northwestern Pacific typhoon intensity controlled by changes in ocean temperatures. *Sci. Adv.* **2015**, *1*, e1500014. [CrossRef] [PubMed]
18. Bin, W.; Elsberry, R.L.; Yuqing, W.; Liguang, W. Dynamics in Tropical Cyclone Motion: A Review. *Chin. J. Atmos. Sci.* **1998**, *22*, 416–434.
19. Sharmila, S.; Walsh, K.J.E. Impact of large-scale dynamic versus thermodynamic climate conditions on contrasting tropical cyclone genesis frequency. *J. Clim.* **2017**, *30*, 8865–8883. [CrossRef]
20. Deshpande, M.S.; Pattnaik, S.; Salvekar, P.S. Impact of cloud parameterization on the numerical simulation of a super cyclone. *Ann. Geophys.* **2012**, *30*, 775–795. [CrossRef]
21. Sandeep, C.P.R.; Krishnamoorthy, C.; Balaji, C. Impact of Cloud Parameterization Schemes on The Simulation of Cyclone *Vardah* using the WRF Model. *Curr. Sci.* **2018**, *115*, 1143. [CrossRef]
22. Raju, P.V.S.; Potty, J.; Mohanty, U.C. Prediction of severe tropical cyclones over the Bay of Bengal during 2007-2010 using high-resolution mesoscale model. *Nat. Hazards* **2012**, *63*, 1361–1374. [CrossRef]
23. Jandaghian, Z.; Touchaei, A.G.; Akbari, H. Sensitivity analysis of physical parameterizations in WRF for urban climate simulations and heat island mitigation in Montreal. *Urban Clim.* **2018**, *24*, 577–599. [CrossRef]
24. Wang, W.; Seaman, N.L. A comparison study of convective parameterization schemes in a mesoscale model. *Mon. Weather Rev.* **1997**, *125*, 252–278. [CrossRef]
25. Chandrasekar, R.; Balaji, C. Sensitivity of tropical cyclone Jal simulations to physics parameterizations. *J. Earth Syst. Sci.* **2012**, *121*, 923–946. [CrossRef]
26. Raju, P.V.S.; Potty, J.; Mohanty, U.C. Sensitivity of physical parameterizations on prediction of tropical cyclone Nargis over the Bay of Bengal using WRF model. *Meteorol. Atmos. Phys.* **2011**, *113*, 125–137. [CrossRef]
27. Kanase, R.D.; Salvekar, P.S. Impact of physical parameterization schemes on track and intensity of severe cyclonic storms in Bay of Bengal. *Meteorol. Atmos. Phys.* **2015**, *127*, 537–559. [CrossRef]
28. Srinivas, C.V.; Bhaskar Rao, D.V.; Yesubabu, V.; Baskaran, R.; Venkatraman, B. Tropical cyclone predictions over the bay of bengal using the high-resolution advanced research weather research and forecasting (ARW) model. *Q. J. R. Meteorol. Soc.* **2013**, *139*, 1810–1825. [CrossRef]
29. Lakshmi, D.D.; Annapurnaiah, K. Impact of Microphysics Schemes in the Simulation of Cyclone Hudhud using WRF-ARW model. *Int. J. Ocean. Oceanogr.* **2016**, *10*, 49–59.
30. Choudhury, D.; Das, S. The sensitivity to the microphysical schemes on the skill of forecasting the track and intensity of tropical cyclones using WRF-ARW model. *J. Earth Syst. Sci.* **2017**, *126*, 57. [CrossRef]

31. Reddy, M.V.; Prasad, S.B.S.; Krishna, U.V.M.; Reddy, K.K. Effect of cumulus and microphysical parameterizations on the JAL cyclone prediction. *Indian J. Radio Space Phys.* **2014**, *43*, 103–123.
32. Osuri, K.K.; Mohanty, U.C.; Routray, A.; Kulkarni, M.A.; Mohapatra, M. Customization of WRF-ARW model with physical parameterization schemes for the simulation of tropical cyclones over North Indian Ocean. *Nat. Hazards* **2012**, *63*, 1337–1359. [CrossRef]
33. Mahala, B.K.; Mohanty, P.K.; Nayak, B.K. Impact of Microphysics Schemes in the Simulation of Cyclone Phailin using WRF Model. *Procedia Eng.* **2015**, *116*, 655–662. [CrossRef]
34. Skamarock, W.C.; Klemp, J.B.; Dudhia, J.; Gill, D.O.; Zhiquan, L.; Berner, J.; Wang, W.; Powers, J.G.; Duda, M.G.; Barker, D.M.; et al. A Description of the Advanced Research WRF Model Version 4. In *NCAR Technical Note NCAR/TN-475+STR*; NCAR: Boulder, CO, USA, 2019; p. 145.
35. Sridhar, V. Tracking the influence of irrigation on land surface fluxes and boundary layer climatology. *J. Contemp. Water Res. Educ.* **2013**, *152*, 79–93. [CrossRef]
36. Sridhar, V.; Anderson, K.A. Human-induced modifications to boundary layer fluxes and their water management implications in a changing climate. *Agric. For. Meteorol.* **2017**, *234*, 66–79. [CrossRef]
37. Lin, Y.-L.; Farley, R.D.; Orville, H.D. Bulk Parameterization of the Snow Field in a Cloud Model. *J. Clim. Appl. Meteorol.* **1983**, *22*, 1065–1092. [CrossRef]
38. Rogers, E.; Black, T.; Ferrier, B.; Lin, Y.; Parrish, D.; DiMego, G. NCEP Meso Eta Analysis and Forecast System: Increase in resolution, new cloud microphysics, modified precipitation assimilation, modified 3DVAR analysis. *NWS Tech. Proced. Bull.* **2001**, *488*, 15.
39. Thompson, G.; Rasmussen, R.M.; Manning, K. Explicit Forecasts of Winter Precipitation Using an Improved Bulk Microphysics Scheme. Part I: Description and Sensitivity Analysis. *Mon. Weather Rev.* **2004**, *132*, 519–542. [CrossRef]
40. Castro, D.M.; Kumar, S.; Rojas, J.L.F.; Alvarez, A.M.; Prado, J.M.V.; Puma, E.V.; Velarde, C.D.C.; Vidal, Y.S. The Impact of Microphysics Parameterization in the Simulation of Two Convective Rainfall Events over the Central Andes of Peru Using WRF-ARW. *Atmosphere* **2019**, *10*, 442. [CrossRef]
41. Morrison, H.; Thompson, G.; Tatarskii, V. Impact of Cloud Microphysics on the Development of Trailing Stratiform Precipitation in a Simulated Squall Line: Comparison of One- and Two-Moment Schemes. *Mon. Weather Rev.* **2009**, *137*, 991–1007. [CrossRef]
42. Hong, S.-Y.; Dudhia, J.; Chen, S.-H. A Revised Approach to Ice Microphysical Processes for the Bulk Parameterization of Clouds and Precipitation. *Mon. Weather Rev.* **2004**, *132*, 103–120. [CrossRef]
43. Hong, S.-Y.; Lim, J.-O.J. The WRF single–moment 6–class microphysics scheme (WSM6). *J. Korean Meteorol. Soc.* **2006**, *42*, 129–151.
44. Li, J.; Ding, C.; Li, F.; Chen, Y. Effects of single- and double-moment microphysics schemes on the intensity of super typhoon Sarika (2016). *Atmos. Res.* **2020**, *238*, 104894. [CrossRef]
45. Park, J.; Cha, D.; Lee, M.K.; Moon, J.; Hahm, S.; Noh, K.; Chan, J.C.L.; Bell, M. Impact of Cloud Microphysics Schemes on Tropical Cyclone Forecast Over the Western North Pacific. *J. Geophys. Res. Atmos.* **2020**, *125*, e2019JD032288. [CrossRef]
46. Mukhopadhyay, P.; Taraphdar, S.; Goswami, B.N. Influence of moist processes on track and intensity forecast of cyclones over the north Indian Ocean. *J. Geophys. Res. Atmos.* **2011**, *116*, 1–21. [CrossRef]
47. Kanase, R.D.; Mukhopadhyay, P.; Salvekar, P.S. Understanding the Role of Cloud and Convective Processes in Simulating the Weaker Tropical Cyclones over Indian Seas. *Pure Appl. Geophys.* **2014**, *172*, 1751–1779. [CrossRef]
48. Maw, K.W.; Min, J. Impacts of Microphysics Schemes and Topography on the Prediction of the Heavy Rainfall in Western Myanmar Associated with Tropical Cyclone ROANU (2016). *Adv. Meteorol.* **2017**, *2017*, 1–22. [CrossRef]
49. Gilmore, M.S.; Straka, J.M.; Rasmussen, E.N. Precipitation uncertainty due to variations in precipitation particle parameters within a simple microphysics scheme. *Mon. Weather Rev.* **2004**, *132*, 2610–2627. [CrossRef]

Publisher's Note: MDPI stays neutral with regard to jurisdictional claims in published maps and institutional affiliations.

atmosphere

MDPI

Article

Future Changes in the Free Tropospheric Freezing Level and Rain–Snow Limit: The Case of Central Chile

Piero Mardones [1,2,3,*] and René D. Garreaud [1,2]

[1] Department of Geophysics, University of Chile, Santiago 8370415, Chile; rgarreau@dgf.uchile.cl
[2] Center for Climate and Resilience Research, University of Chile, Santiago 8370415, Chile
[3] Department of Oceanography, University of Concepción, Concepción 4030000, Chile
* Correspondence: pmardones@dgf.uchile.cl

Received: 14 October 2020; Accepted: 16 November 2020; Published: 23 November 2020

check for updates

Abstract: The freezing level in the free troposphere often intercepts the terrain of the world's major mountain ranges, creating a rain–snow limit. In this work, we use the free tropospheric height of the 0 °C isotherm (H_0) as a proxy of both levels and study its distribution along the western slope of the subtropical Andes (30°–38° S) in present climate and during the rest of the 21st century. This portion of the Andes corresponds to central Chile, a highly populated region where warm winter storms have produced devastating landslides and widespread flooding in the recent past. Our analysis is based on the frequency distribution of H_0 derived from radiosonde and surface observations, atmospheric reanalysis and climate simulations. The future projections primarily employ a scenario of heavy greenhouse gasses emissions (RCP8.5), but we also examine the more benign RCP4.5 scenario. The current H_0 distribution along the central Chile coast shows a gradual decrease southward, with mean heights close to 2600 m ASL (above sea level) at 30 °C S to 2000 m ASL at 38° S for days with precipitation, about 800 m lower than during dry days. The mean value under wet conditions toward the end of the century (under RCP8.5) is close to, or higher than, the upper quartile of the H_0 distribution in the current climate. More worrisome, H_0 values that currently occur only 5% of the time will be exceeded in about a quarter of the rainy days by the end of the century. Under RCP8.5, even moderate daily precipitation can increase river flow to levels that are considered hazardous for central Chile.

Keywords: freezing level; climate change; central Chile; CMIP5; CFSR; flooding

1. Introduction

Most of the precipitation falling on the ground can be traced back to ice crystal formation in the subfreezing environment of the middle and upper troposphere when enough moisture is provided by tropical or extratropical weather systems [1]. As the ice crystals grow, they descend and eventually cross the 0 °C isotherm at a height H_0 above sea level (ASL) and begin to melt. The melting layer depth is highly variable, depending on the air temperature profile and the hydrometeors population (size, type, density), but it generally ranges between 100 and 300 m [2,3]. Over the terrain, the delimitation of sectors receiving rain or snow during a storm is also related to the position of the near surface 0 °C isotherm, although rain can occur with temperatures as cold as −1.5 °C and snow can occur with temperatures as high as +1.5 °C (e.g., [4,5]). The height of the 0 °C level over the terrain is closely tied to the nearby free tropospheric 0 °C isotherm height (H_0) during precipitation events, but the surface level tends to be several hundred meters below its free tropospheric counterpart [6,7]. This offset is driven by the forced ascent of moisture-laden air parcels over the mountainous terrain that caused both adiabatic and diabatic cooling (the latter is due to the greater precipitation rates) as described in [6–8].

Although cloud microphysics and airflow-induced localized cooling can depress the elevation of the rain–snow limit over the terrain relative to the free tropospheric H_0 during a precipitation event, both heights are closely tied between storms (e.g., [4,5,9,10]). Furthermore, because the processes behind such offset are only weakly dependent on air temperature [7], the free tropospheric H_0 suffices to track the variability and change of the rain–snow limit over terrain in the large-scale climate change context addressed in this work.

Significant changes in the rain–snow limit elevation may occur in those regions where H_0 is close to the terrain at height H_G (relative to sea level). Figure 1a shows the long-term mean distribution of the freezing level ($\overline{H_0}$) when precipitation occurs, based on 32 years of reanalysis data (see details in Section 3). Despite the underlying continental distribution, the mean field is mostly symmetric in the zonal direction, with values above 4500 m ASL in the tropics and a gradual decrease down to <500 m ASL poleward of ∼60° in both hemispheres. The condition $H_0 < H_G$, leading to snow as the dominant precipitation type, occurs at high latitudes but also over the world's major mountain ranges: the Tibetan Plateau, the Rocky Mountains, the Alps and the extratropical Andes. There, variations of the freezing level within precipitation events and between storms can produce substantial differences in the pluvial area—the expanse of terrain receiving rainfall—and runoff generation. Lundquist et al. [4] found that the coastal basins in the western United States of America are highly sensitive to H_0 because their pluvial area can change from 25% to 100% of the total area within the range of observed freezing level, thus modulating river flow during winter storms [11]. The freezing level control on surface runoff has also been noted in the Swiss Alps ([12] and reference therein). A similar condition occurs in the Andes cordillera, both in the Peruvian sector [13] and in central Chile [9] where high H_0 events can lead to destructive flash floods [14]. Liu et al. [15] described a catastrophic flooding event in Alberta (Canada) in which, among several factors, the high freezing level played a role not only by increasing the pluvial area but also because of the occurrence of rain-over-snow (ROS). ROS events are a serious hydrological threat most common at high latitudes in the Northern Hemisphere continents [16] but also in midlatitudes mountain areas (e.g., [4,17]).

Given the prospect of global warming during the rest of the 21st century, one may expect a widespread increase in the mean freezing level. Quantifying the impacts of climate change upon the distribution of H_0 thus informs us on potential shifts in the aforementioned hydrometeorological risks in mountainous regions (e.g., [18]) as well as changes in the water stored in the seasonal snowpack [19,20]. Changes in freezing level during precipitation events act in concert with those of precipitation intensity to alter the frequency distribution of runoff and river discharge, but, in a first approximation, they can be treated independently.

In this contribution, we use the tropospheric H_0 distribution to analyze projected changes of the freezing level and rain–snow limit changes over the western side of the subtropical Andes (central Chile). Occurrence of high-impact hydrometeorological events in this region is already high [21,22]; it is derived, at least partially, from high freezing level events [14]. To determine how specific our regional results are, we provide some large-scale context using global maps of selected variables. A brief climate context of the subtropical Andes is provided next in Section 2. The various datasets and methodology are presented in Section 3. Our main results are described in Sections 4.1 (H_0 in the present climate) and Section 4.2 (projected changes under RCP8.5 and RCP4.5). Changes in the H_0 distribution have effects ranging from ecology (e.g., plant distribution) and hydrology (flooding and water availability) to tourism (e.g., snow-related activities) and mining (e.g., open-pit operations). Assessing those impacts is beyond the scope of this paper, requiring hydrological or earth-system modeling, but in Section 4.3, we include a brief discussion on, with simple estimates of, the projected changes in the occurrence of flooding and snowpack water storage, two of the most direct effects of the changing freezing level. Section 5 summarizes the key findings of this work.

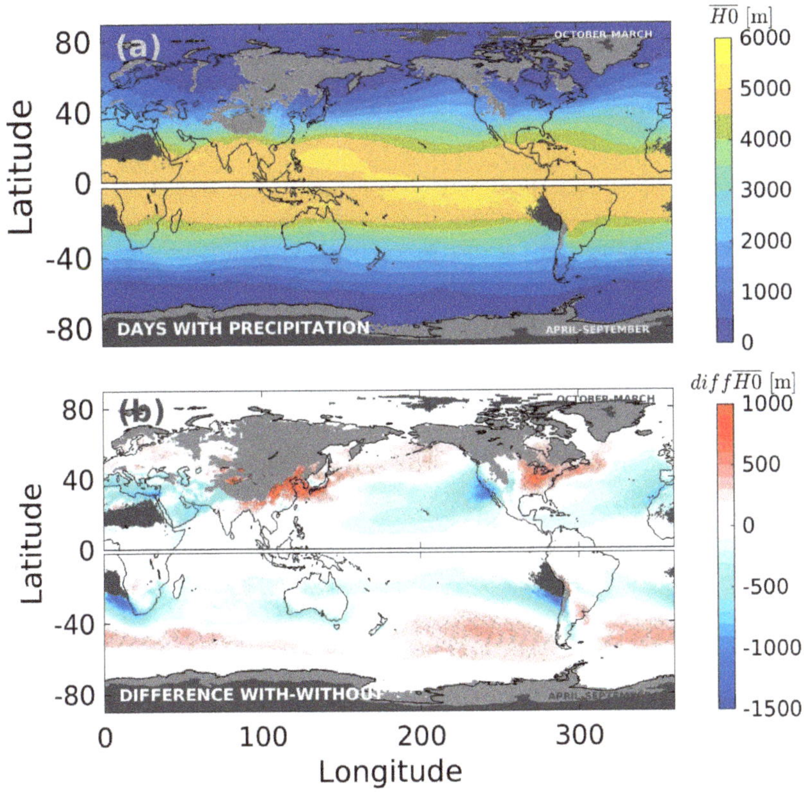

Figure 1. (**a**) Global distribution of mean H_0 (m above sea level (ASL)) calculated at any given grid box for days with precipitation (>5 mm/day in the grid box) during winter months (October–March for the Northern Hemisphere and April–September for the Southern Hemisphere). Data from the Climate Forecast System Reanalysis (CFSR) (see details in Section 3). (**b**) Difference of the mean H_0 between days with and without precipitation. Light gray areas represent those regions where mean H_0 intersects the topography. Dark gray areas indicate arid regions where the low number of rainy days precludes a robust calculation of H_0 under wet conditions.

2. Study Region

The Andes cordillera runs close to the western side of South America from north of the equator (10° N) down to Tierra del Fuego (53° S). Along its subtropical portion (30°–38° S), its crest level reaches more than 5000 m ASL, well above the mean freezing level during winter storms (~2400 m ASL; see Figure 1a). Central Chile, the narrow strip of land to the west of the subtropical Andes, is home to more than 12 million inhabitants with a large concentration (>7 million) in the metropolitan area of Santiago that sits right at the Andean foothills. Annual mean precipitation in this region varies between 200–2000 mm (Figure 2), depending on latitude and height [23].

Precipitation is largely concentrated in the austral winter (May–September) and is mostly caused by cold fronts [24]. The Andes cordillera enhances precipitation by a factor of 2–3 between the upstream lowlands and the upper part of the mountains [23]. Between 10 to 20 precipitation events occur every winter lasting 1–3 days, with storm accumulations that are highly correlated with the amount of water vapor impinging against the Andes [24]. Indeed, extreme events (>50 mm per day) occur when intense atmospheric rivers ahead of cold fronts make landfall in this region [25,26]. Major storms generally result in marked, sudden increases in the flow of the rivers descending from the Andes [27], leading to flooding in central Chile. These events are a significant threat in central Chile, damaging infrastructure and causing loss of lives with a recurrence of 5–10 years [28]. Although the magnitude of the flooding is primarily controlled by the amount of the precipitation falling over the region, the elevation of the freezing level during the storms modulates the hydrological response [9]. During winter storms, $\overline{H_0}$ ~2200 m ASL over the upper Maipo river basin (area: 5500 km^2) east of Santiago, but some precipitation events feature an H_0 in excess of 3500 m ASL, tripling the pluvial area relative to mean conditions, and thus increasing the risk of hydrometeorological hazards [9]. This was the case in 3 May 1993 when a moderate precipitation event occurred under warm conditions (H_0 ~4000 m ASL), resulting in multiple landslides and downstream flooding on the city of Santiago that caused more than 80 fatalities and major damage in public and private infrastructure [14,29,30].

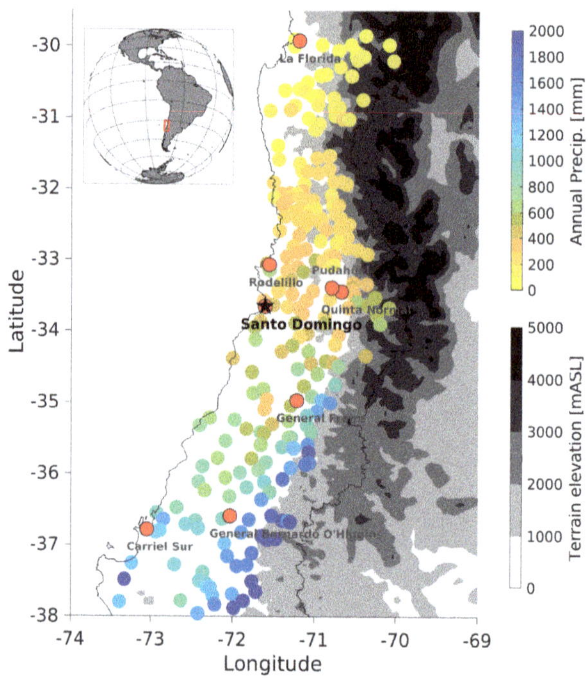

Figure 2. Station-based annual mean precipitation over Central Chile for the period 1981–2010, colored according to the scale to the right. Gray shading indicates topography levels from 0 to 5000 m ASL (every 1000 m). The black star shows the location of the Santo Domingo upper air station and red circles show selected stations with concurrent air temperature and precipitation records used to infer the freezing level during rainy conditions.

3. Data and Models

3.1. Observations

Quality-controlled daily rainfall and average temperature is available from 1999 to 2017 for eight surface stations across central Chile (30°–38° S, Figure 2) operated by the National Weather Service (DMC). In each station, we define a wet day as one with more than 5 mm accumulation and dry days as those with no precipitation. In this work, we only consider the winter semester (May to September) for central Chile and the rest of the Southern Hemisphere (SH). Santo Domingo, a coastal site at 33.65° S, 71.61° W (75 m ASL), is the only radiosonde station operated by DMC in this region (black start in Figure 2), with launches twice daily at 12:00 and 00:00 UTC. In all cases, there is only one level in which the air temperature profile crosses the 0 °C, even if there is an inversion in dry days because they are warm and low [10], so the free tropospheric H_0 was obtained unambiguously from direct interpolation using the temperature and geopotential height of the levels just above and below 0 °C. On a given day, a mean value of H_0 was calculated using the 00:00 UTC (8:00 PM of the previous day), 12:00 UTC (8:00 AM of the current day) and 00:00 UTC (8:00 PM of the current day) values, and then pooled into the wet or dry groups according to the concurrent rainfall data at Santo Domingo. In stations with surface data only (surface air temperature, SAT), we estimated the freezing level during rainy days using a moist adiabatic lapse rate ($\Gamma_{moist} \approx 6.5°C/km$) as $H_{0sfc} = SAT/\Gamma_{moist} + H_G$, which proved to be a good approximation in this region [10,31]. In Section 4.3, the ETOPO2v2 elevation data [32] was used to determine hypsometric curves (basin area below a given height) of Andean selected basins in central Chile. Catchment boundaries are obtained from the CAMELS-CL dataset [33], where basin outlets are defined according to the location of available streamflow gauges and following topographic-driven limits.

3.2. Reanalysis

To supplement the reduced number of stations from where we can derive H_0 in central Chile and obtain a global perspective, we also employed the Climate Forecast System Reanalysis (CFSR, version 1) described in detail by Saha et al. [34]. Atmospheric variables from this state-of-the-art reanalysis system are available from 1979 to 2010 for every 6 hr on a global 0.5° × 0.5° lat-lon grid. As with the Santo Domingo sounding, the vertical profiles of air temperature and geopotential height were used to obtain H_0 in each grid point and time step. We then calculate the daily mean H_0 from the five values centered at 12:00 UTC (00:00, 06:00, 12:00, 18:00, 00:00 UTC) for every day. A comparison between observed H_0 at Santo Domingo and those obtained from CFSR is shown on Figure 3. There is a notable correspondence between daily averages of H_0 for both wet and dry days, with correlation coefficients exceeding 0.9. However, the relative error in the percentiles of the distribution does not exceed 3% in both cases. In Section 4.1, we provide further evidence that H_0 derived from CFSR is a good approximation of the actual H_0 distribution all along central Chile.

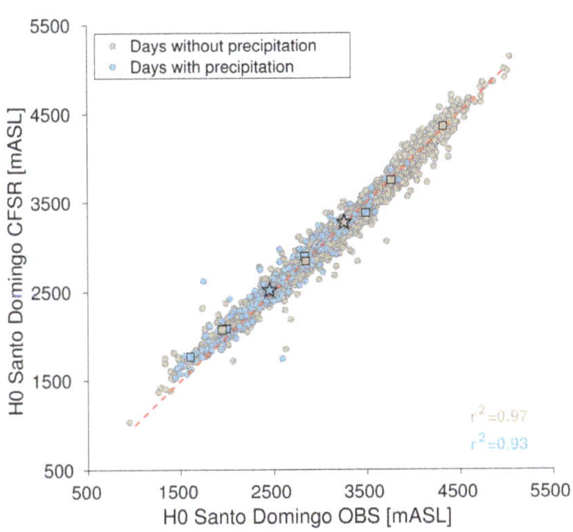

Figure 3. Scatter plot between daily mean values of H_0 observed in Santo Domingo (33.65° S–71.61° W, sounding station) and H_0 from the closest grid point in the CFSR reanalysis (33.50 ° S–71.50° W) for days with (blue dots) and without (brown dots) precipitation at Santo Domingo. Stars and squares represent the mean and percentiles (5th, 25th, 75th and 95th) for each distribution. The dashed red line is the 1:1 line. Data from 1999 to 2010.

3.3. Models

To project the freezing level in the future, we use results from five models included in the Coupled Model Intercomparison Project Phase 5 (CMIP5, [35]) listed in Table 1. The variables used were daily averages of geopotential height, temperature (both at standard isobaric levels) and total precipitation at the surface. This allows us to obtain the daily freezing level grouped into wet and dry days at any given location for their respective winter season (May–September in the Southern Hemisphere and October–April in the Northern Hemisphere). Although the CMIP5 database includes results from more than 50 GCM, differing in their subgrid schemes (parameterizations) and spatial resolution, our analysis is restricted to those models with available surface and upper-air daily data and a grid spacing of ~2° × 2° (or finer) to obtain at least three grid points in latitude and a minimum representation of the Andes cordillera. The five selected models (from independent research centers) in the group conform to a good representation of the central Chile climate [36] and have resolution ranging from 0.75° to 1.86°. Even so, the position of the coastline and Andes ridge of each model can differ substantially from their actual locations, so particular caution was taken in considering the topographic features of each model.

Table 1. Set of Coupled Model Intercomparison Project Phase 5 (CMIP5) models [35] used to characterize the future condition of the H_0 distribution. The variables T and Z correspond to temperature and geopotential height at isobaric levels. The variables pr and sftlf represent precipitation rate and fraction of land area, respectively.

Model	Institution	Variables	Spatial Resolution (lat × lon)	Isobaric Levels (hPa)	Periods
MPI-ESM-LR	Max Planck Institute for Meteorology, Germany	T, Z, pr, sftlf	1.865° × 1.875°	1000, 850, 700, 500, 250, 150, 100, 70, 50, 30, 10, 3, 1, 0.3, 0.1 (15)	1976–2005 (HIS) 2011–2040 (RCPs) 2041–2070 (RCPs) 2071–2100 (RCPs)
MIROC5	Atmosphere and Ocean Research Institute (University of Tokyo)	T, Z, pr, sftlf	1.400° × 1.406°	1000, 850, 700, 500, 250, 100, 50, 10 (8)	
CNRM-CM5	National Center of Meteorological Research, France	T, Z, pr, sftlf	1.400° × 1.406°	1000, 850, 700, 500, 250, 100, 50, 10 (8)	
MRI-CGCM3	Meteorological Research Institute, Japan	T, Z, pr, sftlf	1.121° × 1.125°	1000, 850, 700, 500, 250, 100, 50, 10 (8)	
CMCC-CM	Centro Euro-Mediterraneo sui Cambiamenti Climatici, Italy	T, Z, pr, sftlf	0.748° × 0.750°	1000, 850, 700, 500, 250, 100, 50, 10 (8)	

The present climate distribution of H_0 considers the period between 1976 and 2005 from the historical runs. Based on the previous works [37], the CMCC-CM model has the least errors representing H_0 in central Chile in the present climate, being used as our reference model. For the rest of the 21st century, we use primarily the simulations under the RCP8.5 scenario [38], representing a negative prospect in terms of greenhouse gases emission and atmospheric concentrations [39] in which CO_2-equivalent reaches about 1000 ppm by the end of the century. In this sense, our work illustrates that changes that could occur in a worst-case scenario. Some key analysis, however, were repeated with climate simulations under the more benign RCP4.5 scenario (CO_2-equivalent ~500 ppm by the end of the century) to illustrate the sensitivity of the predicted changes in H_0 to the greenhouse gases concentrations.

4. Results and Discussion

4.1. The Freezing Level in Present Climate

Given the nearly two-dimensional nature of the Andes cordillera, extending almost straight north–south along its subtropical portion, and its proximity to the Pacific shoreline (Figure 2), most of the subsequent analyses are performed using the along-coast (latitudinal) profile of H_0. We acknowledge that the coastal, free tropospheric value of H_0 during a particular storm can differ from the actual snow line over the western slope of the Andes [7,10] but any coast-to-Andes difference is likely to exist both in present and future climates, so it will not preclude exploring the climate change impact upon the freezing level.

Figure 4 shows the latitudinal variation of the mean freezing level ($\overline{H_0}$) along the coast of central Chile using the CFSR values for wet and dry days during the winter semester (May–September). In the case of the wet days, we also included the surface-based $\overline{H_0}$ and the Santo Domingo H_0 distribution. There is a good agreement between the mean values derived from CSFR and the observations along the full transect, except for a ~100 m offset in the mean value that was subsequently removed from the whole latitudinal profile. The mean freezing level during wet days ($\overline{H_0}_{wet}$) gradually decreases southward, from about 2600 m ASL at 30° S to 2000 m ASL at 38° S. Except in the southernmost part of our study region, the mean freezing level is below the Andean crest, and at the latitude of Santiago and Santo Domingo (33° S), $\overline{H_0}_{wet}$ ~2400 m ASL, about half of the altitude of the mountain peaks that reach ~6000 m ASL. The interquartile range of H_0 during wet days is close to 800 m all along central Chile, and extreme high values (95% percentile) can reach up to 3500 m ASL, showing a high variability of the freezing level among winter storms, consistent with previous findings in Garreaud [9].

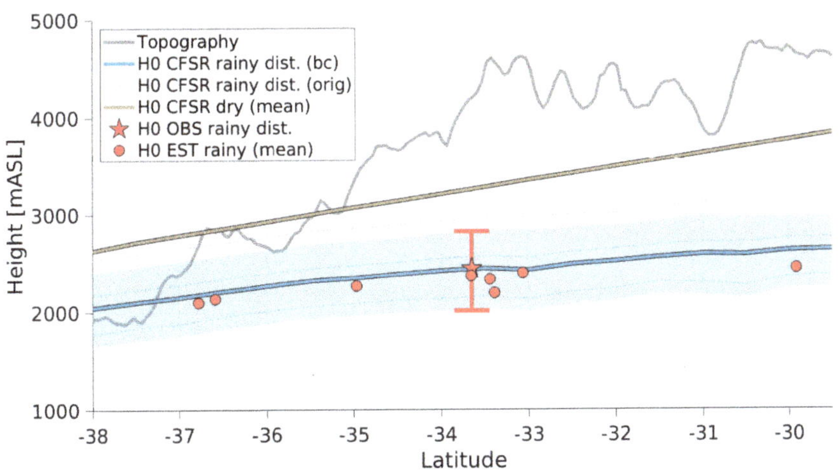

Figure 4. Latitudinal variation of the mean H_0 along the coast of Central Chile during days with (blue line) and without (brown line) precipitation, based on CFSR data. Blue shading indicates the interquartile range of H_0 during wet days. Gray line indicates the 97.5th percentile of terrain elevation, signaling the Andes crest level. Red dots are mean H_0 estimations based on surface data at selected National Weather Service (DMC) stations and the red whiskers indicate the interquartile range of H_0 using the observed values from the Santo Domingo soundings.

The mean freezing level for dry days during winter months (about 90% of the total) also decreases from north to south along the Chilean coast, and it is 800–1000 m higher than $\overline{H_0}_{wet}$. The standard deviation of H_0 under dry conditions is about 700 m, so the freezing level distribution during rainy and dry days have little overlap (see Figure 3). Since both samples are obtained from the winter semester, the lower H_0 values during wet days reflect an actual drop of the air temperature in the lower and middle troposphere in connection with the postfrontal nature of precipitation in central Chile. There, the bulk of the precipitation falls in the first 12–24 h after the front passage cold front in the majority of the winter storms [9,24,37]. The H_0 depression during wet days—relative to dry conditions—also occurs near the west coast of other continents as evidenced in the reanalysis maps (Figure 1b) and documented over the Sierra Nevada in the United States of America [11] and the Iberian Peninsula [40]. Notably, $\overline{H_0}$ tends to rise during winter storms over the east side of the continents (most markedly in North America and East Asia) and the midlatitude oceans.

4.2. Future Changes

Let us begin our description of the future change in the freezing level by considering the frequency distribution of H_0 during winter wet days over Santo Domingo (33° S) from the CMCC-CM model (Figure 5a). For present climate, the simulated H_0 distribution fits well with the observations in terms of the central value and spread. By the end of the century (2071–2100), under the RCP8.5 scenario, the shape of the distribution is preserved, but there is a substantial shift toward higher values (~600 m in the mean), consistent with the expected tropospheric warming. The shift in the H_0 distribution under the RCP4.5 scenario is similar to that in RCP8.5, but the increase in the mean value is about ~400 m. Figure 5b synthesizes the north–south distribution of H_0 (mean value and interquartile range) during wet days for different decades in the 21st century, showing a progressive rise as time progresses. Note that toward the end of the century, $\overline{H_0}_{wet}$ is close to, or higher than, the upper quartile of the H_0 distribution in the current period, and the increase in H_0 seems to be slightly greater in the north. Near the southern limit of our domain (37°–38° S), $\overline{H_0}_{wet}$ in the current climate nearly coincides with the Andean ridge level, so the upper part of the mountains receive snow in about half of the winter storms, allowing the formation of a seasonal snowpack. By the end of the century, however, the mean freezing level during wet days is expected to be several hundred meters above the top of the southern Andes, so snowfall might be quite uncommon even over the highest terrain with a detrimental impact on water availability during the summer months [41]. The projected rise of H_0 toward the end of the century under the RCP4.5 scenario is about 70% of its RCP8.5 counterpart across the whole region.

To place changes throughout central Chile in a global context, Figure 6 shows the change in the mean freezing level ($\Delta\overline{H_0}$) for wet days between the end of century (2071–2010, under RCP8.5) and the historical period (1976–2005) using the CMCC-CM model. There is an increase in $\overline{H_0}$ worldwide, most marked over the subtropical and tropical oceans ($\Delta\overline{H_0}$ close to 1000 m) but rather small at higher latitudes in both hemispheres. The pattern and magnitude of $\Delta\overline{H_0}$ is similar during dry days, but the projected rise tends to be ~100-200 m higher across much of the subtropics and midlatitudes (not shown). In central Chile, for instance, $\Delta\overline{H_0}$ is close to 600 m ASL and 750 m ASL for wet and dry days, respectively.

Next, we examine the changes in freezing level in central Chile using results from the 5 GCM with high spatial resolution and daily data (Section 3.3). Except for CMCC-CM, the simulated H_0 distribution in current climate exhibits biases as large as ±10% [37]; therefore, we use the so-called delta approach here to analyze future changes [42,43]. For each model, we calculate the future (end-of-the-century under RCP8.5) minus present difference in H_0 (mean value and selected percentiles), and then we average the five results. The multimodel mean $\Delta\overline{H_0}$ is then added to the observed (CFSR) profile of $\overline{H_0}$ in the current climate as shown in Figure 7a for the wet and dry winter day groups, along with an indication of the model spread. For the wet days, $\Delta\overline{H_0}$ is about 400 m (250–600 m range among the models), while for days without precipitation the mean change is close to 600 m (400–800 m range). The values of $\Delta\overline{H_0}$ have little variation along the profile. Similar increases were found when considering the median and the lower and upper quartiles, suggesting an overall shift of the H_0 frequency distribution while preserving its shape, as we showed for the specific case of

CMCC-CM (Figure 5a). The latitudinal profiles of $\Delta\overline{H_0}$ for wet and dry days under the RCP4.5 scenario are also included in Figure 7a. When considering the multimodel mean, the rise in mean freezing level is about half of that obtained under RCP8.5, suggesting a linear behavior of the projected changes of the free tropospheric temperature in central Chile, in line with results from Zazulie et al. [44] for surface temperatures over the subtropical Andes.

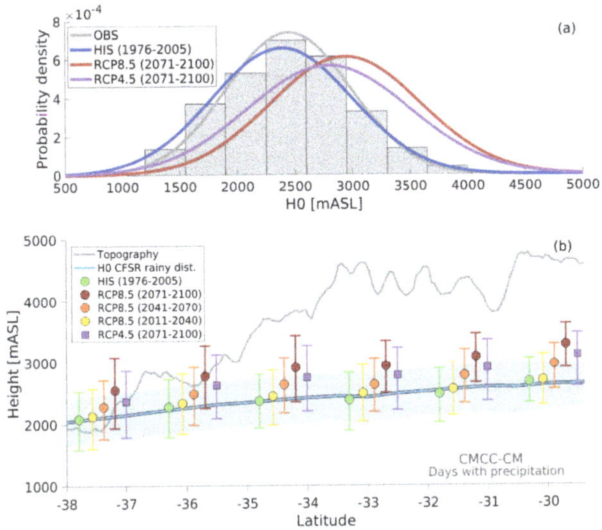

Figure 5. (**a**) The gray bars show the histogram of the observed H_0 at Santo Domingo during wet days and the gray line is the fitted normal distribution. The blue, red and purple lines are the fitted normal distribution of H_0 (during wet days) for present and future climate (2071–2100 under RCP8.5 and RCP4.5), respectively, simulated by the CMCC-CM model and interpolated to Santo Domingo. (**b**) Evolution of the CMCC-CM simulated H_0 distribution for days with precipitation, during different decades in the 21st century along the coast of central Chile. Circles indicate the mean of distribution and whiskers the upper and lower quartiles. The blue line and light blue shading are the present climate mean and interquartile range of H_0 (wet days) based on CFSR data. The gray line indicates the 97.5th percentile of terrain elevation.

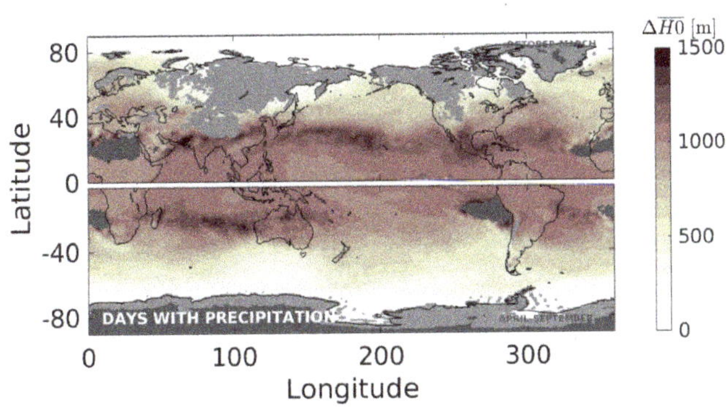

Figure 6. Global distribution of the difference in mean H_0 between the end of century (2071–2100 under scenario RCP8.5) and the historical period (1976–2005) simulated by the model CMCC-CM for winter days with precipitation. The winter months are October–March in the Northern Hemisphere and April–September in the Southern Hemisphere. Light gray areas represent those regions where mean H_0 intersects the topography. Dark gray areas indicate arid regions where the low number of rainy days precludes a robust calculation of H_0 under wet conditions.

As noted in the introduction, winter storms with higher than average H_0 greatly increase the risk of flooding and landslides along central Chile, so particular attention must be placed in the change of extreme events. To gauge those changes, here we consider the variation in the 95% percentile value under wet conditions (H_{095wet}). At the latitude of Santiago, H_{095wet} ~3300 m ASL in the present climate, nearly 1 km above the mean value. For each model, we obtained H_{095wet} in the present-climate simulation, as a function of latitude, and then we calculated the frequency of the time in which this value will be surpassed in the last decades of the 21st century. The multimodel mean change in frequency is shown in Figure 7b. The mean frequency in which future freezing levels (during wet days) will equal or exceed the present climate H_{095wet} is about 20% between 38°–34° S and up to 30% in the northern part of the domain. Thus, what is now labeled as an extreme event will be a condition 4–6 times more frequent by the end of the century (under RCP8.5 scenario), a worrisome projection whose hydrological consequences will be discussed in the next section. Even in the case of the more benign RCP4.5 climate scenario, the frequency of extreme events (5% of the time in the current climate) could double (Figure 7b).

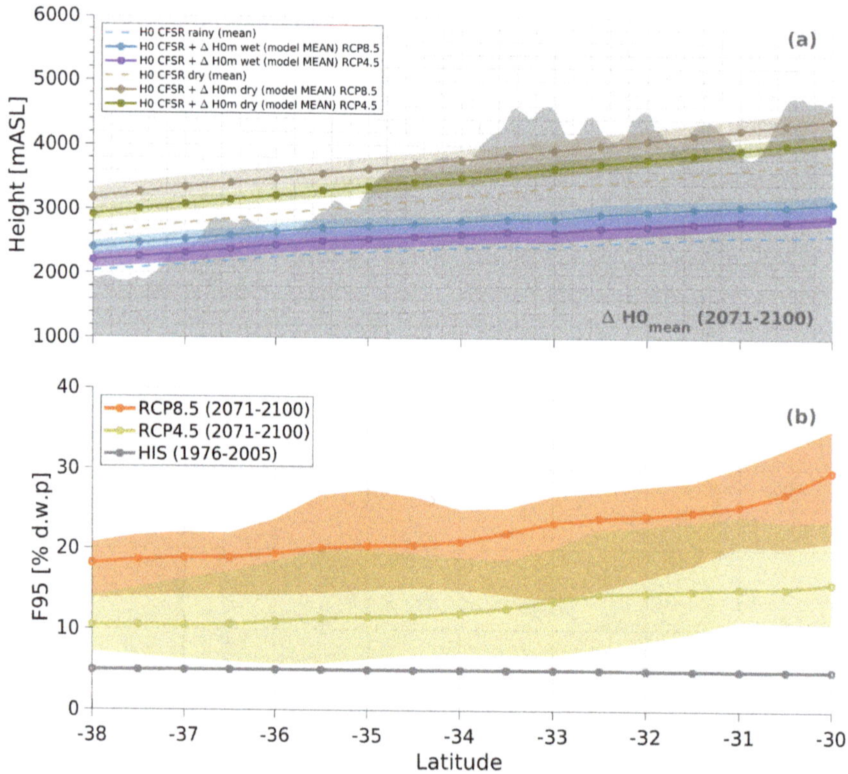

Figure 7. (**a**) Dashed lines indicate the profile of the mean H_0 along the coast of central Chile in present climate based on CFSR data for wet days (blue) and dry days (brown). The solid lines with circles are the multimodel mean H_0 for the end of 21st century under RCP8.5/RCP4.5 and the shading indicates the projection range (maximum-minimum). Gray shaded area indicate the 97.5th percentile of terrain elevation. (**b**) End of the century (2071–2100) multimodel mean frequency of the 95th percentile H_0 value in the present climate. The line with circles and the shaded area represent the average and range (minimum and maximum) projected by the models.

4.3. Hydrological Impact

Here, we provide a rough estimate of the hydrological impact of the H_0 changes during wet days over central Chile contingent of the occurrence of the RCP8.5 heavy emission scenario. Such impact varies with latitude since both $\overline{H_0}_{wet}$ and the terrain elevation decrease toward the south. Indeed, one may expect less acute impacts in the southern part of the domain because even in the current climate, the Andean basin mostly receives rainfall ($\overline{H_0}_{wet} \sim H_G$). Figure 8a shows the area of the terrain above 1000 m ASL and below the present climate mean freezing level (during wet days) for selected Andean basins of central Chile (A_p). The baseline was chosen close to the base of the Andes foothills. We then recalculated the A_p, keeping the baseline but changing the upper limit to $\overline{H_0}_{wet}$ in the future (2071–2100; RCP8.5) and H_{095wet} in the present and future climates. For selected basins, Figure 8b shows the areal increment factors, defined as the ratio of A_p obtained with the new upper limits to the area defined with the present $\overline{H_0}_{wet}$. Considering the rise of the mean freezing level during winter storms, the most affected basins would those between 33° S and 35° S. Particularly, the pluvial area in the upper Maipo River (that drains into the central valley just south of Santiago) almost doubles because of the projected rise in $\overline{H_0}_{wet}$ from present to future. Further south A_p would be almost unaffected since most of the increase in H_0 occurs over the maximum height of the terrain. However, the basins north of 33° S would be less affected since their average slope is gentler. In present climate, the pluvial area also increases markedly when considering the 95% percentile H_0 relative to the mean value. Continuing our focus on the Maipo basin, we found that the pluvial area increases by a factor of 3 when considering the 95% percentile of H_0 in the present and by a factor of 4.5 when considering the 95% percentile events in the future. The increment factor in other basins is lower, but future warm storms might have a pluvial area 2–3 times larger than average storms in the present climate.

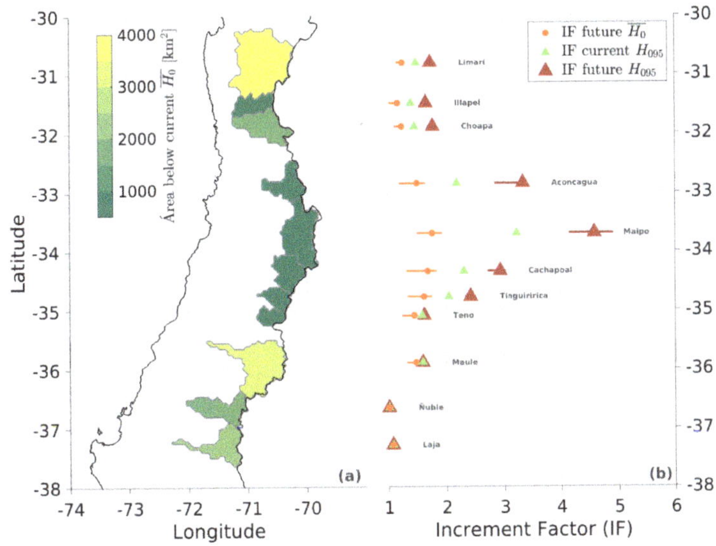

Figure 8. (**a**) Area between 1000 m ASL and below the mean of H_0 in current climate for selected basins in central Chile basins. (**b**) Increment factor of the area considering future H_0 values (mean and 95th percentile). See Section 4.3 for details. Lines in panel (**b**) indicate the range of models for each basin.

The large increase in the projected pluvial area for the Maipo basin, along with its closeness to the large city of Santiago, calls for further analysis. In Figure 9, we present contours of H_0 for mean and extreme cases under wet conditions, superimposed on a topographic map. Note, for instance, that nearly the entire basin would receive rainfall in the 5% of warmer winter storms of the future, a conditions that in the present climate has a probability of approximately less than 0.5%. Of course, the river flow on a given storm is also dependent on the amount of precipitation and soil moisture (which dictate the infiltration rate), but an upper bound of the river flow can be easily obtained as $Q_{max} = PB \times A_p$, where PB is the basin averaged rainfall and A_p depends directly from H_0 [24]. The right panel of Figure 9 shows Q_{max} as a function of the precipitation for selected values of H_0. We have marked the value $Q_{max} = 500$ m^3/s, which is considered dangerous to this area based on the work of Bustos [45], and also include the frequency distribution of the daily precipitation for this basin.

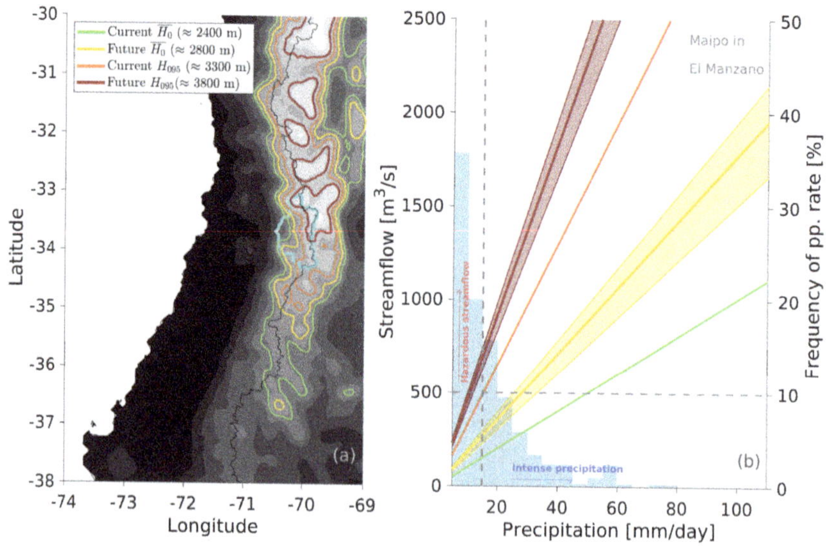

Figure 9. (**a**) Central Chile topography with present and future H_0 contours (mean and 95th percentile value). The upper Maipo river basin is delimited by the cyan line. (**b**) Estimated stream flow at the outlet of the upper Maipo river basin as a function of precipitation for different values of H_0 (mean and 95th percentile value) in present and end-of-the-century climate conditions (2071–2100 under RCP8.5). The histogram in the background shows the distribution of daily accumulated precipitation over 5 mm/day for station Quinta Normal in Santiago close to the outlet of the upper Maipo river basin.

For the mean freezing level in the present, at least 50 mm/day is required for the Maipo River to reach a dangerous level, a daily accumulation that happens in about 5% of the time (Figure 9b). In the future, however, flooding might be caused by a 25 mm/day storm, considering the mean freezing level, and by just 10 mm/day if the storm happens under warm conditions (95th percentile of H_0). This last value is even lower than the average daily accumulation, so such a hazardous combination of precipitation and freezing level leading to flooding may occur rather frequently despite the annual mean precipitation decrease over central Chile predicted for the rest of the 21st century by numerical models [46]. Indeed, projections based on the CMCC-CM model show that daily precipitation rate would change to lower values (Figure 10) and the number of wet days (≥5 mm/day) nearly halves from the current condition to the end of the century. Nonetheless, the number of days in which the combination of H_0 and precipitation results in a upper Maipo river flow ≥500 m^3/s decreases marginally (from 107 to 92 days per 30 years), thus doubling the frequency in which hazardous condition occurs if precipitation is taking place. The increase in such conditional frequency is even more striking considering higher river flow values. For $Q_{max} \geq 2000$ m^3/s, not only does the conditional frequency triple but the total number of days increases toward the end of the century under the RCP8.5 scenario, considering the CMCC-CM model results and the simple hydrological model described before.

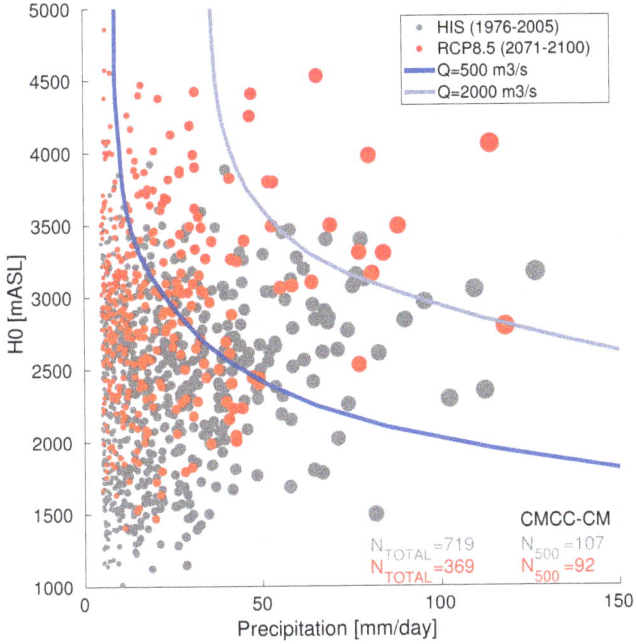

Figure 10. Scatter plot between CMCC-CM simulated daily precipitation and H_0 for wet days in Santiago (model closest point) considering the historical period (1976–2005, gray circles) and the end of century (2071–2100 under RCP8.5, red circles). The light blue line represents the combination of precipitation and H_0 values resulting in a discharge of 500 m^3/s (2000 m^3/s) in at the outlet of the upper Maipo river basin using the simplest hydrological model (Section 4.3). The number of total rainy days (N_{total}) and those resulting in more than 500 m^3/s (N_{500}) are shown.

The projected increase in H_0 during wet days also implies a reduced amount of water stored in the Andean snowpack that forms every winter and releases fresh water during spring/summer when this resource is most needed for agriculture and other uses in central Chile [41]. As before, we provide a rough estimate of these changes here by considering the partition between liquid and solid precipitation during winter in the Maipo river basin. For each storm, the liquid and solid volumes were calculated by multiplying the basin mean daily precipitation for the pluvial and snow areas, defined by the part of the basin below or above the freezing level, respectively. We then added all those volumes during the present (1979–2005) and future (2070–2100, under RCP8.5) periods. Solid precipitation (forming the snow pack) accounts for about 75% of the total in current climate but such contribution decrease to 57% in the far future. The total precipitation over the Maipo River basin also decreases by about 35% (relative to present day), considering the CMCC-CM model results [37]; that figure that is in line with multimodel estimates [46]. The combined effects of warmer storms and less precipitation thus result in future (end of the century, RCP8.5) winter snow accumulation halving present climate values, ensuing a dramatic decrease in summer–fall river flow, an effect documented in other basins in central Chile by Vicuña et al. [47].

5. Conclusions

The distribution of H_0 along central Chile (30°–38° S) was obtained from the CFSR reanalysis for the present climate and five climate models for the rest of the 21st century under the RCP8.5 and RCP4.5 scenarios. In this highly populated region, variations of the freezing level within precipitation events and between storms can produce substantial differences in the pluvial area, affecting runoff generation and the discharge of rivers draining the Andes cordillera.

Relative to present day conditions, the mean H_0 in the future (RCP8.5) would be approximately 400 m higher for days with precipitation and 600 m for days without precipitation. In general, we found a progressive increase of the H_0 distribution while preserving its shape. Toward the end of the century, the mean value under wet conditions is close to or higher than the upper quartile of the H_0 distribution in the current climate, and the increase in H_0 seems to be slightly greater in the north. By the end of the century, however, the mean freezing level during wet days could be several hundred meters above the top of the southern Andes, so snowfall will likely become quite uncommon even over the highest terrain with detrimental impact for the water availability during the summer months. The more benign RCP4.5 scenario also results in a shift of the H_0 distribution toward higher values but with an amplitude about half of its RCP8.5 counterpart.

Regarding the occurrence of particularly warm storms, H_0 values that currently occur in only 5% of the days with precipitation (H_0 > 3300 m ASL) might be concurrent with nearly 25% (10%) of the future winter storms throughout central Chile under RCP8.5 (RCP4.5). The projected changes in H_0 translate into an increase in the pluvial area and, therefore, in the volume of water available for all the basins of central Chile during storms, except in the southernmost basins where the current freezing level is generally above the Andes ridge, at the expense of the water stored in the seasonal snowpack. The basins most affected would be those around 34° S. The upper Maipo River, draining just to the south of the city of Santiago, might experience an increase in its pluvial area by a factor 4 to 5 times the pluvial area during future warm storms relative to mean current conditions. Keep in mind that our inferences are rough estimates of the hydrological response to changing freezing level and contingent to the negative RCP8.5 climate scenario. In this condition, even moderate daily precipitation could increase the river flow to levels that are considered hazardous for central Chile. Thus, even under the prospect of drying along central Chile, warmer winter storms in the future possess a substantial risk of landslides, flashfloods and widespread flooding along the foothills of the subtropical Andes, calling for more comprehensive studies in this subject.

Author Contributions: Conceptualization, P.M. and R.D.G.; data curation, P.M.; formal analysis, P.M. and R.D.G.; funding acquisition, P.M. and R.D.G.; methodology, P.M. and R.D.G.; resources, R.D.G.; writing—original draft, P.M. and R.D.G.; writing—review and editing, P.M. and R.D.G. All authors have read and agreed to the published version of the manuscript.

Funding: This work was funded by the National Agency for Research and Development (ANID)/PFCHA/MAGÍSTER NACIONAL/2017-22170369. The APC was funded by the Center for Climate and Resilience Research (CR2, CONICYT/FONDAP/15110009).

Acknowledgments: We thank the Center for Climate and Resilience Research (CR2, CONICYT/FONDAP/15110009) for its support.

Conflicts of Interest: The authors declare no conflict of interest.

References

1. Hobbs, P.V.; Easter, R.C.; Fraser, A.B. A theoretical study of the flow of air and fallout of solid precipitation over mountainous terrain: Part II. Microphysics. *J. Atmos. Sci.* **1973**, *30*, 813–823. [CrossRef]
2. White, A.B.; Gottas, D.J.; Strem, E.T.; Ralph, F.M.; Neiman, P.J. An automated brightband height detection algorithm for use with Doppler radar spectral moments. *J. Atmos. Ocean. Technol.* **2002**, *19*, 687–697. [CrossRef]
3. Fabry, F.; Zawadzki, I. Long-term radar observations of the melting layer of precipitation and their interpretation. *J. Atmos. Sci.* **1995**, *52*, 838–851. [CrossRef]
4. Lundquist, J.D.; Neiman, P.J.; Martner, B.; White, A.B.; Gottas, D.J.; Ralph, F.M. Rain versus snow in the Sierra Nevada, California: Comparing Doppler profiling radar and surface observations of melting level. *J. Hydrometeorol.* **2008**, *9*, 194–211. [CrossRef]

5. Cui, G.; Bales, R.; Rice, R.; Anderson, M.; Avanzi, F.; Hartsough, P.; Conklin, M. Detecting Rain-Snow-Transition Elevations in Mountain Basins Using Wireless Sensor Networks. *J. Hydrometeorol.* **2020**, *21*, 2061–2081. [CrossRef]
6. Medina, S.; Smull, B.F.; Houze, R.A., Jr.; Steiner, M. Cross-barrier flow during orographic precipitation events: Results from MAP and IMPROVE. *J. Atmos. Sci.* **2005**, *62*, 3580–3598. [CrossRef]
7. Minder, J.R.; Durran, D.R.; Roe, G.H. Mesoscale controls on the mountainside snow line. *J. Atmos. Sci.* **2011**, *68*, 2107–2127. [CrossRef]
8. Marwitz, J.D. Deep orographic storms over the Sierra Nevada. Part I: Thermodynamic and kinematic structure. *J. Atmos. Sci.* **1987**, *44*, 159–173. [CrossRef]
9. Garreaud, R. Warm winter storms in Central Chile. *J. Hydrometeorol.* **2013**, *14*, 1515–1534. [CrossRef]
10. Ibañez, M.; Gironás, J.; Oberli, C.; Chadwick, C.; Garreaud, R.D. Daily and seasonal variation of the surface temperature lapse rate and 0 °C isotherm height in the western subtropical Andes. *Int. J. Climatol.* **2020**. [CrossRef]
11. Sumargo, E.; Cannon, F.; Ralph, F.M.; Henn, B. Freezing level forecast error can consume reservoir flood control storage: Potentials for Lake Oroville and New Bullards Bar reservoirs in California. *Water Resour. Res.* **2020**, *56*, e2020WR027072. [CrossRef]
12. Fehlmann, M.; Gascón, E.; Rohrer, M.; Schwarb, M.; Stoffel, M. Estimating the snowfall limit in alpine and pre-alpine valleys: A local evaluation of operational approaches. *Atmos. Res.* **2018**, *204*, 136–148. [CrossRef]
13. Schauwecker, S.; Rohrer, M.; Huggel, C.; Endries, J.; Montoya, N.; Neukom, R.; Perry, B.; Salzmann, N.; Schwarb, M.; Suarez, W. The freezing level in the tropical Andes, Peru: An indicator for present and future glacier extents. *J. Geophys. Res. Atmos.* **2017**, *122*, 5172–5189. [CrossRef]
14. Garreaud, R.; Rutllant, J. Análisis meteorológico de los aluviones de Antofagasta y Santiago de Chile en el periodo 1991–1993. *Atmósfera* **1996**, *9*, 251–271.
15. Liu, A.; Mooney, C.; Szeto, K.; Thériault, J.; Kochtubajda, B.; Stewart, R.; Boodoo, S.; Goodson, R.; Li, Y.; Pomeroy, J. The June 2013 Alberta catastrophic flooding event: Part 1—Climatological aspects and hydrometeorological features. *Hydrol. Process.* **2016**, *30*, 4899–4916. [CrossRef]
16. Cohen, J.; Ye, H.; Jones, J. Trends and variability in rain-on-snow events. *Geophys. Res. Lett.* **2015**, *42*, 7115–7122. [CrossRef]
17. Grote, T. A synoptic climatology of rain-on-snow flooding in Mid-Atlantic region using NCEP/NCAR Re-Analysis. *Phys. Geogr.* **2020**, 1–20. [CrossRef]
18. Musselman, K.N.; Lehner, F.; Ikeda, K.; Clark, M.P.; Prein, A.F.; Liu, C.; Barlage, M.; Rasmussen, R. Projected increases and shifts in rain-on-snow flood risk over western North America. *Nat. Clim. Chang.* **2018**, *8*, 808–812. [CrossRef]
19. Barnett, T.P.; Adam, J.C.; Lettenmaier, D.P. Potential impacts of a warming climate on water availability in snow-dominated regions. *Nature* **2005**, *438*, 303. [CrossRef]
20. Hatchett, B.J.; Daudert, B.; Garner, C.B.; Oakley, N.S.; Putnam, A.E.; White, A.B. Winter snow level rise in the northern Sierra Nevada from 2008 to 2017. *Water* **2017**, *9*, 899. [CrossRef]
21. Sepúlveda, S.A.; Padilla, C. Rain-induced debris and mudflow triggering factors assessment in the Santiago cordilleran foothills, Central Chile. *Nat. Hazards* **2008**, *47*, 201–215. [CrossRef]
22. Poveda, G.; Espinoza, J.C.; Zuluaga, M.D.; Solman, S.A.; Garreaud Salazar, R.; van Oevelen, P.J. High impact weather events in the Andes. *Front. Earth Sci.* **2020**. [CrossRef]
23. Viale, M.; Garreaud, R. Orographic effects of the subtropical and extratropical Andes on upwind precipitating clouds. *J. Geophys. Res. Atmos.* **2015**, *120*, 4962–4974. [CrossRef]
24. Falvey, M.; Garreaud, R. Wintertime precipitation episodes in central Chile: Associated meteorological conditions and orographic influences. *J. Hydrometeorol.* **2007**, *8*, 171–193. [CrossRef]
25. Viale, M.; Valenzuela, R.; Garreaud, R.D.; Ralph, F.M. Impacts of atmospheric rivers on precipitation in southern South America. *J. Hydrometeorol.* **2018**, *19*, 1671–1687. [CrossRef]
26. Valenzuela, R.A.; Garreaud, R.D. Extreme daily rainfall in central-southern Chile and its relationship with low-level horizontal water vapor fluxes. *J. Hydrometeorol.* **2019**, *20*, 1829–1850. [CrossRef]
27. Cortés, G.; Vargas, X.; McPhee, J. Climatic sensitivity of streamflow timing in the extratropical western Andes Cordillera. *J. Hydrol.* **2011**, *405*, 93–109. [CrossRef]
28. Rojas, O.; Mardones, M.; Arumí, J.L.; Aguayo, M. Una revisión de inundaciones fluviales en Chile, período 1574–2012: Causas, recurrencia y efectos geográficos. *Rev. Geogr. Norte Gd.* **2014**, 177–192. [CrossRef]

29. Ayala, L.; López, A.; Tamburrino, A.; Vera, G. Aspectos hidrometeorológicos e hidrodinámicos de algunos eventos aluvionales recientes en Chile. *XVI Congr. Latinoam. Hidrául.* **1994**, *3*, 39–51.
30. Sepúlveda, S.A.; Rebolledo, S.; Vargas, G. Recent catastrophic debris flows in Chile: Geological hazard, climatic relationships and human response. *Quat. Int.* **2006**, *158*, 83–95. [CrossRef]
31. González, S.; Garreaud, R. Spatial variability of near-surface temperature over the coastal mountains in southern Chile (38 S). *Meteorol. Atmos. Phys.* **2019**, *131*, 89–104. [CrossRef]
32. National Geophysical Data Center. 2-minute Gridded Global Relief Data (ETOPO2) v2. National Geophysical Data Center, NOAA. Available online: https://www.ngdc.noaa.gov/mgg/global/etopo2.html (accessed on 25 May 2019).
33. Alvarez-Garreton, C.; Mendoza, P.A.; Boisier, J.P.; Addor, N.; Galleguillos, M.; Zambrano-Bigiarini, M.; Lara, A.; Cortes, G.; Garreaud, R.; McPhee, J.; et al. The CAMELS-CL dataset: catchment attributes and meteorology for large sample studies-Chile dataset. *Hydrol. Earth Syst. Sci.* **2018**, *22*, 5817–5846. [CrossRef]
34. Saha, S.; Moorthi, S.; Pan, H.L.; Wu, X.; Wang, J.; Nadiga, S.; Tripp, P.; Kistler, R.; Woollen, J.; Behringer, D.; et al. The NCEP climate forecast system reanalysis. *Bull. Am. Meteorol. Soc.* **2010**, *91*, 1015–1058. [CrossRef]
35. Taylor, K.E.; Stouffer, R.J.; Meehl, G.A. An overview of CMIP5 and the experiment design. *Bull. Am. Meteorol. Soc.* **2012**, *93*, 485–498. [CrossRef]
36. Center for Climate and Resilience Research, (CR)2 (FONDAP 15110009). Regional Climate Simulations. Available online: http://simulaciones.cr2.cl/ (accessed on 15 May 2019).
37. Mardones Bascuñan, P.B. Impactos del Cambio Climático en la Altura de la Isoterma 0 °C Sobre Chile Central. Master's Thesis, University of Chile, Santiago, Chile, 2019.
38. Riahi, K.; Rao, S.; Krey, V.; Cho, C.; Chirkov, V.; Fischer, G.; Kindermann, G.; Nakicenovic, N.; Rafaj, P. RCP 8.5—A scenario of comparatively high greenhouse gas emissions. *Clim. Chang.* **2011**, *109*, 33. [CrossRef]
39. Moss, R.H.; Edmonds, J.A.; Hibbard, K.A.; Manning, M.R.; Rose, S.K.; Van Vuuren, D.P.; Carter, T.R.; Emori, S.; Kainuma, M.; Kram, T.; et al. The next generation of scenarios for climate change research and assessment. *Nature* **2010**, *463*, 747–756. [CrossRef] [PubMed]
40. Benarroch, A.; Siles, G.A.; Riera, J.M.; Pérez-Peña, S. Heights of the 0 °C Isotherm and the Bright Band in Madrid: Comparison and Variability. In Proceedings of the 2020 14th European Conference on Antennas and Propagation (EuCAP), Copenhagen, Denmark, 15–20 March 2020; pp. 1–5.
41. Masiokas, M.H.; Villalba, R.; Luckman, B.H.; Le Quesne, C.; Aravena, J.C. Snowpack variations in the central Andes of Argentina and Chile, 1951–2005: Large-scale atmospheric influences and implications for water resources in the region. *J. Clim.* **2006**, *19*, 6334–6352. [CrossRef]
42. Lenderink, G.; Buishand, A.; Van Deursen, W. Estimates of future discharges of the river Rhine using two scenario methodologies: direct versus delta approach. *Hydrol. Earth Syst. Sci.* **2007**, *11*, 1145–1159. [CrossRef]
43. Ekström, M.; Jones, P.; Fowler, H.; Lenderink, G.; Buishand, T.; Conway, D. Regional climate model data used within the SWURVE project 1: projected changes in seasonal patterns and estimation of PET. *Hydrol. Earth Syst. Sci.* **2007**, *11*, 1069–1083. [CrossRef]
44. Zazulie, N.; Rusticucci, M.; Raga, G.B. Regional climate of the Subtropical Central Andes using high-resolution CMIP5 models. Part II: future projections for the twenty-first century. *Clim. Dyn.* **2018**, *51*, 2913–2925. [CrossRef]
45. Bustos Cavada, D. Cambio climático y eventos de emergencia en el suministro de agua potable en el gran Santiago. Bachelor's Thesis, University of Chile, Santiago, Chile, 2011.
46. Bozkurt, D.; Rojas, M.; Boisier, J.P.; Valdivieso, J. Projected hydroclimate changes over Andean basins in central Chile from downscaled CMIP5 models under the low and high emission scenarios. *Clim. Chang.* **2018**, *150*, 131–147. [CrossRef]
47. Vicuña, S.; Garreaud, R.D.; McPhee, J. Climate change impacts on the hydrology of a snowmelt driven basin in semiarid Chile. *Clim. Chang.* **2011**, *105*, 469–488. [CrossRef]

Publisher's Note: MDPI stays neutral with regard to jurisdictional claims in published maps and institutional affiliations.

MDPI
St. Alban-Anlage 66
4052 Basel
Switzerland
Tel. +41 61 683 77 34
Fax +41 61 302 89 18
www.mdpi.com

Atmosphere Editorial Office
E-mail: atmosphere@mdpi.com
www.mdpi.com/journal/atmosphere

www.ingramcontent.com/pod-product-compliance
Lightning Source LLC
Chambersburg PA
CBHW040803250926
51850CB00007BA/1066
* 9 7 8 3 0 3 6 5 2 9 5 3 0 *